APPLIED VIROLOGY RESEARCH

Volume 1

New Vaccines and Chemotherapy

APPLIED VIROLOGY RESEARCH

Editor-in-Chief: Edouard Kurstak, *University of Montreal, Montreal, Quebec, Canada*

Series Editors: R. G. Marusyk, *University of Alberta, Edmonton, Alberta, Canada*
F. A. Murphy, *Centers for Disease Control, Atlanta, Georgia*
M. H. V. Van Regenmortel, *Institute of Molecular and Cellular Biology, Strasbourg, France*

Volume 1 NEW VACCINES AND CHEMOTHERAPY
Edited by Edouard Kurstak, R. G. Marusyk, F. A. Murphy, and M. H. V. Van Regenmortel

---

APPLIED VIROLOGY RESEARCH

Volume 1

New Vaccines and Chemotherapy

Edited by

Edouard Kurstak

University of Montreal
Montreal, Quebec, Canada

R. G. Marusyk

University of Alberta
Edmonton, Alberta, Canada

F. A. Murphy

Centers for Disease Control
Atlanta, Georgia

and

M. H. V. Van Regenmortel

Institute of Molecular and Cellular Biology
Strasbourg, France

Springer Science+Business Media, LLC

Library of Congress Cataloging in Publication Data

New vaccines and chemotherapy.

(Applied virology research; v. 1)
Includes bibliographies and index.
1. Viral vaccines. 2. Antiviral agents. 3. Virus diseases—Chemotherapy. I. Kurstak, Edouard. II. Series. [DNLM: 1. Antigens, Viral—immunology. 2. Viral Vaccines—immunology. 3. Viral Vaccines—therapeutic use. 4. Virus Diseases—prevention & control. QV 268.5 N532]
QR189.5.V5N49 1988 574.2′34 87-32835
ISBN 978-1-4757-9270-6

ISBN 978-1-4757-9270-6 ISBN 978-1-4757-9268-3 (eBook)
DOI 10.1007/978-1-4757-9268-3

Originally published by Plenum Publishing Corporation New York in 1998
Softcover reprint of the hardcover 1st edition 1998

Contributors

Shinobu Abe • Japan Poliomyelitis Research Institute, Higashimurayama, Tokyo 189, Japan

F. E. André • Biological Division, Smith Kline RIT, B-1330 Rixensart, Belgium

Mineo Arita • Department of Enteroviruses, National Institute of Health, Musashimurayama, Tokyo 190-12, Japan

L. A. Babiuk • Veterinary Infectious Disease Organization, Saskatoon, Saskatchewan S7N 0W0, Canada

Phillip W. Berman • Department of Molecular Biology, Genentech, Inc., South San Francisco, California 94080

D. W. Boucher • Bureau of Biologics, Health Protection Branch, Department of National Health and Welfare, Ottawa, Ontario K1A 0L2, Canada

Y. Brossard • Center of Perinatal Blood Biology, 75571 Paris, Cedex 12, France

F. Brown • Wellcome Biotechnology Ltd., Beckenham, Kent BR3 3BS, England

P. G. Canonico • United States Army Medical Research Institute for Infectious Diseases, Fort Detrick, Frederick, Maryland 21701

B. Carlsson • Department of Clinical Immunology, Guldhedsgatan 10, S-413 46 Göteborg, Sweden; and Department of Pediatrics, East Hospital, S-9 41685 Göteborg, Sweden

Kuo-Chi Cheng • Department of Pediatrics, Cornell University Medical College, New York, New York 10021

G. Contreras • Bureau of Biologics, Health Protection Branch, Department of National Health and Welfare, Ottawa, Ontario K1A 0L2, Canada

Erik De Clercq • Rega Institute for Medical Research, The Catholic University of Louvain, B-3000 Louvain, Belgium

A. Delem • Biological Division, Smith Kline RIT, B-1330 Rixensart, Belgium

Donald Dowbenko • Department of Molecular Biology, Genentech, Inc., South San Francisco, California 94080

F. DuBois • Institute of Virology, 37032 Tours, Cedex, France

D. A. Eppstein • Institute of Bio-Organic Chemistry, Syntex Research, Palo Alto, California 94304

Morag Ferguson • National Institute for Biological Standards and Control, Potters Bar, Herts. EN6 3QG, England

A. Paul Fiddian • Clinical Antiviral Therapy Section, Wellcome Research Laboratories, Beckenham, Kent BR3 3BS, England

P. Frenchick • Veterinary Infectious Disease Organization, Saskatoon, Saskatchewan S7N 0W0, Canada

J. Furesz • Bureau of Biologics, Health Protection Branch, Department of National Health and Welfare, Ottawa, Ontario K1A 0L2, Canada

D. J. Gangemi • School of Medicine, University of South Carolina, Columbia, South Carolina 29425

A. Godefroy • Center of Blood Transfusion, 36000 Châteauroux, France

A. Goudeau • Institute of Virology, 37032 Tours, Cedex, France

Timothy Gregory • Department of Process Sciences, Genentech, Inc., South San Francisco, California 94080

F. Gyorkey • Department of Virology and Epidemiology, Baylor College of Medicine, Houston, Texas 77030

A. J. Hansen • Agriculture Canada, Research Station, Summerland, British Columbia V0H 1Z0, Canada

L. Å. Hanson • Department of Clinical Immunology, Guldhedsgatan 10, S-413 46 Göteborg, Sweden; and Department of Pediatrics, East Hospital, S-9 41685 Göteborg, Sweden

S. E. Hasnain • Department of Medicine, University of Alberta, Edmonton, Alberta T6G 2G3, Canada

J. Huchet • Center of Perinatal Blood Biology, 75571 Paris, Cedex 12, France

M. K. Ijaz • Veterinary Infectious Disease Organization, Saskatoon, Saskatchewan S7N 0W0, Canada

E. Isolauri • Department of Clinical Sciences, University of Tampere, SF-33520 Tampere, Finland

Heihachi Itoh† • Japan Poliomyelitis Research Institute, Higashimurayama, Tokyo 189, Japan

F. Jalil • Department of Social and Preventive Pediatrics, King Edward Medical College, Lahore, Pakistan

M. Kende • United States Army Medical Research Institute for Infectious Diseases, Fort Detrick, Frederick, Maryland 21701

†Deceased.

S. B. H. Kent • Division of Biology, California Institute of Technology, Pasadena, California 91125

M. P. Kieny • Transgene S.A., Strasbourg 67000, France

J. Klein • Center of Blood Transfusion, 3600 Châteauroux, France

Michinori Kohara • Department of Microbiology, Faculty of Medicine, University of Tokyo, Hongo, Bunkyo-ku, Tokyo 113, Japan; and Japan Poliomyelitis Research Institute, Higashimurayama, Tokyo 189, Japan

Toshihiko Komatsu • Department of Enteroviruses, National Institute of Health, Musashimurayma, Tokyo 190-12, Japan

J. Kreuter • Institute of Pharmaceutical Technology, Johann Wolfgang Goethe University, D-6000 Frankfurt, Federal Republic of Germany

Shusuke Kuge • Department of Microbiology, Faculty of Medicine, University of Tokyo, Hongo, Bunkyo-ku, Tokyo 113, Japan

W. Lange • Robert Koch Institute, Federal Health Office, D-1000 West Berlin, Germany

Lawrence A. Lasky • Department of Molecular Biology, Genentech, Inc., South San Francisco, California 94080

R. Lathe • LGME–CNRS and U184–INSERM, Strasbourg 67085, France

M. F. K. Leung • Department of Medicine, University of Alberta, Edmonton, Alberta T6G 2G3, Canada

W. C. Leung • Department of Medicine, University of Alberta, Edmonton, Alberta T6G 2G3, Canada

C. Louq • Pasteur-Vaccins, 92430 Marnes-la-Coquette, France

D. I. Magrath • Division of Virology, National Institute of Biological Standards and Control, Potters Bar, Herts. EN6 3QG, England

E. K. Manavathu • Department of Medicine, University of Alberta, Edmonton, Alberta T6G 2G3, Canada

L. Mellander • Department of Clinical Immunology, Guldhedsgatan 10, S-413 46 Göteborg, Sweden; and Department of Pediatrics, East Hospital, S-9 41685 Göteborg, Sweden

Joseph L. Melnick • Department of Virology and Epidemiology, Baylor College of Medicine, Houston, Texas 77030

P. D. Minor • Division of Virology, National Institute of Biological Standards and Control, Potters Bar, Herts. EN6 3QG, England.

Bernard Moss • Laboratory of Viral Diseases, National Institute of Allergy and Infectious Diseases, National Institutes of Health, Bethesda, Maryland 20892

Jennifer Mumford • Equine Virology Unit, Animal Health Trust, Lanwades Park, Kennett, Newmarket, Suffolk CB8 7DW, England

Michael G. Murray • Department of Microbiology, School of Medicine, Health Sciences Center, State University of New York at Stony Brook, Stony Brook, New York 11794

A. R. Neurath • The Lindsley F. Kimball Research Institute of the New York Blood Center, New York, New York 10021

Akio Nomoto • Department of Microbiology, Faculty of Medicine, University of Tokyo, Hongo, Bunkyo-ku, Tokyo 113, Japan

K. Parker • Division of Biology, California Institute of Technology, Pasadena, California 91125

T. Ruuska • Department of Clinical Sciences, University of Tampere, SF-33520 Tampere, Finland

M. I. Sabara • Veterinary Infectious Disease Organization, Saskatoon, Saskatchewan S7N 0W0, Canada

G. C. Schild • National Institute for Biological Standards and Control, Potters Bar, Herts. EN6 3QG, England.

G. Schuster • Department of Biosciences, Karl-Marx-University of Leipzig, DDR 7010 Leipzig, German Democratic Republic

Valerie Seagroatt • National Institute for Biological Standards and Control, Potters Bar, Herts. EN6 3QG, England

Bert L. Semler • Department of Microbiology and Molecular Genetics, College of Medicine, University of California, Irvine, California 92717

Geoffrey L. Smith • Department of Pathology, University of Cambridge, Cambridge CB2 1QP, England

J. C. Soulié • Center of Perinatal Blood Biology, 75571 Paris, Cedex 12, France

N. Strick • The Lindsley F. Kimball Research Institute of The New York Blood Center, New York, New York 10021

K. Suryanarayana • Department of Medicine, University of Alberta, Edmonton, Alberta T6G 2G3, Canada

David Y. Thomas • Biotechnology Research Institute, National Research Council of Canada, Montreal, Quebec H4P 2R2, Canada

F. Tron • Pasteur-Vaccins, 92430 Marnes-la-Coquette, France

A. L. van Wezel† • National Institute for Public Health and Environmental Hygiene, Bilthoven, The Netherlands

Thierry Vernet • Biotechnology Research Institute, National Research Council of Canada, Montreal, Quebec H4P 2R2, Canada

†Deceased.

T. Vesikari • Department of Clinical Sciences, University of Tampere, SF-33520 Tampere, Finland

R. G. Webster • Department of Virology and Molecular Biology, St. Jude Children's Research Hospital, Memphis, Tennessee 38101

T. J. Wiktor† • The Wistar Institute, Philadelphia, Pennsylvania 19104

Eckard Wimmer • Department of Microbiology, School of Medicine, Health Sciences Center, State University of New York at Stony Brook, Stony Brook, New York 11794

J. M. Wood • Division of Virology, National Institute of Biological Standards and Control, Potters Bar, Herts. EN6 3QG, England

C. Yong Kang • Department of Microbiology and Immunology, University of Ottawa, School of Medicine, Ottawa, Ontario K1H 8M5, Canada

S. Zaman • Department of Social and Preventive Pediatrics, King Edward Medical College, Lahore, Pakistan

J. Zwaagstra • Department of Medicine, University of Alberta, Edmonton, Alberta T6G 2G3, Canada

†Deceased.

Preface to the Series

Viral diseases contribute significantly to human morbidity and mortality and cause severe economic losses by affecting livestock and crops in all countries. Even with the preventive measures taken in the United States, losses caused by viral diseases annually exceed billions of dollars. Five million people worldwide die every year from acute gastroenteritis, mainly of rotavirus origin, and more than one million children die annually from measles. In addition, rabies and viral hepatitis continue to be diseases of major public health concern in many countries of the Third World, where more than 200 million people are chronically infected with hepatitis B virus. The recent discovery of acquired immunodeficiency syndrome (AIDS), which is caused by a retrovirus, mobilized health services and enormous resources. This virus infection and its epidemic development clearly demonstrate the importance of applied virology research and the limits of our understanding of molecular mechanisms of viral pathogenicity and immunogenicity.

The limitations of our knowledge and understanding of viral diseases extend to the production of safe and reliable vaccines, particularly for genetically unstable viruses, and to antiviral chemotherapy. The number of antiviral drugs currently available is still rather limited, despite extensive research efforts. The main problem is finding compounds that selectively inhibit virus replication without producing toxic effects on cells. Indeed, the experimental efficacy of several drugs, for example, new nucleoside derivatives, some of which are analogues of acyclovir, makes it clear that antiviral chemotherapy must come of age because many new compounds show promise as antiviral agents. In the field of antiviral vaccine production, molecular biologists are using a wide variety of new techniques and tools, such as genetic engineering technology, to refine our understanding of molecular pathogenicity of viruses and of genetic sequences responsible for virulence. Identification of genes that induce virulence is vital to the construction of improved antiviral vaccines.

Novel types of vaccines are presently receiving particular attention. For example, the protein that carries the protective epitopes of hepatitis B virus, which is produced by expressing the appropriate viral gene in yeast or in mammalian cell systems, is now available.

Another group of new vaccines are produced by using vaccinia virus as a vector for the expression of genes of several viruses. Vaccines for rabies, influenza, respiratory syncytial disease, hepatitis B, herpes infection, and AIDS, which are based on greatly enhanced expression of the viral genes in vaccinia virus, are being tested. Also of interest is baculovirus, an insect cell vector system now used in the development of recombinant DNA vaccines for a variety of important human and animal virus diseases. This system yields very large quantities of properly processed and folded proteins from the rabies, hepatitis B, AIDS, and Epstein-Barr viruses.

The synthetic peptides, which act as specific immunogens, have also received attention as new antiviral vaccines. The recent experimental performance of new synthetic peptides of foot-and-mouth disease virus, as well as peptide-based vaccines for poliovirus, rotavirus, hepatitis B, and Venezuelan equine encephalitis virus, gives strong support for this group of specific immunogens. However, testing of these synthetic peptide vaccines is in the early stages and future research will have to answer several questions about their safety, efficacy, and immune responses.

Current attempts at developing synthetic vaccines are based either on recombinant DNA technology or on chemical-peptide synthesis. Several virus proteins have been produced in bacterial, yeast, or animal cells through the use of recombinant DNA technology, while live vaccines have been produced by introducing relevant genes into the genome of vaccinia virus. By using solid-phase peptide synthesis, it has been possible to obtain peptides that mimic the antigenic determinants of viral proteins, that elicit a protective immunity against several viruses. Both the chemical and the recombinant approaches have led to the development of experimental vaccines. It should become clear within a few years which approach will lead to vaccines superior to the ones in use today.

The recent development of monoclonal antibody production techniques and enzyme immunoassays permits their application in virology research, diagnosis of viral diseases, and vaccine assessment and standardization. These techniques are useful at different stages in the development of vaccines, mainly in the antigenic characterization of infectious agents with monoclonal antibodies, in assessment schemes in research and clinical assays, and in production.

This new series, entitled *Applied Virology Research,* is intended to promote the publication of overviews on new virology research data, which will include within their scope such subjects as vaccine production, antiviral chemotherapy, diagnosis kits, reagent production, and instrumentation for automation interfaced with computers for rapid and accurate data processing.

We sincerely hope that *Applied Virology Research* will serve a large audience of virologists, immunologists, geneticists, biochemists, chemists, and molecular biologists, as well as specialists of vaccine production and experts of health services involved in the control and treatment of viral diseases of plants, animals, and man. This series will also be of interest to all diagnostic laboratories, specialists, and physicians dealing with infectious diseases.

Edouard Kurstak

Montreal, Canada

Preface to Volume 1

In the first volume of *Applied Virology Research,* particular emphasis is given to the development of new vaccines and antiviral chemical compounds. Among the timely research areas presented in this volume are (1) the development of a new generation of virus vaccines and their immunogenicity; (2) vaccination with synthetic peptides; (3) quality control, potency, and standardization of viral vaccines; and (4) antiviral chemotherapy.

Three novel types of vaccines to control viral diseases are given special attention. The first group of new vaccines is based on genetically engineered viral proteins. Some of these proteins are ready for use as human and animal vaccines. For example, proteins carrying the protective epitopes of hepatitis B virus, produced by expressing the appropriate viral gene in mammalian or other cell systems, may prove to be a cost-effective source of vaccine. Similar approaches may lead to the production of viral proteins for vaccines against the acquired immunodeficiency syndrome (AIDS), herpes simplex virus, poliovirus, rotaviruses, and foot-and-mouth disease, among others. Of interest is the eukaryotic expression system, the baculovirus-driven production of proteins from other viruses in insect cells or tissues. This system has yielded very large quantities of properly processed and folded proteins from hepatitis B and Epstein–Barr viruses.

The second group of new vaccines described in this volume are those dependent on the use of vaccinia virus as a vector for the expression of genes of several viruses at once. New vaccines for rabies, hepatitis B, influenza, respiratory syncytial disease, and human immunodeficiency virus that are based on greatly enhanced expression of the viral genes in vaccinia virus are being investigated and tested.

The relative ease with which immunogenic viral proteins can now be amplified or expressed by cloning in biologic vectors is obviously of major importance for the control of viral diseases. Many of the problems previously encountered in the development of cloned vaccine immunogens have now been overcome through the use of expression systems that appear to mimic the normal host processing of viral glycoproteins.

The introduction to the research laboratory of automated peptide synthesizers now permits the rapid production of synthetic peptides based on the known or predicted sequence of antigenic domains on the surface of neutralizing antibody-inducing viral components. Four chapters are devoted to the third group of vaccines—those based on such synthetic peptides. A number of these peptides have been prepared from a variety of viruses, but their efficacious use in vaccines has, unfortunately, not progressed as rapidly. Good levels of neutralizing antibodies can be obtained after synthetic peptide products are administered, thus leading to protection of the host against challenges by virulent viruses. Most testing of these synthetic peptide vaccines is in the early *in vitro* stages, although vaccines for foot-and-mouth disease, polio, and hepatitis B virus have recently begun to be tested *in vivo*. Ongoing research promises to answer detailed questions regarding efficacy, safety, dosage, proper use of adjuvants, induction of cell-mediated immunity, and practical delivery vehicles for delayed booster immunization. Several chapters also describe recent developments in the quality-control testing of virus vaccines and in the methodology for potency and stability assays.

Research directed toward the chemotherapeutic control of virus infections has recently undergone a resurgence, largely due to the search for a chemical agent to be used in the treatment of AIDS, such as the recently approved azidothymidine. Interferons and several antiviral drugs of proven efficacy, i.e., idoxuridine, trifluridine, amantadine, rimantadine, vidarabine, and acyclovir, have already been used in the treatment of influenza A, varicella zoster, herpes simplex, respiratory syncytial virus, and arenavirus infections. The clinical potential of such antiviral chemical agents is recognized for the chemotherapy or prophylaxis of numerous virus diseases.

This volume also includes detailed descriptions of recent developments in the use of nucleoside analogues in the chemotherapy of virus diseases and of new analogues to acyclovir with increased potency to herpes viruses. Particular attention is given to antiphytoviral substances and to the chemotherapy of plant virus infections, as well as to carrier-mediated drug-delivery systems for controlled release and targeting of chemical compounds to the affected cells.

The contributors to this volume are leading experts in viral vaccines, biotechnology, and antiviral chemotherapy. The editors wish to express their sincere gratitude to all of them for the effort and care with which they prepared their manuscripts. We are convinced that this series—which is designed for all research workers involved in the production of viral vaccines and antiviral chemical compounds, for medical and veterinary clinicians, and for plant pathologists—will be a very useful tool for all those concerned with the control of viral diseases.

The editors' thanks are also addressed to the staff of Plenum Publishing Corporation for their efforts in the production of this series.

E. Kurstak
Montreal, Quebec, Canada
R. G. Marusyk
Edmonton, Alberta, Canada
F. A. Murphy
Atlanta, Georgia
M. H. V. Van Regenmortel
Strasbourg, France

Contents

IV. ANTIVIRAL CHEMOTHERAPY

Introduction

Viral Vaccines

Joseph L. Melnick

As with history in general, the history of vaccines needs to be reexamined and updated. My task is to look back to see what has been successful and to look forward to see what remains to be accomplished in the prevention of viral diseases by vaccines. Also, I shall refer to the pertinent material discussed at two recent conferences of the Institute of Medicine, National Academy of Sciences, on virus vaccines under development and their target populations in the United States (1985*b*) and in developing countries (1986). These reports, plus a third on Vaccine Supply and Innovation (1985*a*), should be required reading for all those in both the public and the private sector who have a responsibility or interest in vaccines for the prevention of human disease.

It has been through the development and use of vaccines that many viral diseases have been brought under control. The vaccines consist either of infectious living attenuated viruses or of noninfectious killed viruses or subviral antigens. When we look at the record, it is the live vaccines that have given the great successes in controlling diseases around the world. Examples are smallpox, yellow fever, poliomyelitis, measles, mumps, and rubella.

The historic landmarks in the development of vaccines for virus diseases of humans are shown in Table 1. One of the greatest triumphs of mankind has been the purposeful eradication of smallpox from the world's population. Almost 200 years ago, Jenner noted that cowpox conferred immunity against smallpox. He inoculated material from the cowpox lesions as a means of protecting persons against smallpox, and the first successful vaccine came into use. Jenner showed that a mild disease produced by design with one virus could protect against more virulent disease caused by a related agent. Even before Jenner's time, it was known that some smallpox cases were mild, and material from the lesions of such patients was used to protect others against the severe form of smallpox, but the risk of this procedure was high. Other dramatic results of widespread vaccination can be seen in the rapid decreases, in the United States, of cases of four childhood viral diseases: poliomyelitis, measles, rubella, and mumps (Fig. 1).

The viral vaccines now in use are listed in Table 2: those recommended for the

Joseph L. Melnick • Department of Virology and Epidemiology, Baylor College of Medicine, Houston, Texas 77030.

Table 1. Virus Vaccines

Year	Immunization introduced
1721	Variolation
1798	Smallpox attenuated
1885	Rabies attenuated and inactivated
1936	Yellow fever attenuated
1940s	Influenza inactivated
1954	Poliomyelitis inactivated
1961	Poliomyelitis attenuated
1960s	Measles (inactivated) attenuated
	Mumps attenuated
	Rubella attenuated
1982	Hepatitis B subviral particles

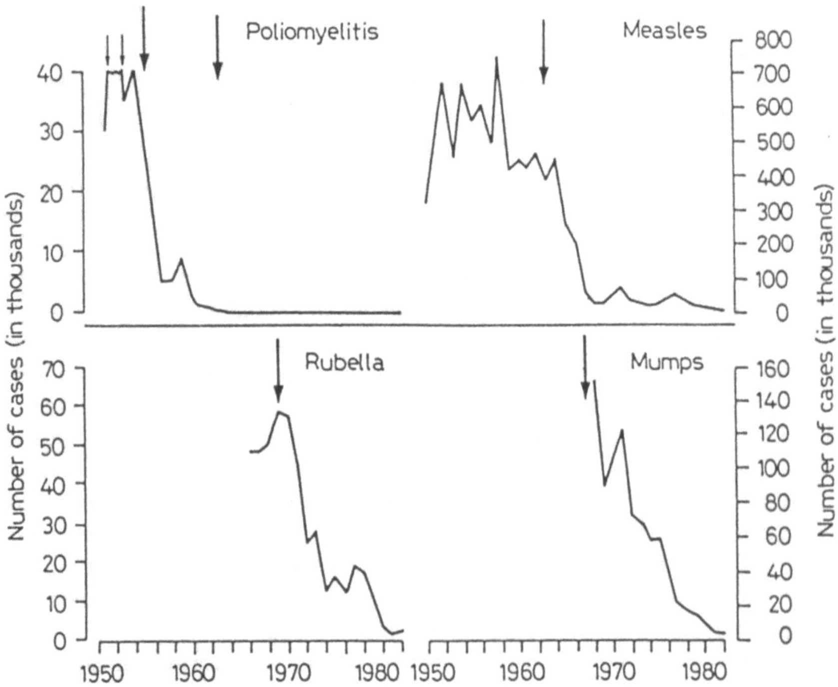

Figure 1. Record of the decrease in cases of poliomyelitis, rubella, measles, and mumps in the United States following widespread administration of vaccines. Numbers of cases are indicated in thousands. (Note that the scale differs in the different sectors of the illustration.)

Table 2. Principal Vaccines Used in Prevention of Virus Diseases of Humans

Disease	Source of vaccine	Condition of virus	Route of administration
	Recommended immunization for general public		
Poliomyelitis[a]	Tissue culture (human diploid cell line, monkey kidney)	Live attenuated	Oral
		Killed	Subcutaneous
Measles[a]	Tissue culture (chick embryo)	Live attenuated[b]	Subcutaneous[c]
Mumps[a]	Tissue culture (chick embryo)	Live attenuated	Subcutaneous
Rubella[a,d]	Tissue culture (duck embryo, rabbit, or human diploid)	Live attenuated	Subcutaneous
	Immunization recommended only under certain conditions (epidemics, exposure, travel, military)		
Smallpox[e]	Lymph from calf or sheep (glycerolated, lyophilized) Chorioallantois, tissue cultures (lyophilized)	Live vaccinia	Intradermal (multiple pressure, multiple puncture)
Yellow fever	Tissue cultures and eggs (17D strain)	Live attenuated	Subcutaneous or intradermal
Hepatitis type B	Purified HB_sAg from ''healthy'' carriers	Killed	Subcutaneous
	Purified immunogen from yeast expressing a recombinant HB_sAg gene	Viral protein	Subcutaneous
Influenza	Highly purified or subunit forms of chick embryo allantoic fluid (formalinized or UV-irradiated)	Killed	Subcutaneous or intradermal
Rabies	Duck embryo or human diploid cells	Killed	Subcutaneous
Adenovirus[f]	Human diploid cell cultures	Live attenuated	Oral, by enteric-coated capsule
Japanese B encephalitis[g]	Mouse brain (formalinized), tissue culture	Killed	Subcutaneous

[a] Available also as combined vaccines.

[b] Killed measles vaccine was available for a short period. However, a serious hypersensitivity reaction often occurs when children who have received primary immunization with killed measles vaccine are later exposed to live measles virus. Because of this complication, killed measles vaccine is no longer recommended.

[c] With less attenuated strains, immune globulin USP is given in another limb at the time of vaccination.

[d] Neither monovalent rubella vaccine nor combination vaccines incorporating rubella virus should be administered to a postpubertal susceptible woman unless she is not pregnant and understands that it is imperative not to become pregnant for at least 3 months after vaccination. (The time immediately postpartum has been suggested as a safe period for vaccination.)

[e] Since smallpox virus seems to have been totally eradicated from the world, vaccination is no longer recommended. However, stocks of vaccine are held in depots for use in the event that cases should reappear.

[f] Recently licensed but recommended only for military populations in which epidemic respiratory disease caused by adenovirus is a frequent occurrence.

[g] Not available in the United States except for the Armed Forces or for investigative purposes.

general public and those recommended only for special populations. However, some general principles apply to most virus vaccines used in the prevention of human disease. Vaccination (or recovery from natural infection) does not always result in total immunity against a subsequent exposure to the wild virus. This situation holds true for diseases for which successful control measures are available, including polio, smallpox, influenza, rubella, measles, mumps, and adenovirus infections. But the replication of the wild virus, if it occurs, is markedly restricted. Disease can be controlled by vaccination even if the vaccine-induced resistance only limits the multiplication of virulent virus, thus preventing its spread to target organs where the pathologic damage is done (e.g., polio and measles viruses prevented from invading the brain and spinal cord; rubella virus prevented from infecting the embryo).

Killed-virus vaccines prepared from whole virions generally stimulate the development of circulating antibody against the coat proteins of the virus, conferring some degree of resistance. For some diseases, killed-virus vaccines are currently the only ones licensed, but there are some caveats associated with them:

1. Since virulent strains often have been the ones chosen as killed-virus vaccine strains, extreme care is required in the manufacture of these vaccines to make certain that no residual live virulent virus is present in the vaccine.
2. The immunity conferred is often brief and must be boosted. It not only involves the logistic problem of repeatedly reaching the persons in need of immunization but has also caused concern about the possible effects (hypersensitivity reactions) of repeated administration of foreign proteins in the vaccine.
3. Some killed-virus vaccines (such as the early measles vaccine) have induced hypersensitivity to subsequent infection by wild virus.
4. Sometimes protection conferred by parenteral administration of killed-virus vaccine is limited. In these instances, even when the vaccine stimulates circulating antibody (IgM, IgG) to satisfactory levels, local resistance (IgA) is not induced adequately at the natural portal of entry or the primary site of multiplication of the wild virus infection (e.g., nasopharynx for respiratory viruses, alimentary tract for poliovirus) (Fig. 2).

Attenuated live-virus vaccines have the advantage of behaving like the natural infection in their effect on immunity. They multiply in the host and tend to stimulate production of longer-lasting humoral antibody and also to induce cellular immunity and resistance at the portal of entry (Fig. 2). However, some disadvantages are associated with live attenuated vaccines.

1. There may be a risk that the vaccine virus could revert to greater virulence during its multiplication in the vaccinee. Although reversion of licensed vaccines has not proved a major problem in practice, its potential exists. Monitoring of vaccine programs should be required.
2. Unrecognized adventitious agents latently infecting the culture substrate (eggs, primary cell cultures) may enter the vaccine stocks. Viruses found in vaccines have included avian leukosis virus, simian papovavirus SV40, and simian

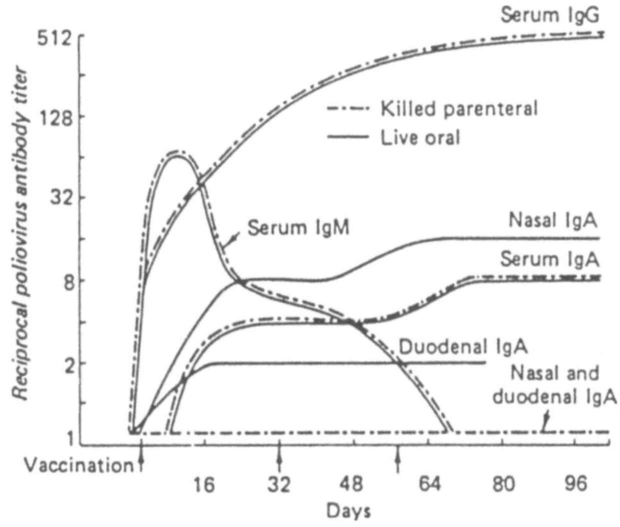

Figure 2. Serum and secretory antibody response to orally administered live attenuated poliovaccine and to intramuscular inoculation of killed poliovaccine. (From Ogra *et al.*, 1980.)

cytomegalovirus. The problem of adventitious contaminants may be circumvented through the use of normal cells serially propagated in culture (e.g., human diploid cell lines) as substrates for cultivation of vaccine viruses. Vaccines prepared in such cultures have been in use for years and have been safely administered to many millions of persons.

3. The storage constraints and limited shelf-life of live attenuated vaccines present problems, but these can be overcome in some cases by extraordinary efforts to maintain the cold chain of refrigeration, even under difficult field conditions, and also, more advantageously, by the use of viral stabilizers (e.g., molar $MgCl_2$ for poliovaccine).

Until recently, virus strains suitable for live-virus vaccines were developed chiefly by selecting naturally attenuated strains or by cultivating the virus serially in various hosts and cultures in the hope of deriving an attenuated strain. The search for such strains is now being approached by laboratory manipulations aimed at specific genetic alterations in the virus. These include the development of host-range mutants, temperature-sensitive mutants, cold-adapted mutants, deletion mutants, reassortants, and genetic recombinants.

A theoretical possibility has been raised that antibody responses might be diminished or that interference might occur if two or more live-virus vaccines were given at the same time. In practice, however, simultaneous administration of live vaccines can be safe and effective. Trivalent live oral poliovaccine (when given in three doses), and the combined live measles, mumps, and rubella vaccine, given by injection, are effective. Antibody response to each component of these combination vaccines is comparable to the antibody response elicited by the individual vaccine given separately.

One fact cannot be overemphasized: an effective vaccine does not protect against disease until it is administered at the proper time and in the proper dosage to susceptible individuals. The failure to reach all children with complete courses of immunization is reflected in many countries by the continued occurrence of paralytic poliomyelitis and measles in those who have not been vaccinated, or have been incompletely vaccinated.

Several new approaches to vaccines are promising. Efforts are under way to provide for administration of vaccine at the portal of entry. Intranasally administered aerosol vaccines are being developed, particularly for respiratory disease viruses and for measles virus, to stimulate local antibody.

Considerable success has been associated with a new live varicella vaccine. However, some concern has been raised about the possibility of vaccinated subjects contracting zoster in later life. Vaccines are under development for cytomegalovirus (CMV), Epstein–Barr virus (EBV), and herpes simplex virus (HSV), but, as with the varicella–zoster virus (VZV) vaccine, there is concern about the long-term effects. Attenuated vaccines are also under development for rotaviral gastroenteritis, viral hepatitis A, dengue, and respiratory syncytial disease.

Purification is being used more widely to eliminate nonviral proteins and thus reduce the possibility of adverse reactions to the vaccine. In some instances, purified material can also be administered in more concentrated form, thus containing greatly increased amounts of the specifically desired antigen. An improved inactivated poliovirus vaccine that has been developed and used for several years in the Netherlands is becoming available to other countries.

Subviral components are being obtained by breaking the virion apart so as to include in the vaccine only those components needed to stimulate protective antibody. The current influenza vaccines in the United States are made in this way. Presenting the subunit vaccine in micelles has been found to greatly enhance the immunogenic response to a hepatitis B polypeptide vaccine, not only in experimental animals but in humans as well.

Genetic manipulation is being utilized to produce recombinants or temperature-sensitive mutants that can then serve as attenuated live-virus vaccines. The use of vaccinia virus as a vector for cloned immunizing genes of other viruses is an important new development. This method may have wide application in developing countries because the vaccine is easily administered, and needed vaccines such as that for hepatitis B—a vaccine that has heretofore been expensive and in short supply—should now be available at low cost. It is possible to insert genes for immunogens of several viruses, so that one vaccination could simultaneously protect against a number of diseases. However, many persons are already immune to vaccinia, and vaccinia virus is not totally risk-free. It should be kept in mind that before smallpox was eradicated worldwide, some nations, including the United States, discontinued general vaccination against smallpox when the risk of imported disease—if suitable precautions were continued with travelers—became less than the risk of serious vaccinia-induced complications of vaccination. The live oral adenovaccine might prove an alternative vector. The possibilities seem endless, as witness the recent experiments in which the gene for hepatitis B surface antigen (HB_sAg) has been inserted into HSV and is readily expressed in cells in which the herpes virus replicates. This approach may prove particularly useful in obtaining gene products of highly virulent viruses that are extraordinarily dangerous to handle in the laboratory. The glycoprotein antigens of human immunodeficiency virus have recently been expressed after insertion of the corresponding viral gene into vaccinia virus.

Other developments also are related to recombinant DNA. The genetic material coding for hepatitis B virus immunizing antigen has been introduced into bacteria, yeast, and continuous mammalian cell lines genetically engineered to yield the antigen in quantity. This has led to the licensure of the first vaccine manufactured by recombinant DNA technology—hepatitis B vaccine prepared from yeast.

Mammalian cell lines have important advantages as substrates for production of vaccines made by means of recombinant DNA. Since these cells are the natural hosts of mammalian viruses, the viral proteins would be expected to be folded into the proper conformation, otherwise processed and modified in a manner similar to that of the naturally occurring antigen, and more likely to yield the native antigenic structure of the viral protein. In turn, purification should be more rapid and more economical. HB_sAg produced by recombinant DNA in mammalian cells proved more potent (in tests in mice and chimpanzees) than corresponding material prepared from serum or yeast. Another subunit vaccine (HSV glycoprotein D) produced in mammalian cells has shown excellent immunogenicity and efficacy in experimental animals.

For foot-and-mouth disease virus, another step was necessary. Since the virus has an RNA genome, complementary DNA (cDNA) first had to be prepared, for insertion into the bacterial DNA. Immunizing antigen has been obtained from bacteria containing this recombinant DNA. Similar approaches are being made with cDNA clones of different segments of the rotavirus genome.

The possibility of even more precise tailoring of vaccines to have the desired properties, and only those properties, is envisioned in the field of synthetic vaccines. Antigenically active polypeptides have been synthesized for hepatitis B, influenza, and polioviruses. They have induced neutralizing antibodies in animals. The possibility of producing synthetic immunizing antigens for human vaccination is being actively explored. However, the synthetic peptides have been weak antigens, and a search for potent and safe adjuvants continues in several laboratories. A synthetic vaccine containing three different antigens of foot-and-mouth disease virus has been highly immunogenic in small animals and is being tested in cattle.

Synthetic peptides may play another role in immunology. Some peptides have failed to induce neutralizing antibodies in experimental animals but have apparently primed the animals' immune systems so that a single challenge dose of whole virus produced a powerful antibody response. Synthetic vaccine may be made more antigenic in a synergistic fashion if two epitopes of the viral coat protein are combined into a single peptide.

One problem with synthetic peptides is that a number of antigenic determinants are not continuous sets of amino acids, but rather are amino acids brought together by folding of the proteins. Such sequences cannot be readily mimicked by synthetic peptides. It seems that in some instances conformation may be even more important than amino acid composition and sequence as far as antigenic determinants are concerned. Cyclized peptides have proved more immunogenic than their linear counterparts. Conformation must be the reason why anti-idiotype antibodies are able to (1) enhance antibody formation upon subsequent injection of a subimmunogenic dose of the natural antigen, and (2) even induce an antibody response (albeit often a weak one) by themselves.

Potential viral vaccines that, according to the Institute of Medicine (1985*b*, 1986), may reasonably be expected to become available in the near future are listed in Tables 3 and 4, for the United States and for developing countries, respectively. The type of vaccine under development for each virus disease is listed, as is the corresponding target

Table 3. Potential Virus Vaccines for the United States: 1985[a]

Pathogen	Vaccine	Target population	Persons entering target population each year
Influenza viruses A and B	Subunit vaccine (purified hemagglutinin/neuraminidase proteins)	High-risk population	48 million
	Attenuated live virus	High-risk population	
Parainfluenza viruses	Trivalent, subunit vaccine (must contain fusion protein)	Infants	Birth cohort (3.8 million)
Respiratory syncytial virus	Glycoprotein produced by recombinant DNA technology	Infants	Birth cohort (3.8 million)
Hepatitis A virus	Attenuated live virus	Susceptibles of all ages (routine for children)	Birth cohort (3.8 million)
	Subunit	Susceptibles of all ages (routine for children)	
Hepatitis B virus	Glycoprotein produced by recombinant DNA technology	High-risk groups (e.g., health professionals, homosexuals, IV drug users)	350,000
Herpes simplex viruses 1 and 2	Glycoprotein produced by recombinant DNA technology	Children up to age 12 and older susceptibles	Birth cohort (3.8 million)
	Attenuated live virus	Children up to age 12 and older susceptibles	
Varicella virus	Attenuated live virus	Recipients of organ and bone marrow transplants and persons under age 25 with lymphomas and leukemias	13,000
		Normal susceptibles, routine for children (booster for adults)	Birth cohort (3.8 million)
Cytomegalovirus	Attenuated live virus	Seronegative (SN) recipients of bone marrow and organ transplants and SN persons with leukemias and lymphomas	Approx. 1500
		Nonpregnant adolescent females	Females, 14 years of age (1.8 million)
	Glycoprotein produced by recombinant DNA technology	All children	Birth cohort (3.8 million)
Rotavirus	Attenuated live bovine virus	Infants	Birth cohort (3.8 million)
	Attenuated live human or reassortant virus	Infants	

[a] From the Institute of Medicine, National Academy of Sciences (1985*b*).

Table 4. Potential Virus Vaccines for Developing Countries: 1986[a]

Pathogen	Vaccine	Target population	Persons entering target population each year
Parainfluenza viruses	Trivalent, subunit vaccine (which must contain fusion proteins)	Infants	115 million
Respiratory syncytial virus	Polypeptides produced by recombinant DNA technology	Infants at earliest possible age	115 million
	Attenuated live virus	Infants at earliest possible age	
Hepatitis A virus	Attenuated live virus	Susceptibles of all ages; routine for preschool children	115 million
	Polypeptide recombinant vaccine produced in yeast	Susceptibles of all ages; routine for preschool children	
Hepatitis B virus	Polypeptide produced by recombinant DNA technology	Areas with high perinatal infection; all infants at birth (if possible). Other areas: all infants, simultaneous with other vaccinations, at earliest possible age	115 million
Dengue virus	Attenuated live vector virus containing gene for broadly cross-reacting protective antigen	Infants and children in endemic areas; travelers to endemic areas	48 million
Japanese encephalitis virus	Inactivated virus produced in cell culture	Children in epidemic and endemic areas; foreign visitors to epidemic regions	65 million
Yellow fever virus	Attenuated live virus produced in cell culture	Young children	25 million
Rabies virus	Vero cell-derived vaccine	Individuals at high risk, plus postexposure prophylaxis	5M (postexposure)
	Glycoprotein produced by rDNA technology in mammalian cells	Individuals at high risk, plus postexposure prophylaxis	5M (postexposure)
	Attenuated live vector virus containing gene for protective glycoprotein antigen	Birth cohort in areas of high risk	53 million
Rotavirus	Attenuated live bovine rotavirus	Infants at earliest possible age	115 million
	Rhesus monkey rotavirus	Infants at earliest possible age	
	Attenuated live human or reassortant virus	Infants at earliest possible age	

[a] From the Institute of Medicine, National Academy of Sciences (1986).

population. Also listed for each vaccine is the number of persons entering the target population each year.

Local recommendations for the administration of vaccines should be made for specific circumstances. For example, Table 2 refers to currently available vaccines recommended for the general population of the United States, including pediatric immunizations; also presented are the vaccines used in the United States only for persons in certain special circumstances, such as those at risk because of epidemics and special occupational and other opportunities for exposure (e.g., travel to endemic areas or living in large groups such as military recruits, in whom adenovirus respiratory disease epidemics occur frequently). Table 5 summarizes recommendations of the Advisory Committee on Immunization Practices (ACIP) for adult immunizations, published by the Centers for Disease Control (CDC), for those in special occupations, life-styles, and environmental circumstances in the United States. Table 6 lists the recommendations for travelers leaving the United States for other parts of the world and for foreign students and immigrants entering the United States.

Even when effective vaccines are available, they are not effective until they reach the target population of susceptible persons. A quotation from a recent issue of *The Lancet* deserves serious attention (Editorial, 1985). If present conditions persist, "of every 1000 children born into the world, 5 will grow up crippled by poliomyelitis, 10 die of neonatal tetanus, 20 die of whooping-cough, and 30 or more die of measles or its complications." The total of preventable childhood deaths worldwide has been estimated at five million per year. To meet these problems, the World Health Assembly proposed that immunization against the six target diseases be extended to all the world's children. This proposal led to the establishment of the WHO Expanded Programme on Immunization (EPI). There have been impressive successes but, as shown in Tables 7 and 8, much remains to be accomplished.

Table 7 indicates that as of 1985 there was wide variation in vaccine coverage among

Table 5. Vaccines Recommended for Special Occupations, Life-styles, and Environmental Circumstances

Indication	Vaccine
Occupation	
Hospital, laboratory, and other health care personnel	Hepatitis B
	Polio
	Influenza
Staff of institutions for the mentally retarded	Hepatitis B
Veterinarians and animal handlers	Rabies
Selected field workers	Plague
Life-styles	
Homosexual males	Hepatitis B
Illicit drug users	Hepatitis B
Environmental situation	
Inmates of long-term correctional facilities	Hepatitis B
Residents of institutions for the mentally retarded	Hepatitis B
High-school and college students who as children were unvaccinated or who received faulty vaccine or dosage schedules now recognized as suboptimal	Measles Rubella

Table 6. Vaccines Recommended for Travelers, Foreign Students, and Immigrants

Indication	Vaccine
Travel	Measles
	Rubella
	Polio
	Yellow fever
	Hepatitis B
	Rabies
	Meningococcal polysaccharide
	Typhoid
	Cholera
	Plague
	Immunoglobulin
Foreign students, immigrants	Measles
	Rubella
	Diphtheria
	Tetanus

the developing countries participating in the Expanded Programme on Immunization. In 1974 when the Programme was launched, WHO estimated that, in developing countries, the percentage of infants under 1 year of age who had received DPT (diphtheria, pertussis, and tetanus) and poliomyelitis vaccines was less than 5%. As of 1985, 36% of the children in the developing countries (excluding China) receive the full course of three doses of oral polio vaccine, and 21% receive measles vaccine. For China, 78% receive oral polio vaccine and 74% measles vaccine (see Table 8). WHO estimates that immunization against the six childhood diseases shown in Tables 7 and 8 is now saving the lives of about 800,000 children in the developing countries each year. But there is ample room for improvement, as summarized in Table 8, which includes developed countries as well. As of 1985, more than 250,000 cases of paralytic poliomyelitis and more than 3 million deaths caused by measles, pertussis, and neonatal tetanus continue to occur annually.

The conditions under which immunizations are conducted vary geographically. Whereas in some of the developed industrialized countries vaccination of infants can be safely postponed until the ideal age for the maximum immune response, waiting until the maternal antibodies and immaturity of the immune system no longer limit the vaccine's effectiveness, in other parts of the world infants are heavily exposed to disease agents almost from the moment of birth. In areas in which the EPI target diseases cause numerous deaths in the first year of life, a new recommendation is that poliovaccine in these countries should now be given at birth. Although the serologic response is less than it is in later age groups, nonetheless 70% or more of the infants acquire local immunity in the intestinal tract, 30–50% also acquire antibodies against at least one poliovirus type, and others become immunologically primed. Where there is early and heavy exposure, DPT could also advantageously be offered earlier in life. A suggested immunization schedule for these circumstances is (1) at birth, trivalent oral polio vaccine and bacille Calmette-Guérin (BCG); (2) at 6, 10, and 14 weeks, trivalent oral poliovaccine and DPT; and (3) at 9 months, measles vaccine.

Table 7. Vaccination Coverage (Percentage Vaccinated, among Children under 1 Year of Age) in Developing Countries Ranked by Surviving Infants: 1985

Country	Infants <1 year old (in millions)	Vaccination coverage			
		BCG	DPT (3 doses)	Polio (3 doses)	Measles
India	21.74	65	51	37	*
Indonesia	4.59	56	6	7	7
Nigeria	4.11	*	*	*	*
Pakistan	3.46	73	64	65	80
Bangladesh	2.95	2	2	1	1
Brazil	2.83	79	67	99	80
Mexico	2.46	24	26	91	30
Iran	2.02	10	68	65	69
Viet Nam	1.66	5	5	2	4
Philippines	1.56	76	61	58	30
Egypt	1.54	53	57	67	41
Ethiopia	1.45	16	9	9	16
Turkey	1.40	65	56	59	30
Zaire	1.29	34	16	18	20
Burma	1.29	24	8	1	*
South Africa	1.11	*	*	*	*
Thailand	1.05	81	57	56	7
Kenya	0.98	*	*	*	*
Tanzania	0.95	84	9	56	82
South Korea	0.92	84	69	78	*
Morocco	0.91	70	48	48	42
Colombia	0.86	72	41	60	42
Sudan	0.85	7	4	4	3
Algeria	0.81	59	33	30	17
Argentina	0.70	64	65	64	62

* Data not available.

Unfortunately, one of the reasons children remain unprotected against the six EPI target diseases is that the vaccinating teams are reluctant to vaccinate children who have even mild malaise. It needs to be emphasized that low-grade fever, mild respiratory infection, diarrhea, or other minor illnesses should not be regarded as contraindications to immunization. Because the risk of encountering the wild virus is so great, and the probability of the child not being brought back for immunization at another time is also great, health service workers are being urged not to withhold vaccine except for very serious illness. Thus, if a child is not so ill that hospitalization is required, the vaccine should be administered.

BENEFIT–COST CONSIDERATIONS FOR VIRAL VACCINES

It is important to note that once a country has eliminated indigenous cases of a viral disease it must still maintain careful surveillance and containment measures and appropriate vaccination programs until the disease is completely controlled everywhere or until the agent is eradicated from the world. In the United States, for example, the last known case of smallpox occurred in 1949, but barriers against importation had to be maintained for more than 25 years, until global eradication was accomplished. Even apart from the elimination of human suffering and death that the disease had entailed in many parts of the world, and despite the known risks of the vaccine itself in a limited proportion of vaccinees, the eradication program yielded remarkable benefits in purely economic terms. It has been estimated that $200 million, split evenly between international and national sources, was invested in the worldwide smallpox eradication program over a 13-year period. Five times this amount is now saved annually by the nations of the world because protective activities have been discontinued. For the United States, the total amount ($489,782,640) contributed to WHO's regular budget between 1948 and 1979—when WHO certified that smallpox had been eradicated—was regained in just 3 years.

The case for the economic benefits of controlling and ultimately eliminating paralytic poliomyelitis is well known. With regard to measles, it has been estimated that for the period 1963–1972 in the United States, an annual net savings of $130 million was achieved by vaccination. Currently, the annual saving is about $500 million. This means a benefit–cost ratio of 10 : 1. In the developing world, with the higher rates of disease and death from measles, the returns are even greater, in some instances 20 : 1 or higher. The combined savings in the United States as a result of the use of vaccines against polio, measles, rubella, mumps, and hepatitis total more than $1 billion annually.

The so-called high-technology vaccines might at first glance be considered irrelevant to those involved in health problems of developing countries in which the entire health system may labor under great shortage of funds and personnel. "High technology" should ultimately result not only in better vaccines, but also in vaccines that are far less expensive to produce and simpler to distribute and administer. They should induce long-

Table 8. Vaccination Coverage (Percentage Vaccinated, among Children Under 1 Year of Age) in Developing and Developed Countries: 1985

Group or country	Infants <1 year old (millions)	Vaccination coverage[a]			
		BCG	DPT (3 doses)	Polio (3 doses)	Measles
Developing countries					
25 large	63.49	49	37	37	19
Other (small)	17.36	43	33	32	28
Subtotal					
(excl. China)	80.85	48	36	36	21
China	19.23	50	63	78	74
Total in developing countries	100.08	48	41	44	33
Total in developed countries	17.31	54	62	66	73
Global total	117.39	49	45	47	39

[a] The price per dose of vaccine used in the Expanded Programme of Immunization was as follows: bacille Calmette-Guérin (BCG), $0.055; diphtheria, pertussis, and tetanus (DPT), $0.018; oral polio vaccine (OPV), $0.017; and measles, $0.069.

lasting immunity with relatively few doses of vaccine. Thus, although the early steps toward some of the new vaccines may be expensive, it is anticipated that they will soon become cost effective, both in terms of disease prevention and as regards costs of administration. It is for these reasons that the WHO has taken a leading role in sharing new information and resources on the development, production, and use of vaccines.

One of the truly dangerous problems hampering the improvement of current vaccines and the development of new ones is that of liability. Obviously some compensation should be awarded to those who, in the course of a vaccine program that greatly benefits the population generally, are themselves harmed by the rare adverse reactions to the vaccine. In effect, these few persons, in their own vaccine-induced problems, pay the cost for the vaccine's benefits to the society as a whole. But some of the awards made in the courts are shocking in their size and rationale: one manufacturer was ordered to pay a victim of a vaccine $10 million, of which $8 million was punitive damages. A number of manufacturers have already abandoned the vaccine field, at least in part for this reason, thereby limiting the supply of vaccines. In the United States, there is now only one manufacturer of measles vaccine, one of oral polio vaccine, one of hepatitis B vaccine, one of vaccines for mumps and rubella, and one of pertussis vaccine. Even the purchase and use in the United States of vaccines prepared in other countries is hampered by reluctance of foreign manufacturers to become involved in potential liability questions.

Progress in the development of a live-virus measles vaccine to be administered as an aerosol by the respiratory route is likely to be slow, for reasons unrelated to the merits or drawbacks of the vaccine itself. A problem to be considered is the probability that a manufacturer would be virtually swamped in lawsuits. One can only imagine how much "guilt by temporal association" and costly litigation would beset manufacturers and health service providers if a vaccine were given by the respiratory route and an unrelated serious respiratory disease occurred within the next few weeks.

The U.S. administrative and legislative branches of government are beginning to come to grips with the problem, whereas in some other nations legislation dealing with assistance to those experiencing vaccine damage is already in place. In most cases, the solutions have involved some form of governmental compensation, with some definitions of the compensation for various levels of injury.

REFERENCES

Editorial (1985). *Lancet* **1,** 438–439 (Feb. 23).

Institute of Medicine, National Academy of Sciences (1985*a*). *Vaccine Supply and Innovation.* National Academy Press, Washington, D.C.

Institute of Medicine, National Academy of Sciences (1985*b*). *New Vaccine Development: Establishing Priorities. Vol. 1: Diseases of Importance in the United States,* National Academy Press, Washington, D.C.

Institute of Medicine, National Academy of Sciences (1986). *New Vaccine Development: Establishing Priorities. Vol. 2: Diseases of Importance in the Developing World,* National Academy Press, Washington, D.C.

Ogra, P. L., Fishaut, M., and Gallagher, M. R. (1980). *Rev. Infect. Dis.* **2,** 352–369.

I

The New Generation of Virus Vaccines and Their Immunogenicity

1

Production of Viral Glycoproteins in Genetically Engineered Mammalian Cell Lines for Use as Vaccines against Herpes Simplex Virus and the Acquired Immune Deficiency Syndrome Retrovirus

Phillip W. Berman, Timothy Gregory, Donald Dowbenko, and Laurence A. Lasky

I. INTRODUCTION

One promise of recombinant DNA technology is the possibility of developing new and improved vaccines to prevent the transmission of infectious diseases. Using these techniques, it is possible to produce virtually unlimited quantities of the surface antigens of pathogenic organisms without resorting to the large-scale culture of the pathogen itself. Conventional methods of vaccine production (i.e., live attenuated viruses, killed viruses, or extracts of killed viruses) are limited by the fact that some pathogens are difficult or uneconomical to grow. In addition, society is still haunted by fears that such vaccine preparations could be contaminated by unattenuated or inadequately inactivated virus. Even when the most stringent production methods are applied, manufacturers must be concerned with the possibility that a vaccine could induce latent infections, oncogenic transformation, or autoimmunity. The recombinant DNA approach to vaccine development circumvents these problems by providing a preparation consisting of a single highly purified protein, derived from a safe noninfectious source.

The basic steps in the production of a recombinant vaccine are as follows. First, the molecules on the surface of the virus that induce the formation of neutralizing antibodies must be identified. Next, the viral gene encoding the neutralizing antigen must be cloned and sequenced. The cloned gene then has to be inserted into a vector to permit its expression in the cell selected as the production substrate. Finally, processes have to be

Phillip W. Berman, Donald Dowbenko, and Lawrence A. Lasky • Department of Molecular Biology, Genentech, Inc., South San Francisco, California 94080. *Timothy Gregory* • Department of Process Sciences, Genentech, Inc., South San Francisco, California 94080.

developed for recovery, purification, and formulation of the recombinant protein. Several years ago it seemed that microorganisms such as *Escherichia coli* or yeast would be useful substrates to produce vaccines. Experience with a number of viral antigens, especially those from enveloped viruses, have shown that viral surface proteins produced in *E. coli* or yeast are often unsatisfactory in that the recombinant products are typically denatured and insoluble and lack much of the antigenic structure of the native molecule. Although technologies have been devised to renature enzymes and hormones produced in recombinant microorganisms, these methods have not been particularly successful when applied to viral glycoproteins. This may relate in part to the fact that virus surface antigens may be designed to aggregate to form oligomeric structures such as capsids or spikes. In addition, the surface antigens of enveloped viruses are often glycosylated. A number of studies have demonstrated that the carbohydrate typically found on viral surface antigens is crucial to the folding, solubility, and antigenic characteristics of glycoproteins (Alexander and Elder, 1984). Because bacteria do not possess protein glycosylation pathways, and because yeast has an incomplete glycosylation pathway relative to vertebrates, it is unlikely that these organisms could accurately duplicate the complex structures of viral glycoproteins.

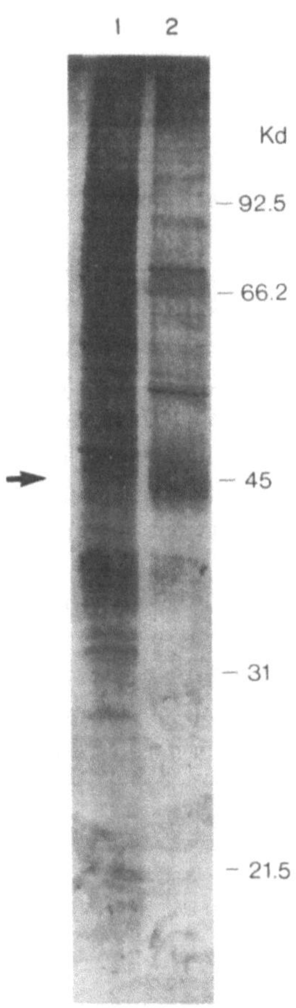

Figure 1. Analysis of cell culture medium conditioned by the growth of Chinese hamster ovary (CHO) cell lines transfected with truncated form of type 1 herpes simplex virus (HSV-1) glycoprotein D (gD-1t). One μg of serum-free culture medium from an unamplified cell line (lane 1) or an amplified cell line (lane 2) was analyzed by sodium dodecyl sulfate-polyacrylamide gel electrophoresis (SDS-PAGE). Proteins were then visualized by silver staining. Arrow indicates the location of the gD-1t protein.

To solve this problem, mammalian cells have been employed as substrates for the production of recombinant vaccines. Mammalian cells have the advantage that they are able to glycosylate and fold recombinant viral glycoproteins into a conformation antigenically similar to that found in the virus. In addition, it has been possible to modify the viral envelope glycoprotein genes by deletion of transmembrane-binding domains so that recombinant molecules can be secreted from transfected mammalian cell lines. Cell-culture medium conditioned by the growth of such cell lines can be harvested as a clear liquid greatly enriched in the recombinant glycoprotein. For example, Fig. 1 illustrates the composition of conditioned cell-culture medium from a cell line producing a secreted form of herpes simplex virus glycoprotein D. It can be seen that after gene-amplification techniques are applied, the resulting culture medium is greatly enriched for the recombinant viral glycoprotein. Purification of the active ingredient from a feedstock of such high quality typically involves less steps than that required for recovery of recombinant proteins from *E. coli* or yeast pastes, resulting in a more economical recovery process.

II. HERPES SIMPLEX VIRUS VACCINE

The first membrane glycoprotein gene truncated to be secreted from mammalian cells for use as a vaccine was glycoprotein D (gD-1) from type 1 herpes simplex virus (HSV-1). Lasky *et al.* (1984) showed that gD-1 could be secreted from transfected cell lines, and that immunization with the recombinant protein protected mice from lethal virus infections by either HSV-1 or HSV-2. More recently, it was shown that this vaccine protect guinea pigs from genital herpes infections (Berman *et al.*, 1985). As can be seen in Table 1, guinea pigs vaccinated with gD-1 in complete Freund's adjuvant were completely protected from any signs of virus infection (Berman *et al.*, 1985). Guinea pigs vaccinated with gD-1 incorporated in alum adjuvants were protected to a lesser yet significant extent.

Table 1. Protection of Guinea Pigs from Genital HSV-2 Infection by Vaccination with Recombinant HSV-1 and HSV-2 gD

Immunogen	Average neutralization[a] titer *in vitro* HSV-1	HSV-2	Number of animals asymptomatic after HSV-2 challenge	Peak mean lesion score[b]
gD-lt in Freund's adjuvant	9.3 ± 0.8	0.8 ± 1.4	15 of 15	0
Freund's adjuvant placebo	<3	<3	1 of 14	3.3 ± 0.4
gD-lt in alum–phosphate	9.2 ± 2.2	5.6 ± 2.2	2 of 9	1.3 ± 0.4
gD-lt in alum–hydroxide	10.9 ± 1.2	6.7 ± 1.5	4 of 9	0.9 ± 0.4
Alum–phosphate–placebo	<3	<3	1 of 10	3.2 ± 0.4
gD-2t in alum–phosphate	ND	ND	8 of 14	0.71 ± 0.22
Alum–phosphate–placebo	ND	ND	1 of 15	3.0 ± 0.35

[a] *In vitro* neutralization assays were performed as described by Lasky, *et al.* (1984). Values represent the highest serum dilution (log_2) that gave 50% plaque reduction.

[b] The clinical symptoms of genital HSV-2 infection in guinea pigs were judged on a scale of 0–4: 0, no symptoms; 1, swelling and erythema; 2, small vesicles; 3, large vesicles; 4, large ulcerated lesions. ND, not done.

Although some animals showed signs of viral infection, the severity of the infections was greatly reduced relative to the controls. It can be seen that of the animals that did show signs of viral infection, the mean score was reduced from 3.8 to <1. When glycoprotein D from HSV-2 (gD-2) was tested, it was found to provide greater protection from a HSV-2 challenge than the HSV-1 protein. Because the animals were challenged with a dose of virus close to the LD_{50} of guinea pigs, the viral infections induced were undoubtedly more severe than those that occur in humans. The next question to be considered was whether the recombinant vaccine was merely providing protection from the symptoms of virus infection or whether it actually inhibited virus infectivity. To answer this question, we recently attempted to recover latent virus from the dorsal root ganglia of protected animals (Kern *et al.*, 1987). In these studies, latent virus could be recovered from 70% of the control animals ($N = 7$); 27% of the animals vaccinated with gD-1 ($N = 11$), and none of the animals vaccinated with gD-2 ($N = 9$). This work provides evidence that the recombinant vaccine provided protection from the establishment of latent infections.

III. AIDS RETROVIRUS VACCINE

In principle, the approach described for the production of an HSV vaccine should be applicable to other viruses. Most recently we have applied this technology toward the production of a vaccine against the retrovirus responsible for acquired immunodeficiency syndrome (AIDS) (Fauci *et al.*, 1984; Staal and Gallo, 1985). AIDS is caused by a retrovirus, variously termed lymphadenopathy virus (LAV), human T-cell lymphotropic virus type III (HTLV-III), AIDS-related virus (ARV), and most recently human immunodeficiency virus (HIV) (Coffin *et al.*, 1986). The virus infects lymphocytes bearing the OKT4 surface antigen, resulting in a selective depletion of this important class of regulatory cells from the immune system. For vaccine development, the most important protein in the virus is that located on the viral surface. Analysis in several laboratories has shown that the AIDS retrovirus envelope protein is synthesized as a 160,000-M_r precursor that is subsequently cleaved at an internal processing site to generate molecules with the apparent molecular weights of 120,000 (gP120) and 41,000 (gP41) (Veronese *et al.*, 1985). Sequence analysis of the viral genome has revealed an open reading frame of 856 amino acids found to correspond to the AIDS retrovirus envelope protein gene (Muesing *et al.*, 1985). Hydropathic analysis (Hopp and Woods, 1981) of the predicted amino acid sequence showed three prominent hydrophobic domains. Near the amino-terminal (residues 1–30) is a hydrophobic domain identified as a signal sequence (Allan *et al.*, 1985). A second hydrophopic domain (residues 515–541) begins immediately following a dibasic protease-processing site (residues 510–511). The third hydrophobic domain beginning at residue 684 and terminating at residue 706 is thought to correspond to a transmembrane-binding domain. Another significant aspect of the envelope proteins structure is the fact that is highly glycosylated and possesses approximately 30 potential *N*-linked glycosylation sites.

The AIDS retrovirus envelope protein is thought to mediate two distinct activities: (1) binding to the T4 antigen of target cells, and (2) promoting the fusion of virus infected cells to form multinucleate giant cells or syncytia (Staal and Gallo, 1985; Dalgleish *et al.*, 1984; Klatzman *et al.*, 1984). The simplest model to account for the orientation of the

AIDS retrovirus envelope protein in the viral surface is shown in Fig. 2. By analogy with other virus fusagenic proteins, it appears that the carboxy-terminal hydrophobic domain anchors the 160,000-M_r precursor to the lipid bilayer. Some time after export to the cell surface or incorporation into the viral envelope, the envelope protein becomes activated by proteolytic cleavage at the internal processing sites (residues 510–511). This cleavage exposes a hydrophobic domain (residues 515–541) that could then promote cell fusion by insertion into the host cell membrane. According to this model, only gP41 is anchored to the lipid bilayer, and gP120 is associated with gP41 by either covalent interactions (disulfide bridges) or by noncovalent interactions. Because this model predicts that approximately 467 residues of gP120 would be exposed on the cell surface while only 174 residues of gP41 would be on a surface, we first attempted to express gp120. In order to obtain a secreted product, the protein was truncated so as to delete the second hydrophobic domain and was fused to an in-phase stop codon. The truncated protein was inserted into an expression vector similar to that described by Lasky *et al.* (1984) and transfected into

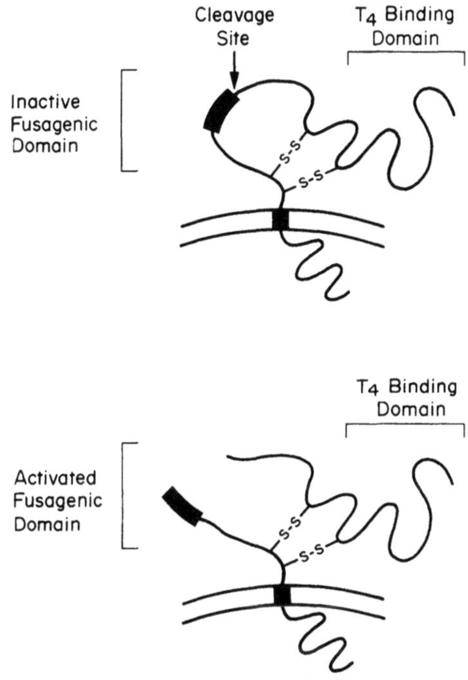

Figure 2. Hypothetical structure for the acquired immunodeficiency syndrome (AIDS) retrovirus envelope glycoprotein. The AIDS retrovirus envelope glycoprotein is thought to possess the following functional domains: (1) an amino-terminal signal sequence ensuring that the protein is synthesized on membrane-bound ribosomes, (2) a dibasic amino acid protease-processing site cleaved to generate gp120 and gp41 from the 16,000-M_r precursor, (3) a domain capable of binding to the OKT4 cell surface antigen, (4) a fusagenic domain that mediates syncytia and multinucleate giant cell formation, (5) a transmembrane-binding domain, and (6) a cytoplasmic tail whose function is yet unknown. The simplest model to account for the orientation of these molecular features is shown above. According to this model, gp160 is inserted into a lipid bilayer such that all gp120 (residues 1–511) and the amino-terminal half of gp41 (residues 512–683) are exposed on the cell surface. As with most other membrane proteins, the signal sequence (residues 1–44) is cleaved cotranslationally, and is not present in the mature protein. The nascent protein is anchored to the lipid bilayer by a single hydrophobic membrane-binding domain (involving residues 684–706). Residues of 707–856 probably do not cross the lipid bilayer and thus constitute the cytoplasmic tail. By analogy with other fusagenic viral proteins, cleavage at the dibasic amino acid processing site (residues 510–511) is necessary to activate the fusagenic activity. This cleavage probably occurs after gp160 has reached the cell surface, since intracellular cleavage would lead to fusion of intracellular membranes. The actual fusagenic domain may involve the cluster of hydrophobic amino acids adjacent to the cleavage site (residues 512–540). After cleavage, it would be expected that this hydrophobic region could insert into the lipid bilayer of a neighboring cell; thus, a intermembrane bridge is formed that facilitates entry of the virus into the target cell and mediates cell fusion in virus-infected cells. After cleavage, gp120 is attached to gp41 by either covalent (e.g., disulfide bridges) or noncovalent interactions. It has been shown that the OKT4-binding site, which serves to target the virus to OKT4-bearing cell, is located on gp120, although the residues involved have yet to be identified.

Chinese Hamster ovary (CHO) cells. It was possible to improve expression of the recombinant protein by exchanging the gD signal sequence for the gp120 signal sequence (Lasky *et al.*, 1986). Cells producing gp120 were identified by immunofluorescence using antibodies obtained from a healthy homosexual male with a high titer of anti-gp120 antibodies. The cells were then analyzed by by radioimmunoprecipitation of ^{35}S-labeled cell lysates. Figure 3 shows that a molecule with a molecular weight of 130,000 could be specifically immunoprecipitated from growth-conditioned culture medium. In order to determine whether the protein was glycosylated, the protein was treated with endoglycosidase H (endo H) and neuraminidase. As can be seen, the secreted molecule was sensitive to neuraminidase and resistant to endo H, demonstrating that it possessed the complex form of N-linked carbohydrate. Inside the cell, a precursor of the secreted protein was identified. This protein had a molecular weight of approximately 96,000 and was resistant to neuraminidase, but sensitive to endo H, indicating that it possessed the simple high mannose form of complex carbohydrate. The fact that endo H treatment reduced the apparent molecular weight from 96,000 to 69,000 demonstrates that the recombinant molecule is highly glycosylated. By means of an affinity column built using

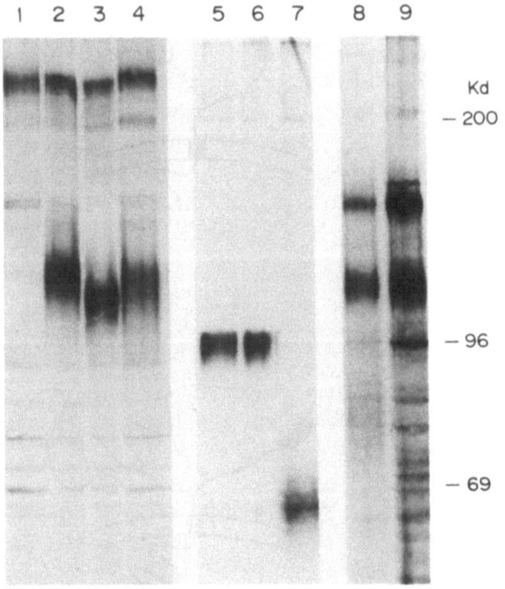

Figure 3. Molecular and immunologic characterization of the intracellular and secreted forms of recombinant acquired immunodeficiency syndrome (AIDS) retrovirus envelope glycoprotein, gp130. The gene encoding the AIDS retrovirus envelope glycoprotein was truncated at a codon corresponding to amino acid residue 531 so as to delete the transmembrane-binding domain. Chinese hamster ovary (CHO) cells were transfected with this truncated gene, and a stable cell line, termed D531, that constitutively secreted the viral glycoprotein was isolated. D531 cells and H9 cells infected with the AIDS retrovirus were metabolically labeled with [^{35}S]methionine and analyzed by radioimmunoprecipitation followed by sodium dodecyl sulfate-polyacrylamide gel electrophoresis (SDS-PAGE) as described by Lasky *et al.* (1984). Lanes 1–4 illustrate results from the analysis of growth-conditioned culture medium from the D531 cell line. Lane 1, control immunoprecipitation using normal human serum; lane 2, immunoprecipitation of a protein with an apparent molecular weight of 130,000 by serum from an individual seropositive for the AIDS retrovirus envelope glycoprotein (AIDS serum); lane 3, neuraminidase treatment of the 130,000-M_r glycoprotein; lane 4, endoglycosidase H treatment of the 130,000-M_r glycoprotein. Lanes 5–7 illustrate results obtained from immunoprecipitation of solubilized D531 cells. Lane 5, immunoprecipitation of a protein with an apparent molecular weight of 96,000 with AIDS serum; lane 6, treatment of the 96,000-M_r protein with neuraminidase; lane 7, treatment of the 96,000-M_r protein with endoglycosidase H. Lanes 8–9 compare the immunoprecipitation of native AIDS retrovirus envelope glycoproteins, gp120 and gp160, by antibodies produced in guinea pigs against recombinant gp130, lane 8; and antibodies from AIDS serum.

antibodies from healthy homosexual male subjects, we have been able to isolate the recombinant protein from growth-conditioned medium. We have immunized a number of animals with this material and have found that they make antibodies that react with the native virus envelope protein. Figure 3 shows that rabbit antibodies to gp120 react with the ^{35}S-labeled virus protein in radioimmunoprecipitation experiments. This result demonstrated that the recombinant protein shares antigenic determinants with both gp160 and gp120. Together, these studies demonstrate that useful quantities of recombinant gp120 can be produced and that it is highly immunogenic. The next issue to resolve is whether the recombinant protein induces the formation of neutralizing antibodies and, if so, whether the antibodies neutralize various clinical isolates.

IV. CONCLUSION

Genes encoding virus surface antigens have been expressed in a variety of cellular substrates. With the notable exceptions of foot and mouth disease virus (Kleid *et al.*, 1981) and hepatitis B surface antigen (Valenzuela *et al.*, 1982; Hitzman *et al.*, 1983), *E. coli* and yeast have not proved particularly useful for the production of recombinant vaccines. Recombinant DNA technology was recently used to engineer live virus vaccines (Smith *et al.*, 1983; Paoletti *et al.*, 1984); however, vaccine products produced from live engineered viruses appear to have many of the same safety problems of conventional vaccines. Of all methods described, it appears that mammalian cells provide the best substrate for the production of recombinant vaccines. They provide a glycosylated molecule that can be easily isolated. Because they consist of a single highly purified protein, their composition can be easily verified and their properties fully investigated.

REFERENCES

Alexander, S., and Elder, J. H. (1984). *Science* **226,** 1328–1330.

Allan, J., Coligan, J., Barin, F., McLane, M., Sodrowski, J., Rosen, C., Haseltine, W., Lee, T., and Essex, M. (1985). *Science* **228,** 1091–1094.

Berman, P. W., Gregory, T., Crase, D., and Lasky, L. A. (1985). *Science* **227,** 1490–1492.

Coffin, J., Haase, A., Levy, J. A., Montagnier, L., Orosziam, S., Teick, N., Temin, H., Toyoshima, K., Varmus, H., Vogt, P., Weiss, R. (1986). *Nature (Lond.)* **321,** 10.

Dalgleish, A. G., Beverly, P. C. L., Clapham, P. R., Crawford, D. H., Greaves, M. F., and Weiss, R. A. (1984). *Nature (Lond.)* **312,** 763–766.

Fauci, A., Macker, A., Longo, D., Lane, H., Rook, A., Masur, H., and Gelmann, E. (1984). *Ann. Med.* **100,** 92–106.

Hitzman, R. A., Chen, C. Y., Hagie, F. E., Patzer, E. J., Liu, C-C., Estell, J. D., Miller, J. V., Yaffe, A., Kleid, D. G., Levinson, A. D., and Opperman, H. (1983). *Nucleic Acids Res.* **11,** 2745–2763.

Hopp, T. P., and Woods, K. R. (1981). *Proc. Natl. Acad. Sci. USA* **78,** 3824–3828.

Kern, E. R., Vogt, P. E., Gregory, T., Lasky, L. A., and Berman, P. W. (1987). *J. Virol.* (in press).

Klatzman, D., Champagne, E., Chamoret, S., Greuest, J., Guefard, D., Hercond, T., Gluckman, J. C., and Montagnier, L. (1984). *Nature (Lond.)* **312,** 767–768.

Kleid, D. G., Yansura, D., Small, B., Dowbenko, D., Moore, D. M., Grubman, M. J., McKercher, P. D., Morgan, D. O., Robertson, B. H., and Bachrach, H. L. (1981). *Science* **214,** 1125–1129.

Lasky, L. A., Dowbenko, D., Simonsen, C. C., and Berman, P. W. (1984). *Bio/Technology* **2,** 527–532.

Lasky, L. A., Groopman, J. E., Fennie, C. W., Benz, P. M., Capon, D. J., Dowbenko, D. J., Nakamura, G. R., Nunes, W. M., Renz, M. E., and Berman, P. W. (1986). *Science* **233,** 209–212.

Muesing, M. A., Smith, D. H., Cabradilla, C. D., Benton, C. V., Lasky, L. A., and Capon, D. J. (1985). *Nature (Lond.)* **313,** 450–458.

Paoletti, E., Lipinskas, B. R., Samsonoff, C., Mercer, S., and Panicali, D. (1984). *Proc. Natl. Acad. Sci. USA* **81,** 193–197.

Smith, G. L., Murphy, B. R., and Moss, B. (1983). *Proc. Natl. Acad. Sci. USA* **80,** 7155–7159.

Staal, F. W., and Gallo, R. C. (1985). *Blood* **65,** 253–263.

Valenzuela, P., Medina, A., Rutter, W. J., Ammerer, G., and Hal, B. D., (1982). *Nature (Lond.)* **298,** 347–350.

Veronese, F. D., DeVico, A. L., Copeland, T. D., Oroszlan, S., Gallo, R. C., Sarnagadharan, M. G. (1985). *Science* **229,** 1402–1405.

2

Development of Fungal and Algal Cells for Expression of Herpes Virus Genes

W. C. Leung, E. K. Manavathu, J. Zwaagstra, K. Suryanarayana, S. E. Hasnain, and M. F. K. Leung

I. INTRODUCTION

The advances of recombinant DNA technology have permitted the routine isolation and characterization of genes of medical and virologic interests. Most of these genes can be re-expressed into functional proteins upon reintroduction into suitable host cells. It is by now quite apparent that the expression of a gene into functional protein depends not only on the nature of the expressed protein, but on the post-translation modifications available in the host cells as well. For example, the expression and assembly of hepatitis B virus surface antigen (HB_sAg) polypeptide into the antigenic and immunogenic 22-nm particle was readily carried out in yeast cells (Valenzuela *et al.*, 1982), but not in *Escherichia coli*. The recent demonstration that chymosin can be expressed and secreted in functional form in the filamentous fungus *Aspergillus*, but not in *E. coli*, serves as another example. More subtle and intricate differences do exist even in closely related host cells. For example, the human growth hormone expressed in *E. coli* retained the N-terminal methionine, while the same polypeptide expressed in *Pseudomonas* was properly processed and even secreted (Gregory *et al.*, 1984). The mammalian cells and insect cells have now been used quite frequently for the expression of human and viral genes, which require glycosylation and protein processing to produce functional gene products. However, they are by no means host cells of universal applicability. There are instances in which genes failed to express, or expressed only in minute amounts (e.g., blood coagulation factor VIII gene), when they were reintroduced back into mammalian cells (Wood *et al.*, 1984).

II. CHOICE OF ADDITIONAL HOST CELLS

In view of the fact that the post-translational modifications available are specific for each host cell, it will then be of great use to develop additional host cells for the proper

W. C. Leung, E. K. Manavathu, J. Zwaagstra, K. Suryanarayana, S. E. Hasnain, and M. F. K. Leung • Department of Medicine, University of Alberta, Edmonton, Alberta T6G 2G3, Canada.

expression of cloned genes and processing of expressed polypeptides. We have devised a gene cloning and expression system for two unusual host cells: one is the multinucleate, filamentous water fungus *Achlya ambisexualis;* the other is the unicellular, eukaryotic green algae *Chlamydomonas reinhardtii.* Both host cells were chosen for the following considerations:

1. They are nonpathogenic. There will not be major biosafety concerns over their use in gene expression and also for large-scale production of bioproducts.
2. Both host cells are known to possess the capacity for post-translational modification of proteins, including glycosylation and protein secretion (Sengupta *et al.*, 1981).
3. Both host cells can be grown on simple culture media at very low cost. There will not be major difficulties in large-scale fermentation. *Achlya* is a particularly fast-growing fungus.
4. The genetic system for *Chlamydomonas* is fairly well defined, and a large number of genetic mutants and variants are already available. A considerable number of cell physiologic studies have been performed in *Achlya.* The steroid hormone receptor for this fungus is virtually identical to that of the mammalian cell (Riehl and Toft, 1984). This observation suggests that this fungus could provide a suitable environment for the proper expression and processing of proteins of mammalian cells and viral origin.

III. CLONING VECTOR USING HETEROLOGOUS COMPONENTS

The availability of selection markers, promoters, and replicons would permit the construction of cloning vectors. However, endogenous plasmids or viruses are not yet

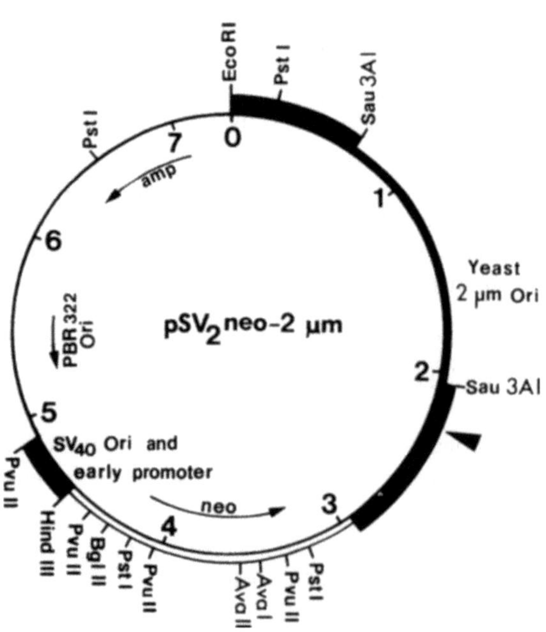

Figure 1. Restriction map of pSV_2neo-2μm. A 1.576-kbar Sau 3A1 fragment from the cosmid pYC1 (Hohn and Hinnen, 1980) containing the yeast 2μm origin of replication sequences was inserted into the unique BamH1 site of pSV2neo (Southern and Berg, 1982). Only selected restriction endonuclease cut sites are shown. (———) pBR322; (▭▭▭) neo-gene; (▬▬) yeast 2μm; (■■) SV40. Solid arrow indicates the location of SV40 T antigen polyadenylation signal sequences.

identified in either *Chlamydomonas* or *Achlya* to allow for ready isolation of these components. In an effort to overcome these difficulties, our approach is to explore the possibility of using highly efficient heterologous promoter and termination sequences, as well as DNA replication origin and drug selection marker for the development of gene cloning and expression vectors in these host cells. We rationalize that the highly efficient nature of these components will permit them to overcome the expected reduction in efficiency in these host cells.

Figure 1 illustrates the cloning vector $pSV_2neo\text{-}2\mu m$ that proved to produce stable transformants in either *Chlamydomonas* or the *Achlya* cells. The animal virus SV40 early promoter and enhancer sequences as well as the termination sequences were used to direct RNA transcription. To allow for autonomous replication of the vector DNA in the transformed cells, the DNA replication origin of yeast 2μm plasmid was employed. A positive selection marker, the bacterial aminoglycoside 3′-phosphotransferase (neo) gene was used to confer resistance to neomycin and its analogue, G-418. The choice of a positive selection scheme overcomes the nonavailability of genetic selection markers, thereby opening the possibility of performing gene cloning and expression studies in host cells of selective advantages (e.g., protease-deficient cells) as well.

IV. DNA-MEDIATED TRANSFORMATION OF CHLAMYDOMONAS REINHARDTII

The DNA-mediated transformation of *Chlamydomonas reinhardtii* was detailed in Hasnain *et al.* (1985, 1986). Briefly, the cell wall-deficient mutant of *C. reinhardtii* was mixed with the vector DNA in the presence of poly-L-orthinine and $ZnSO_4$. The reaction mixture was then plated out in agar media. Colonies were picked and selected in alternate agar media or liquid media with increasing concentration of G-418 or kanamycin. The algal transformants remained resistant for well over 230 cell generations, indicating that it is a stable transformation rather than an abortive transient transformation. Cells transformed by a control plasmid using the endogenous *Chlamydomonas* autonomous replicating sequence failed to survive beyond 50 cell generations in the presence of G-418. This observation suggested that the yeast 2μm DNA contributed to the long-term maintenance of transformation. A quick-blot messenger RNA (mRNA) analysis indicated the presence of neo gene transcript, suggesting that the G-418 resistance was attributable to the transcription of neo-gene, rather than to phenotypic changes of membrane permeability of G-418. Southern blot analyses of the transformant DNA indicated that most of the transforming plasmid DNA were in free episomal form. The vector DNA recovered by transformation of *E. coli* HB101 with total DNA preparations extracted from the *Chlamydomonas* transformants. In essence, our system fulfills all the necessary features of a gene cloning system.

V. DNA-MEDIATED TRANSFORMATION OF ACHLYA AMBISEXUALIS

The experimental protocol for transformation of this fungus was quite similar to that observed for *Chlamydomonas,* but with some differences. After removal of the cell wall by digestion with driselase (Schafrick and Horgen, 1979), the vector $pSV_2neo\text{-}2\mu m$ DNA can be taken up by the fungal protoplast in the presence of polyethylene glycol. A soft agar overlay was used to select for transformants exhibiting resistance to G-418 during cell-wall regeneration. Transformed cultures have grown to more than 200 times original weight by tip growth yet still remained resistant to G-418. Again, vector DNA can be recovered by transforming *E. coli* HB101 with total DNA preparations from the fungal transformants. Preliminary RNA blot analyses indicated the presence of neo-gene transcripts. Preliminary Southern blot analyses indicated the presence of multiple DNA bands specific for the vector DNA, suggesting the presence of vector DNA in episomal form as well as integration into the cell genome.

VI. EXPRESSION OF THYMIDINE KINASE OF HERPES SIMPLEX VIRUS IN FUNGUS

We have obtained a recombinant DNA clone from Dr. C. Lau, Howard Hughes Medical Research Institute, San Francisco, in which the HSV TK gene was driven by the mouse metallothionen gene promoter. This plasmid was ligated to the vector $pSV_2neo2\mu m$ through the EcoR1 site, and the resulting plasmid was used to transform the fungal cell of *Achlya*. One of the transformants exhibited the presence of HSV TK-specific transcripts. HSV TK enzymatic activity with the characteristic Rm 0.5 peak was observed when the cell extract was analyzed by electrophoresis on native gel. This is the first known foreign gene to be expressed successfully in the fungal cell of *Achlya ambisexualis*.

VII. GLYCOPROTEIN gB GENE OF HERPES SIMPLEX VIRUS TYPE 2

Among the foreign genes we are interested in conducting gene expression studies with in *Chlamydomonas* and *Achlya* are human γ-interferon (IF_γ) (Leung *et al.*, 1985) and glycoprotein gB gene of herpes simplex virus type 2 (HSV-2). The gB-2 gene is one of the major neutralization glycoproteins for HSV-2 and is therefore a candidate for the development of a subunit vaccine.

The gB-1 gene of HSV-1 is located on the 7.8-kb BamHI G DNA fragment. This DNA fragment was used as a probe to select, by nucleic acid hybridization, the 6.6-kb BamHI/Bgl II DNA fragment containing the gB-2 gene of HSV-2, strain 333. A 544-base pair (bp) Xho I restriction fragment containing the 5′-end of the gB-1 gene was used to map the location of the 5′-end of the gB-2 gene by Southern blot analysis. A detailed restriction endonuclease map of the gB-2 gene was constructed as well.

Subfragments of the gB-2 gene were cloned into M13 phage vectors for DNA sequencing by the dideoxy chain-termination method, using a universal primer to obtain DNA sequence of the terminal regions of the fragment. Synthetic DNA primers were also chemically synthesized (Jing *et al.*, 1986) to prime for DNA sequences farther along each fragment.

Computer analysis of the gB-2 gene revealed that it exhibited a high degree of homolog with the gB-1 gene sequence of HSV-1 (Bzik *et al.*, 1984). Figure 2 shows the extensive sequence homology between the promoter regions of the gB genes of HSV-1 (strain KOS) and HSV-2 (strain 333). The available open reading frame suggested a

```
                             -100                       -75
HSV-2 (333)     TCGGGCGCTGATTGGGCCGTCAGCGAGTTTCAAAAATTCTACTGTTTTGACGGGT
                     X                    X     X XX  X              XX
HSV-1 (KOS)     TCGGGTGCTGATTGGGCCGTCAGCGAATTTCAGAGGTTTTACTGTTTTGACGGCA

                         -50                        -25
HSV-2 (333)     TTTCCCGAATCACGCCCACCCAGCGCCGCCTGGCGATATATTCGCGAGCTCATTATC

HSV-1 (KOS)     TTTCCGGAATAACGCCCACTCAGCGCCGCCTGGCGATATATTCGCGAGCTGATTATC
                     CAT  SEQUENCE                    TATA  BOX

                  +1                      +25                      +50
HSV-2 (333)     GCCACCACACTCTTTGCCTCGGTGTACCGGTGCGGGGAGCTTGAGTTGCGCCGCCCCGACTGCA
                                                         X              X
HSV-1 (KOS)     GCCACCACACTCTTTGCCTCGGTCTACCGGTGCGGGGAGCTCGAGTTGCGCCGCCCGGACTGCA
                   mRNA START
```

Figure 2. Sequence homology between the gB genes of HSV-1 and HSV-2.

polypeptide of close to 900 amino acids in length. It contained three major domains: an N-terminal domain projecting toward the extracellular environment, a hydrophobic domain spanning the cell membrane, and a positively charged cytoplasmic domain.

VIII. CONCLUSION

Additional host cells were developed for the proper expression and processing of viral genes and cellular genes, including the algal and fungal cells. The heterologous component approach was used to construct a cloning and expression vector that gave rise to stable transformants. The HSV TK is the first nonselected gene to be expressed in the fungal cells of *Achlya*. Further studies on expression of other genes, including human IF_{γ} and the gB gene of HSV were in progress.

ACKNOWLEDGMENTS

We are indebted to Monica Jewell for preparing the manuscript. The studies were supported by operating grants from the Medical Research Council of Canada and the National Cancer Institute of Canada. E.K.M., K.S., and S.E.H. are recipients of AHFMR postdoctoral fellowships. J.Z. is a recipient of an AHFMR predoctoral studentship. W.C.L. is an AHFMR scholar.

REFERENCES

Bzik D. J., Fox, B. A., DeLuca, N. A., Person, S., et al. (1984). *Virology* **133,** 301–314.

Gregory, L. C., McKeown, K. A., Jones, A. J. S., Seeburg, P. H., Heyneker, H. L. (1984). *Bio/Technology* **2,** 161–165.

Hohn, B., and Hinnen, A. (1980). In *Genetic Engineering: Principles and Methods* (J. K. Setlow and A. Hollaender (ed.), Vol. 2, pp. 169–183. Plenum, New York.
Hasnain, S. E., Manavathu, E. K., and Leung, W.-C. (1985). *Mol. Cell. Biol.* **5,** 3647–3650.
Hasnain, S. E., Manavathu, E. K., Suryanarayana, K., and Leung, W.-C. (1986). In: Proteins in Food, Health and Industry, A. Srinivasan (ed.), Oxford IBH Press, New Delhi.
Jing, G. Z., Liu, A. P., and Leung, W.-C. (1986). *Anal. Biochem.* **155,** 376–378.
Leung, W.-C., Liu, A. P., Wu, X. N., Hasnain, S. E., Ritzel, G., and Jing, G. Z. (1985). In *Biology of Interferon System* (H. Hirchener and H. Schellekens, eds.), pp. 419–424, Elsevier, New York.
Riehl, R. M., and Toft, D. O. (1984). *J. Biol. Chem.* **259,** 15324–15330.
Schafrick, M. T., and Horgen, P. A. (1979). *Cytobios* **22,** 97–104.
Sengupta, C., Brandhaust, B. P., and Verma, D. P. (1981). *Biochem. Biophys. Acta.* **674,** 105–117.
Southern, P. J., and Berg, P. (1982). *J. Mol. Appl. Genet.* **1,** 327–341.
Valenzuela, P., Medina, A., Rutter, W. J., Ammerer, G., and Hall, B. G. (1982). *Nature (Lond.)* **298,** 347–350.
Wood, W. I., et al. (1984). *Nature (Lond.)* **312,** 330–337.

3

Expression of Hepatitis B Virus Surface Antigen Gene in Mammalian, Yeast, and Insect Cells

C. Yong Kang, Thierry Vernet, and David Y. Thomas

I. INTRODUCTION

Hepatitis B virus (HBV) infections cause serious liver disease and are a worldwide major public health problem. Vaccines against human HBV have been developed in three different countries using formalin-inactivated 22-nm hepatitis B virus surface antigen (HB_sAg) particles from chronic active hepatitis patients. The hepatitis B virus vaccine has proved effective in terms of producing a protective level of circulating antibodies in the majority of persons vaccined. However, the vaccine supply is limited. The cost of vaccination is among the highest of all vaccines currently available. In order to overcome these problems, recombinant DNA techniques have been applied, and the viral genome has been cloned, sequenced, and expressed in various heterologous expression systems. Recently, a subunit vaccine was developed by expressing the cloned HB_sAg gene in yeast. However, the cost of the vaccine has not yet been reduced significantly. A subunit vaccine developed by recombinant DNA technology must fulfill certain requirements, such as potency, purity, and lower cost of production. Thus, one of the most active areas of recombinant DNA technology in the past several years has been the search for an expression system that will produce large quantities of desired gene products, provide protein modification similar to that of the naturally occurring proteins, and allow for the secretion of the expressed proteins out of cells. This chapter compares the level of expression and antigenicity of HB_sAg expressed in mammalian, yeast, and insect cell systems, using recombinant DNA vector.

C. Yong Kang • Department of Microbiology and Immunology, University of Ottawa, School of Medicine, Ottawa, Ontario K1H 8M5, Canada. *Thierry Vernet and David Y. Thomas* • Biotechnology Research Institute, National Research Council of Canada, Montreal, Quebec H4P 2R2, Canada.

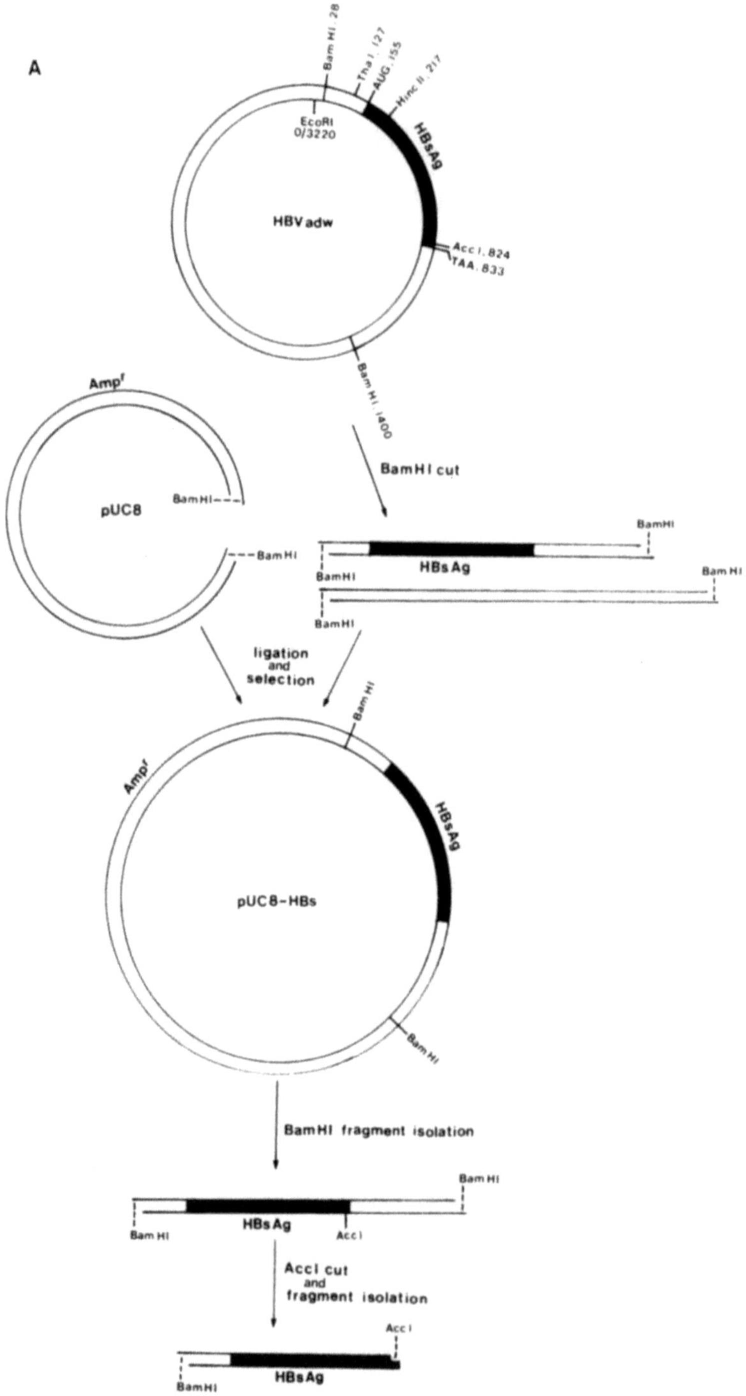

Figure 1. Construction of pJC119–HB$_s$YK5. (A) A pUC8-based plasmid containing the entire coding region of hepatitis B virus surface antigen (HB$_s$Ag) DNA (pUC8-HB$_s$) was constructed using a 1400-base-pair (bp) Bam

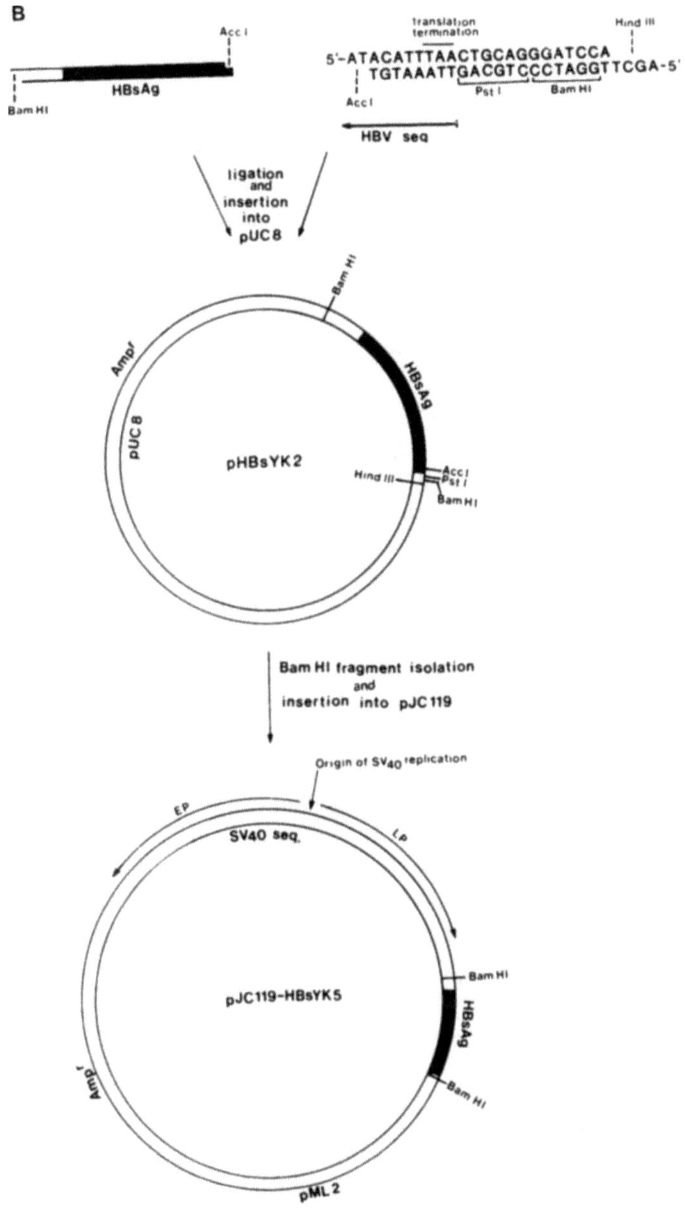

HI fragment of the hepatitis B virus (HBV) genome. The 1400-bp BamHI fragment was isolated from recombinant plasmid pUC8–HB_s and digested with Acc I. (B) The HB_sAg gene having BamHI and Acc I sticky ends was ligated to a synthetic oligonucleotide containing an Acc I sticky end, the translation termination signal of the HB_sAg gene, a Pst I site, a BamHI site, and a *Hind*III sticky end. This DNA construct was inserted between the BamHI and *Hind*III sites of pUC8 (pHB_sYK2). The plasmid pHB_sYK2 was then digested with BamHI, and an 828-bp fragment was isolated by electrophoresis in an agarose gel and ligated into pJC119.

II. EXPRESSION OF THE CLONED HB_sAg GENE

A. Expression of the HB_sAg Gene in Mammalian Cells

We have cloned two BamHI fragments of the *adw* subtype of the HBV genome using standard cloning techniques (Maniatis *et al.*, 1982). We selected a plasmid containing the 1400-base-pair (bp) BamHI fragment, including the coding sequences of the entire HB_sAg gene. The 1400-bp BamHI fragment was modified by digesting with Acc I to eliminate downstream noncoding sequences and by adding a synthetic oligonucleotide that contains the original HB_sAg translation termination codon plus new Pst I, BamHI, and Hind III restriction sites (Fig. 1). The modified HB_sAg gene was inserted into plasmid pJC119, which contains sequences derived from the bacterial plasmid pML2, allowing its replication in *Escherichia coli* as well as sequences from a late region deletion mutant of SV40. The SV40 origin of replication in pJC119 permits replication of the plasmid in mammalian cells in the presence of the large T antigen of SV40. The BamHI or Xho I cloning sites in pJC119 span the SV40 deletion. The deletion removed most of the SV40 VP 1 gene, including the translation initiation codon; however, the sequences required for correct transcriptional initiation, termination, and polyadenylation of late transcripts remain. The translational initiation and termination signals are provided by the coding sequence of the gene cloned into the BamHI or Xho I site. Expression of the cloned gene can be achieved by transfection of the recombinant plasmid into Cos 1 cells. These cells constitutively express the large T antigen of SV40, permitting both the replication of plasmid pJC119 and transcription from the SV40 late promoter. pJC119–HBsYK5 was introduced into Cos 1 cells by the calcium phosphate method of transfection (Graham and van der Eb, 1973). HB_sAg-specific transcripts were analyzed using Northern blots. A single HB_sAg-specific RNA transcript was synthesized in Cos 1 cells. In Cos 1 cells, a 23,000-M_r protein was detected that was not present in Cos 1 cells transfected with pJC119 alone. The extracellular fluid (ECF) culture was monitored for the antigen production using a commercially available radioimmunoassay (RIA) kit (Abbott). Table 1 shows the results of the RIA using cell-free culture supernatants and cell extracts of pJC119–HB_sYK5-transfected Cos 1 cells. One can detect immunologically reactive HB_sAg in both the culture supernatant and the cell extracts. As a control, cells transfected with the HB_sAg gene in the reverse orientation failed to produce a significant level of HB_sAg (Table 1). Electron microscopic examination of ECF showed particles with a morphology identical to that of 22-nm particles isolated from patients with chronic active hepatitis. Although the 22-nm particle of HB_sAg produced in Cos 1 cells appears indistinguishable from the naturally occurring 22-nm particles isolated from chronic active hepatitis patients, the level of expression in Cos 1 cells using the SV40-based transient expression system is not satisfactory. Therefore, we explored the possibility of expressing the HB_sAg in the yeast, *Saccharomyces cerevisiae*.

B. Expression of the HB_sAg Gene in *Saccharomyces cerevisiae*

Several laboratories have shown that the cloned HB_sAg gene can be expressed in yeast, *S. cerevisiae*, under the control of various promoters (Valenzuela *et al.*, 1982; Miyanohara *et al.*, 1983; Hitzeman *et al.*, 1983; Choo *et al.*, 1985). All the expressed

Table 1. Production of HB_sAg from Cos 1 Cells Transformed with Plasmid pJC119-HBsYK5[a,b]

Recombinant plasmid	Cell culture supernatant (cpm)	Cell extracts (cpm)
pJC119-HBsYK5(CO)	21,470	16,100
pJC119-HBsYK6(RO)	270	290
Positive control	14,190	
Negative control	310	

[a] Exponentially growing Cos 1 cells were transformed with pJC119-HBsYK5 or with pJC119-HB_sYK6 plasmids using the calcium phosphate transfection method. After 48 hr, extracellular fluid (ECF) and cell extracts were monitored for HB_sAg production. The cell extracts were prepared as follows: transfected cells were centrifuged, the pellet was washed, cells were disrupted by glass homogenization, and lysates were centrifuged at 5000 rpm for 10 min. The cytoplasmic extracts were used for radioimmunoassay.
[b] CO, correct orientation; RO, reverse orientation of the HB_sAg gene in plasmid pJC119.

antigens in yeast are cell associated, and extra effort is needed to purify 22-nm HB_sAg particles from yeast cells. We have attempted not only to express HB_sAg in yeast but to obtain secretion of the expressed product from yeast cells using the signal polypeptide or the naturally secreted subunits encoded by the yeast killer toxin gene as well. The active secreted form of yeast killer toxin is a 21,000-M_r α/β protein, the extracellular proteolytic cleavage product of a 32,000-M_r intracellular precursor protein (Skipper *et al.*, 1984). We inserted the HB_sAg gene at the Bal I site within the gene coding for the yeast killer preprotoxin coding sequences (Lolle *et al.*, 1984). This construct would be expected to produce a fusion product of the yeast killer toxin protein and the HB_sAg protein, as one of the proteolytic cleavage sites in the yeast killer preprotoxin has been eliminated (Fig. 2). We inserted the fused gene into the BamHI site of the pYT760 yeast plasmid (pSCK–G20), which contains the inducible galactokinase gene promoter and a copy of the yeast *LEU 2* gene (Fig. 3). The plasmid pSCK–G20 was used to transform the *S. cerevisiae* strain SC252, which requires leucine for growth. We selected *S. cerevisiae* transformants that grew on medium lacking leucine. *S. cerevisiae* transformed with plasmid pSCK–G20 and induced with galactose produced 22-nm particles. Electron microscopic examination revealed that the 22-nm particles produced in yeast were morphologically indistinguishable from the 22-nm particles isolated from patients with chronic active hepatitis (Fig. 4). These 22-nm particles, however, failed to react with anti-HB_sAg antibody. This may be attributable to masking of antigenic sites in the chimeric protein by the uncleaved portion of the yeast killer toxin. In an attempt to obtain proteolytic cleavage of the toxin sequences, we reconstructed the junction sequences between the yeast killer toxin sequence and the HB_sAg gene to facilitate proteolytic cleavage between the yeast killer toxin peptide and HB_sAg. It is known that *S. cerevisiae* has an endopeptidase that cleaves after

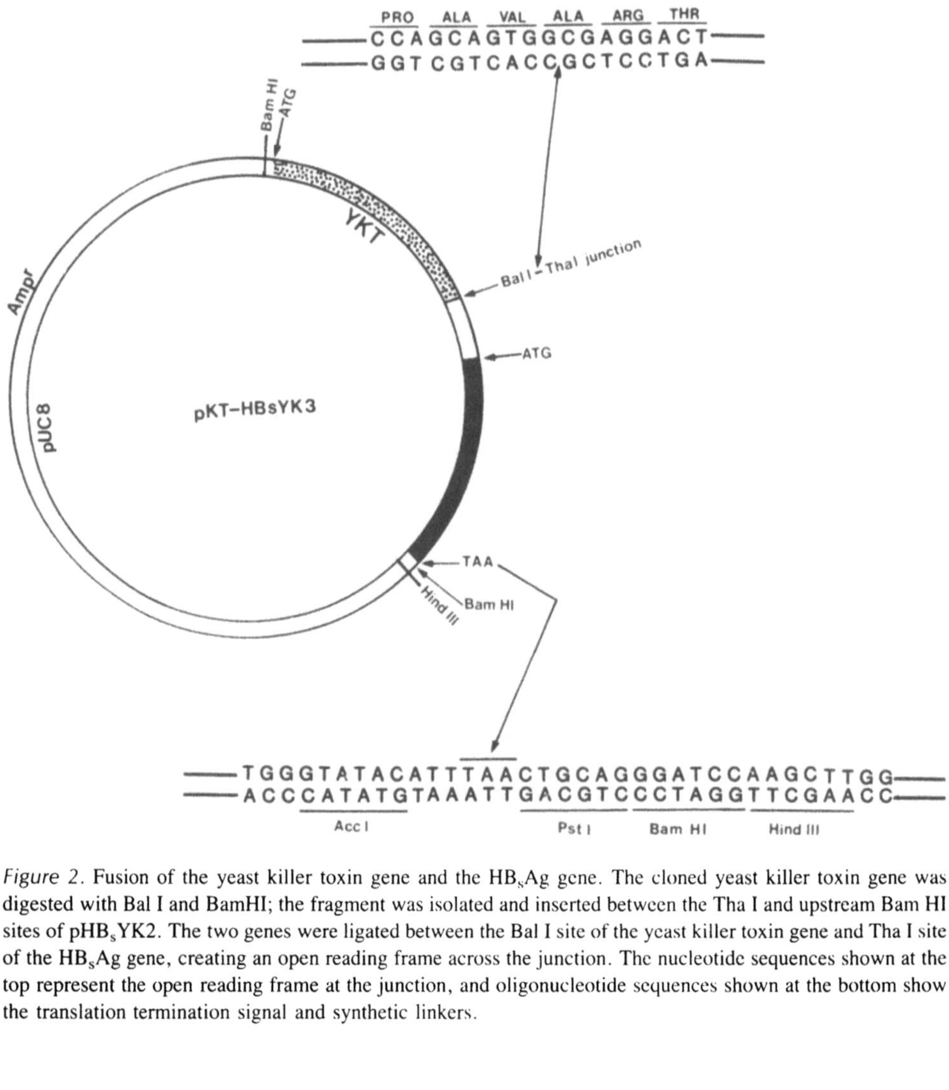

Figure 2. Fusion of the yeast killer toxin gene and the HB_sAg gene. The cloned yeast killer toxin gene was digested with Bal I and BamHI; the fragment was isolated and inserted between the Tha I and upstream Bam HI sites of pHB_sYK2. The two genes were ligated between the Bal I site of the yeast killer toxin gene and Tha I site of the HB_sAg gene, creating an open reading frame across the junction. The nucleotide sequences shown at the top represent the open reading frame at the junction, and oligonucleotide sequences shown at the bottom show the translation termination signal and synthetic linkers.

basic amino acid residues and a dipeptidyl aminopeptidase activity that removes $(glu\text{-}ala)_n$ sequences (Julius *et al.*, 1983); we therefore inserted an oligonucleotide that codes for the $(glu\text{-}ala)_3$ sequence between the yeast killer toxin γ- and β-domains and the HB_sAg gene (Fig. 5). This construct was inserted into the BamHI site of the pYT760 plasmid. The recombinant plasmid was used to transform *S. cerevisiae;* HB_sAg production in the cells and in the ECF was monitored. The HB_sAg expressed in yeast cells transfected with this modified plasmid reacted strongly with anti-HB_sAg antibody but was intracellular. Although the level of expression in yeast was relatively high, it was difficult to purify the cell-associated HB_sAg gene products from the total cellular extracts. Therefore, we searched for a better expression system that would secrete the HB_sAg gene products.

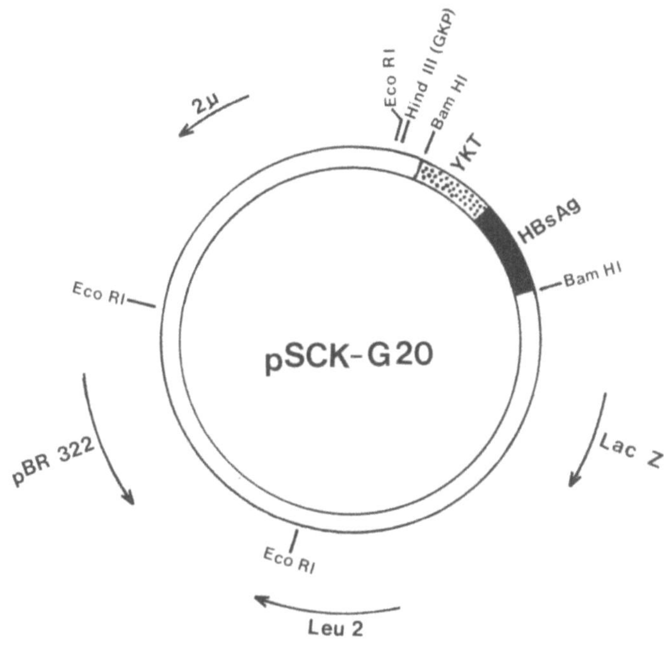

Figure 3. Construction of a yeast expression vector containing the fused yeast killer toxin and HB_sAg genes. The BamHI fragment of pKT-HB_sYK3 was inserted at the BamHI site of the yeast expression vector pYT760. The plasmid pYT760 contains the galactokinase promoter (GKP) inserted at the Hind III site.

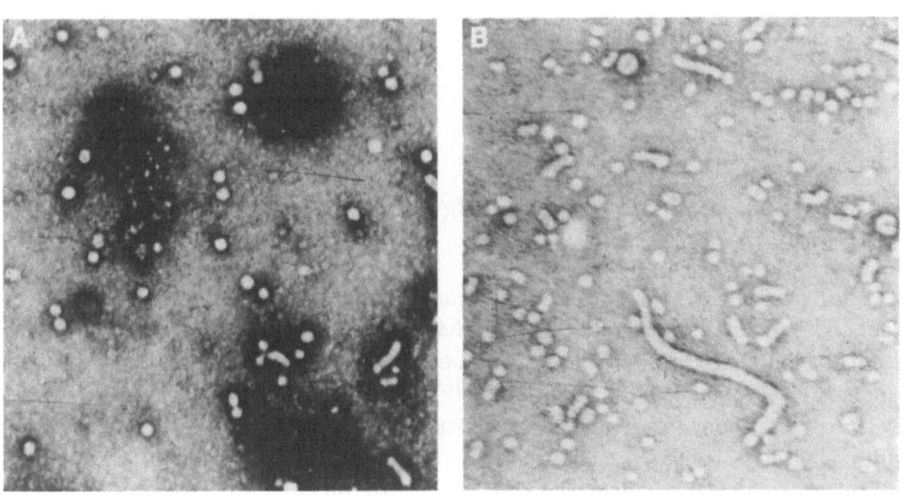

Figure 4. Electron micrographs of 22-nm particles from *Saccharomyces cerevisiae* transformed with pSCK-G20. (A) Particles prepared from yeast transformed with pSCK-G20 after induction with galactose. (B) Three different forms of hepatitis B virus surface antigen (HB, Ag) positive human blood: the Dane particle, the 22-nm particle, and the tubular form from the plasma of a patient with chronic active hepatitis.

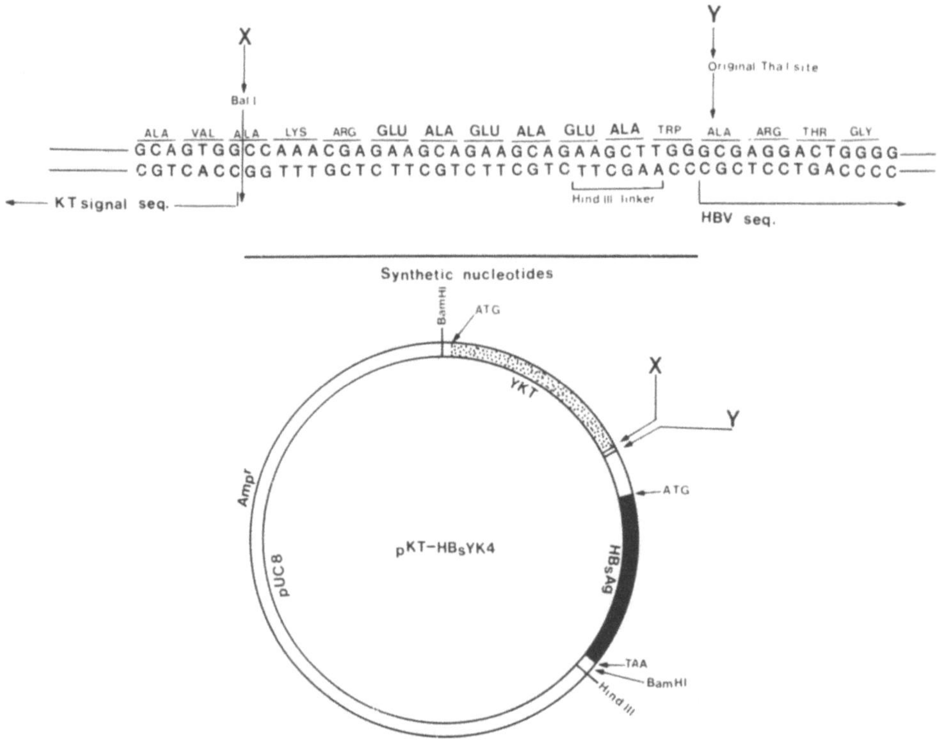

Figure 5. Construction of a chimeric gene containing a proteolytic cleavage site. Synthetic oligonucleotides were inserted between the Bal I site of the yeast killer toxin and Tha I site of hepatitis B virus surface antigen (HB_sAg) gene (X-Y) in pKT–HB_sYK3 to obtain three repeats of a glu-ala coding sequence. This plasmid is designated pKT–HB_sYK4. The BamHI fragment of pKT–HB_sYK4 was inserted into pYT760 for expression in yeast.

C. Expression and Secretion of HB_sAg in *S. frugiperda* Cells Using the Baculovirus Expression Vector

The baculovirus, *Autographa californica* nucleopolyhedrosis virus (AcNPV), is a helper-independent viral expression vector that has been shown to be a suitable expression system for the efficient production of proteins in cultured insect cells. This system has been used to produce human β-interferon (IFN_β) (Smith *et al.*, 1983); to express *E. coli* β-galactosidase (Pennock *et al.*, 1984); to express and secrete human interleukin-2 (IL-2) (Smith *et al.*, 1985); to produce human c-myc protein (Miyamoto *et al.*, 1985); to synthesize, glycosylate, and secrete human α-interferon (IFN_αs (Maeda *et al.*, 1985); to express influenza virus hemagglutinin (Possee, 1986); and for the expression of S RNA-coded genes of the lymphocytic choriomeningitis arenavirus (Matsuura *et al.*, 1986). It has been shown in this system that IFN_α and IFN_β are glycosylated by an N-linked sugar, that the pre-IFN_β signal peptide is removed, and that the protein is secreted efficiently. It has also been shown, using this vector, that human IL-2 was produced and secreted into the ECF and that it stimulated the growth of IL-2-dependent cell lines (Smith *et al.*, 1985).

We inserted BamHI fragments of the HB_sAg gene isolated from plasmid pHBsYK2 (Fig. 1) into the pAcRP6 transfer vector (Matsuura *et al.*, 1986), as shown in Fig. 6. The recombinant plasmid pAcRP6–HB_sYK14 contains the full coding sequence of the HB_sAg

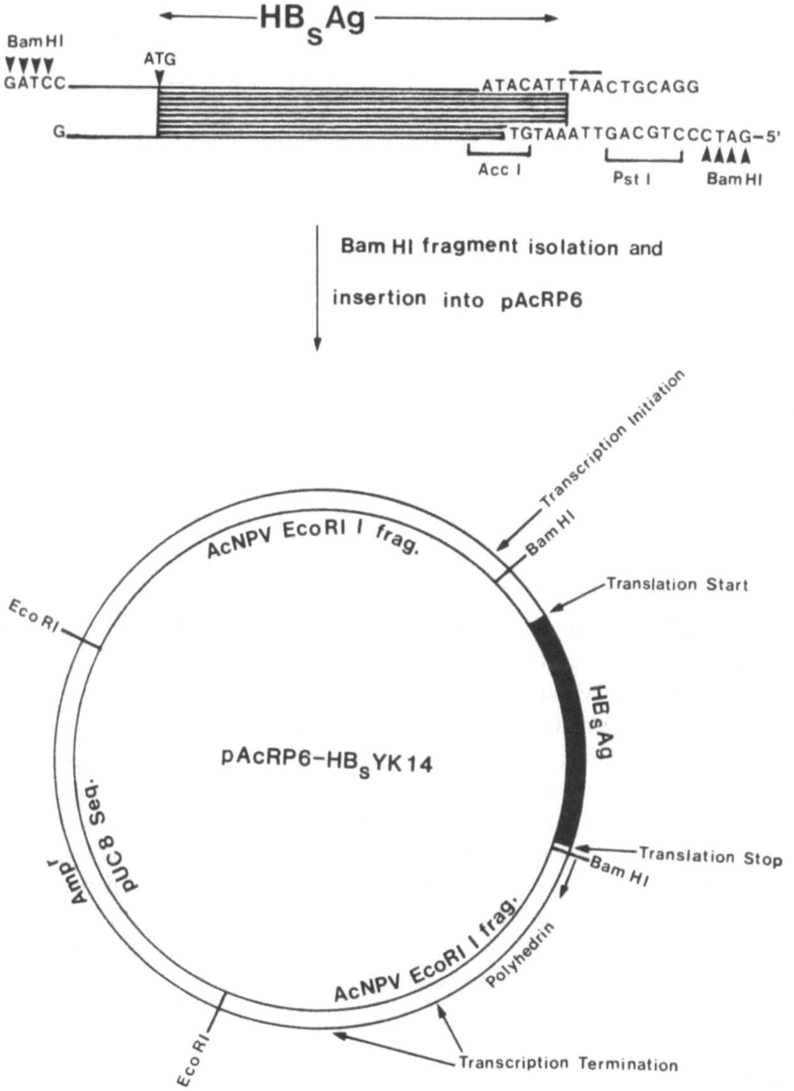

Figure 6. Construction of baculovirus transfer vector pAcRP6–HB_sYK14. The Bam HI fragment of pHB_sYK2 was isolated and inserted into the BamHI site of the transfer vector pAcRP6. All junction sequences were confirmed by the Maxam and Gilbert DNA-sequencing method. The plasmid pAcRP6–HB_sYK14 was used to cotransfect *S. frugiperda* cells with wild-type baculovirus genomic DNA. Polyhedrin-negative AcNPV recombinant virus was isolated and purified.

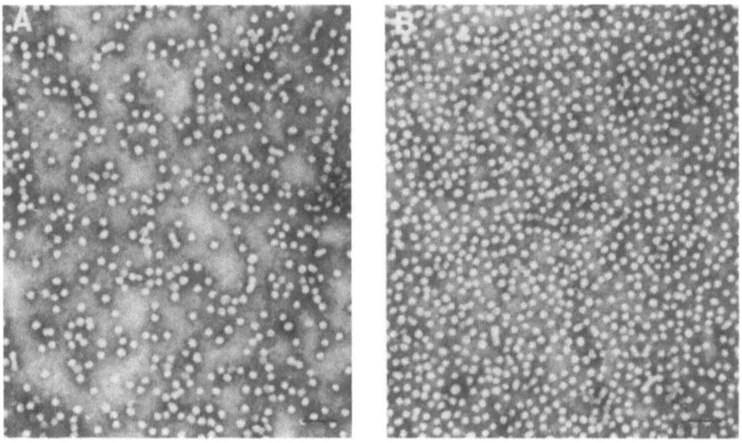

Figure 7. Electron micrographs of hepatitis B virus surface antigen (HB_sAg) particles produced by *Saccharomyces frugiperda* cells infected with recombinant AcNPV–HB_sYK14 virus. The recombinant AcNPV–HB_sYK14 virus was isolated following cotransfection of *S. frugiperda* cells with pAcRP6–HB_sYK14 and wild-type DNA of the baculovirus. The recombinant AcNPV–HB_sYK14 virus-infected cell culture media was harvested 4 days postinfection. Cells and cell debris were removed by centrifugation at 3000 rpm for 10 min, and the supernatant was further centrifuged at 25,000 rpm for 1 hr to remove all baculovirus particles. The supernatant was then concentrated with ammonium sulfate and examined electron microscopically. (A) Particles produced and secreted from *S. frugiperda* cells infected with AcNPV–HB_sYK14 virus. (B) Purified 22-nm particles from HB_sAg-positive human blood.

gene. The recombinant plasmid DNA was purified and cotransfected into *S. frugiperda* cells with wild-type AcNPV DNA. The plaques were screened for recombinant virus, which does not produce polyhedrin. The polyhedrin-negative plaques were isolated and subsequently plaque purified three times in order to obtain recombinant virus (AcNPV–HB_sYK14), which was free of wild-type virus. *S. frugiperda* cells were infected with the AcNPV–HB_sYK14 recombinant virus and culture fluid was monitored for the production of HB_sAg using a commercially available RIA kit (Abbott). The kinetics of 22-nm particle production demonstrated that the particles were formed and secreted into the ECF as early as 12 hr postinfection and continued to be produced until 6 days after infection. Electron microscopic examination revealed that 22-nm particles isolated from the ECF of the recombinant virus-infected *S. frugiperda* cells were indistinguishable from 22-nm particles isolated from the plasma of the chronic active hepatitis patients (Fig. 7). The level of HB_sAg production from the recombinant virus-infected *S. frugiperda* cells was approximately 10 mg/liter of suspension culture.

III. DISCUSSION

A cloned HB_sAg gene has been expressed in mammalian, yeast, and insect cell systems using appropriate expression vectors. RIA and electron microscopic examination of the HB_sAg particles produced from these three systems showed the following results. Cos 1 cells produced 22-nm particles of HB_sAg antigenically and morphologically similar to naturally occurring 22-nm particles. However, the efficiency of Cos 1 cell transformation was less than 10%, and the level of expression was not acceptable for vaccine preparation. Selection of transformants using a plasmid containing the neomycin resistance and a temperature-sensitive variant of Cos 1 cells has also been tried but without success. HB_sAg was also expressed in the form of 22-nm particles in yeast cells, using the galactokinase promoter and a part of the yeast killer toxin. However, these 22-nm particles were poorly antigenic and not readily assayable with the commercially available RIA kit. The introduction of an endopeptidase (lys-arg) and an exopeptidase site, $(glu\text{-}ala)_3$, at the junction of the yeast killer toxin polypeptide and the HB_sAg gene also produced 22-nm particles that were not secreted. These intracellular 22-nm particles were readily detectable with the RIA kit. It is therefore postulated that the yeast killer toxin portion of the chimeric protein may mask antigenic sites of HB_sAg. Consequently, the 22-nm particles were not recognized by the antibodies available in the RIA kit. When the yeast killer toxin portion was probably cleaved from the chimeric protein by introduction of the proteolytic processing sequences, the 22-nm particles were antigenic.

By contrast, a baculovirus expression system provided the expression of the cloned HB_sAg gene in *S. frugiperda* cells and secretion of antigenic 22-nm particles. RIA and electron microscopic examination revealed that 22-nm particles isolated from the ECF of *S. frugiperda* cells were identical to 22-nm particles isolated from chronic active hepatitis patients. The 22-nm particles produced from *S. frugiperda* cells contained predominantly a 22,000-M_r protein (data not shown). The level of 22-nm particle production was approximately 10 μg/ml of *S. frugiperda* cell culture, which is high enough for commercial application. Furthermore, our results showed that the same level of HB_sAg can be produced when serum-free medium is used during infection of *S. frugiperda* cells with recombinant AcNPV.

IV. CONCLUSION

Hepatitis B virus surface antigen expressed in Cos 1 cells under the control of SV40 late promoter secreted antigenically active 22-nm particles. It has been shown that the particles produced in mammalian cells are glycosylated and induce protective immunity in vaccinated chimpanzees. However, the level of expression is not adequate for vaccine production.

Yeast cells have been used to express HB_sAg; however, the particles show relatively low antigenicity. Yeast cells do not appear to glycosylate the particles, although 22-nm particles were formed after the cells were disrupted. The HB_sAg particles produced from yeast cells can induce protective immunity in vaccinated chimpanzees, and a yeast vaccine is currently licensed.

By contrast, HB_sAg produced by the baculovirus expression system is glycosylated and is secreted as 22-nm particles. The level of expression is the highest achieved among all the expression systems examined to date. The particles have morphologic and antigenic properties identical to those of 22-nm particles isolated from the plasma of patients with chronic active hepatitis. Thus, the baculovirus expression system provides the most promising expression system for the production of HBV vaccine and probably other vaccines as well.

REFERENCES

Choo, K.-B., Wu, S.-M., Lee, H.-H., and Lo, S. J. (1985). *Biochem. Biophys. Res. Commun.* **131,** 160–166.

Graham, F. L., and van der Eb, A. J. (1973). *Virology* **52,** 456–467.

Hitzeman, R. A., Chen, C. Y., Hagie, F. E., Patzer, E. J., Liu, C.-C., Estell, H. (1983). *Nucleic Acids Res.* **11,** 2745–2763.

Julius, D., Blair, L., Brake, A., Sprague, G., and Thorner, J. (1983). *Cell* **32,** 839–852.

Lolle, S., Skipper, N., Bussey, H., and Thomas, D. Y. (1984). *EMBO J.* **3,** 1383–1387.

Maeda, S., Kawai, T., Obinata, M., Fujiwara, H., Horiuchi, Y., Saeki, Y., Sato, Y., and Furusawa, M. (1985). *Nature (Lond.)* **315,** 592–594.

Maniatis, T., Fritsch, E. F., and Sambrook, J. (1982). *Molecular Cloning: A Laboratory Manual.* Cold Spring Harbor Laboratory, Cold Spring Harbor, New York.

Matsuura, Y., Possee, R. D., and Bishop, D. H. L. (1986). *J. Gen. Virol.* **67,** 1515–1529.

Miyamoto, C., Smith, G. E., Farrell-Towt, J., Chizzonite, R., Summers, M. D., and Ju, G. (1985). *Mol. Cell Biol.* **5,** 2860–2865.

Miyanohara, A., Toh-e, A., Nozaki, C., Hamada, F., Ohtomo, N., and Matsurbara, K. (1983). *Proc. Natl. Acad. Sci. USA* **80,** 1–5.

Pennock, G. D., Shoemaker, C., and Miller, L. K. (1984). *Mol. Cell Biol.* **4,** 399–406.

Possee, R. D. (1986). *Virus Res.* **5,** 43–59.

Skipper, N., Thomas, D. Y., and Lau, P. (1984). *EMBO J.* **3,** 107–111.

Smith, G. E., Summers, M. D., and Fraser, M. J. (1983). *Mol. Cell Biol.* **3,** 2156–2165.

Smith, G. E., Ju, G., Ericson, B. L., Moschera, J., Lahm, H.-W., Chizzonite, R., and Summers, M. D. (1985). *Proc. Natl. Acad. Sci. USA* **82,** 8404–8408.

Valenzuela, P., Medina, A., Rutter, W. J., Ammerer, G., and Hall, B. D. (1982). *Nature (Lond.)* **298,** 347–350.

4

The Development of New Poliovirus Vaccines Based on Molecular Cloning

Akio Nomoto, Michinori Kohara, Shusuke Kuge, Shinobu Abe, Bert L. Semler, Toshihiko Komatsu, Mineo Arita, and Heihachi Itoh

I. INTRODUCTION

Poliovirus, known since 1908 to be the causative agent of poliomyelitis (Paul, 1971), is a human enterovirus of the Picornaviridae. This virus consists of a single-stranded RNA of plus-strand polarity and 60 copies each of four capsid proteins, VP1, VP2, VP3, and VP4; it occurs in three serologically distinct types, i.e., type 1, type 2, and type 3.

Two strategies have been pursued to control poliomyelitis (Paul, 1971). The first strategy involved the preparation of inactivated virus as vaccines (Salk, 1960). The second strategy involved the isolation of attenuated virus strains; those used by Sabin were found to be most effective as oral live vaccines (Sabin and Boulger, 1973; Melnick, 1984). Although these effective poliovirus vaccines have been available since the mid-1950s, paralytic poliomyelitis remains a serious threat in many countries, especially in developing countries in which effective vaccination against poliomyelitis is not sufficiently inclusive.

The virtue of the oral live polio vaccines has been widely recognized (Melnick, 1984). However, the Sabin vaccine strains similar to other live vaccines have the inherent problem of risk of reversion to virulence upon repeated passages. Indeed, a very small number of cases of paralytic poliomyelitis continues to occur in countries with extensive oral polio vaccine programs (Melnick, 1984). Experimental evidence strongly suggests

Akio Nomoto and Shusuke Kuge • Department of Microbiology, Faculty of Medicine, University of Tokyo Hongo, Bunkyo-ku, Tokyo 113, Japan. *Michinori Kohara* • Department of Microbiology, Faculty of Medicine, University of Tokyo, Hongo, Bunkyo-ku, Tokyo 113, Japan; and Japan Poliomyelitis Research Institute, Higashimurayama, Tokyo 189, Japan. *Shinobu Abe and Heihachi Itoh†* • Japan Poliomyelitis Research Institute, Higashimurayama, Tokyo 189, Japan. *Bert L. Semler* • Department of Microbiology and Molecular Genetics, College of Medicine, University of California, Irvine, California 92717. *Toshihiko Komatsu and Mineo Arita* • Department of Enteroviruses, National Institute of Health, Musashimuryama, Tokyo 190-12, Japan.
†Deceased.

that most of these cases are caused by the vaccines themselves, especially type 2 and type 3 vaccines (Kew *et al.*, 1981; Minor, 1982).

The rate by which spontaneous mutations occur is especially high in single-stranded RNA genome replication (10^{-3}–10^{-4}) as compared with double-stranded DNA replication (10^{-8}–10^{-11}) (Holland *et al.*, 1982). The single-stranded RNA genome of poliovirus is not an exception to this high rate of mutation. As a result, every poliovirus preparation is a mixture of different genotypes, and even the original Sabin vaccine strain cannot be considered completely homogeneous.

The high frequency of mutations that occurs during poliovirus genome replication makes it difficult to maintain the attenuated phenotype of the polio live vaccines. In addition, attempts to develop new poliovirus vaccine strains using methods of plaque purification have proved unsuccessful to date. Because the worldwide eradication of poliovirus may take generations or may be altogether impossible (Chin, 1984; Gregg, 1984), global control of poliomyelitis may eventually be hampered by the limiting amount of available stock vaccine virus (WHO Consultative Group on Oral Poliomyelitis Vaccine, 1977).

This chapter describes our recent work focused on development of new poliovirus vaccines based on molecular cloning. Prospective strategies for *in vitro* construction of oral live polio vaccines are discussed.

II. GENERAL FEATURE OF POLIOVIRUS GENOME

The genome of poliovirus is composed of approximately 7500 nucleotides, polyadenylylated at the 3′-terminal (Yogo and Wimmer, 1972) and covalently attached to a genome-linked protein (VPg) at the 5′-terminal (Wimmer, 1982) (Fig. 1). This RNA is infectious in mammalian cells regardless of whether VPg is attached at the 5′-terminal (Nomoto *et al.*, 1977). In the host cell cytoplasm, the viral RNA is translated into a single continuous polyprotein with a molecular weight of 247,000; the polyprotein is subsequently cleaved by proteases to form viral structural and nonstructural proteins (Kitamura *et al.*, 1981; Hanecak *et al.*, 1984; Toyoda *et al.*, 1986). Studies on the sequences of RNA and of amino acid in viral polypeptides have provided a precise viral protein map (Kitamura *et al.*, 1981) (see Fig. 1).

To date, complementary DNAs (cDNAs) to the genomes of both the virulent and attenuated strains of all three poliovirus serotypes have been cloned using bacterial plasmids, and the total nucleotide sequences of these genomes have been determined. Poliovirus strains whose genome sequences have been elucidated are listed in Table 1.

Comparison of the complete nucleotide sequences of all three poliovirus serotype genomes, carried out by Toyoda *et al.* (1984), revealed the conserved (hence probably essential) genetic informations of poliovirus that included nucleotide sequences of the noncoding regions of the genome, amino acid sequences of the viral nonstructural proteins, and characteristics of viral structural proteins contributing to virion formation. On the basis of these data, the genetic variation of poliovirus, the cleavage signals in polyprotein processing, the initiation site(s) of translation, the variations in the noncoding regions, and the potential antigenic sites involved in neutralization by antibodies have been discussed (see Toyoda *et al.*, 1984 for details).

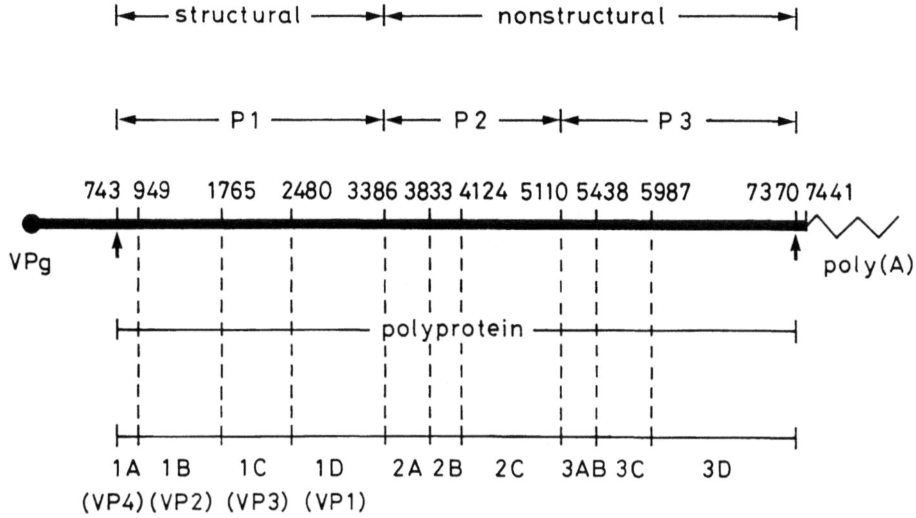

Figure 1. Gene organization of poliovirus. The genome RNA is shown as a solid bar. Numbers above the genome indicate nucleotide positions from the 5′-terminal. Lines indicate the polypeptides and nomenclature of the viral proteins follows the report by Rueckert and Wimmer (1984). Filled arrows indicate the initiation and termination sites of viral protein synthesis.

III. PRODUCTION OF POLIOVIRUS ANTIGENIC PEPTIDES

The antigenic determinants at the surface of a nonenveloped virion such as poliovirus are protrusions formed by the polypeptide chains of the tightly aggregated capsid proteins. Chemical labeling (Wetz and Habermehl, 1979) and cross-linking of monoclonal neutralizing antibodies (Emini *et al.*, 1982) have revealed that VP1 is the predominantly exposed capsid protein and carrier of neutralizing epitopes. Viral polypeptides carrying capsid protein VP1 might therefore be capable of inducing anti-poliovirus neutralizing antibodies.

To test the possibility of developing subunit vaccines, poliovirus cDNA encoding

Table 1. Poliovirus Strains

Serotype		Strain	Reference
Type 1	Virulent	Mahoney	Kitamura *et al.* (1981)
			Racaniello and Baltimore (1981*a*)
	Attenuated	Sabin 1 (LSc,2ab)	Nomoto *et al.* (1982)
Type 2	Virulent	Lansing	LaMonica *et al.* (1986)
	Attenuated	Sabin 2 (p712,Ch,2ab)	Toyoda *et al.* (1984)
Type 3	Virulent	Leon	Stanway *et al.* (1984)
	Attenuated	Sabin 3 (Leon $12a_1b$)	Stanway *et al.* (1983)
			Toyoda *et al.* (1984)

viral polypeptide including capsid protein VP1 was expressed in *Escherichia coli* under control of the penicillinase promoter system (Kuge *et al.*, 1984) (Fig. 2). In the expressing vectors, poliovirus sequences were designed to be read in phase and therefore to be expressed as fusion proteins with the bacterial peptides. In addition, the *E. coli* tryptophan operon promoter–operator system was inserted upstream of the penicillinase system in order to obtain stronger expression of the poliovirus sequences. The antigenic polypeptides produced in the bacteria transformed with the plasmids were detected by immunoprecipitation with antibodies to capsid proteins VP1 or VP3 followed by sodium dodecyl sulfate-polyacrylamide gel electrophoresis (SDS-PAGE) (Fig. 2). Similar work was also done by van der Werf *et al.* (1983).

The antigenic polypeptides, however, displayed no ability to elicit anti-poliovirus neutralizing antibody in rabbits, although the peptides was found capable of priming the immune response in rabbits (Table 2). The priming ability and low immunogenicity of short synthetic peptide corresponding to antigenic determinants of type 1 poliovirus had

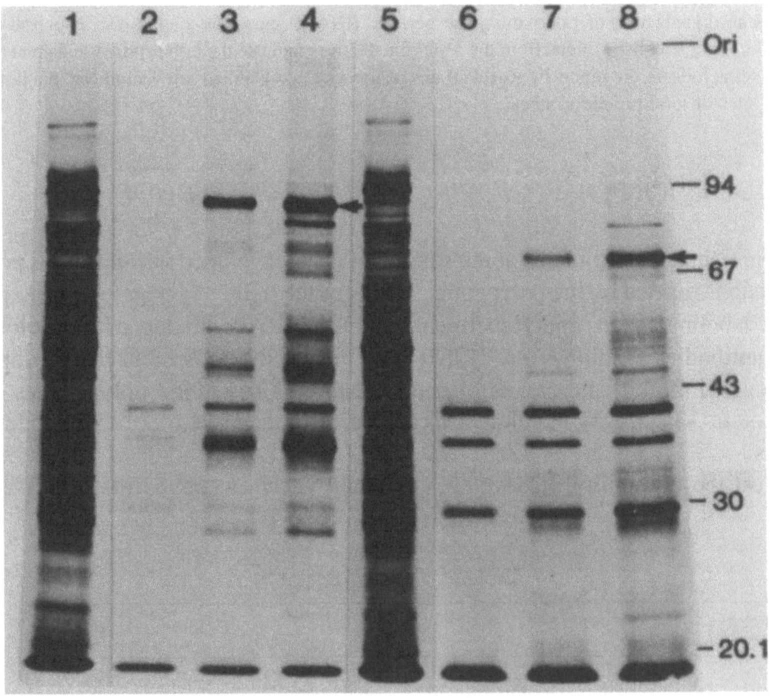

Figure 2. Immunoprecipitated proteins on sodium dodecyl sulfate-polyacrylamide gel. Proteins of *Escherichia coli* carrying plasmids pLT-P12 (lanes 1–4) and pGT–P12 (lanes 5–8). Total proteins of *E. coli* carrying pLT–P12 (lane 1) and pGT–P12 (lane 5). Immunoprecipitated proteins with normal sera (lanes 2 and 6), anti-VP3 IgG (lanes 3 and 7), and anti-VP1 IgG (lanes 4 and 8). Positions of specifically precipitated peptides (P12) are indicated by arrows. Peptides (P12) has amino acid sequence corresponding to those of most VP3 and whole VP1. Molecular weights are shown at the right. Peptides expressed from plasmid pLT–P12 were used in the experiment of Table 2. (From Kuge *et al.*, 1984.)

Table 2. Priming of Anti-Poliovirus Immune Response in Rabbits

Injection			
First[a]	Second[b]	Rabbit	Titer of anti-type 1 sera[c]
P12[d]	P12	A	<1[f]
		B	1
P12	Sabin 1[e]	C	>64
		D	>128
VP1	VP1	E	11
		F	11
VP1	Sabin 1	G	>128
		H	>128
VP3	VP3	I	<1
		J	6
VP3	Sabin 1	K	>128
		L	>128
XXXXXXX	Sabin 1	M	11
		N	4
Sabin 1	Sabin 1	O	>128
		P	>128

[a] Intradermal.
[b] Intramuscular, 8 weeks after the first injection.
[c] Sera were prepared 10 days after the second injection.
[d] Peptide expressed in *Escherichia coli* shown in Fig. 2; 50 μg of peptide, diluted 1 : 1 with complete Freund's adjuvant (CFA) was injected.
[e] 10^6 PFU of poliovirus without CFA was injected.
[f] Serum titer sufficient to neutralize 214 TCD_{50} of type 1 poliovirus.

initially been reported by Emini *et al.* (1983). Our results obtained as above are compatible with those of Emini *et al.* (1983). It is therefore not easy to consider these poliovirus antigenic peptides subunit vaccines.

Recently, Diamond *et al.* (1985) reported that one amino acid change alters the distant (in terms of linear sequence) immunogenic structure of antibody-binding sites of poliovirus. The results indicate that epitope functions expressed in antibody-binding sites are supported by amino acids in other loci of capsid proteins. Thus, expression of immunogenicity of poliovirus appears to depend on specific conformation of capsid proteins formed only in the virion particle. Highly immunogenic subunit vaccines of poliovirus might be designed and produced in the future based on accumulated knowledge concerning functional conformation of neutralizing epitopes of poliovirus. Crystallography on picornaviruses must be a powerful tool for elucidation of the functional epitope structures (Rossman *et al.*, 1985; Hogle *et al.*, 1985).

IV. PRESERVATION OF THE GENETIC INFORMATION OF THE SABIN STRAIN OF TYPE 1 POLIOVIRUS

The high frequency of mutation occurring in poliovirus genome replication suggests that the seed viruses for polio live vaccines will be used up because of their limiting amount (WHO Consultative Group on Oral Poliomyelitis Vaccine, 1977). However, for the control of poliomyelitis in developed and in developing countries, it is desirable to preserve the constancy and quality of the present seed viruses of the Sabin vaccine strains.

A. Construction of Infectious cDNA Clones

One possibility for stabilization of the poliovirus genotypes is to carry out large-scale propagations of the genomes as cDNAs cloned into bacteria plasmids. Racaniello and Baltimore (1981*b*) assembled a full-length clone of the genome of the virulent Mahoney strain of type 1 poliovirus from subgenomic cDNA clones; these workers showed that transfection of plasmid DNA from this clone would produce infectious poliovirus in

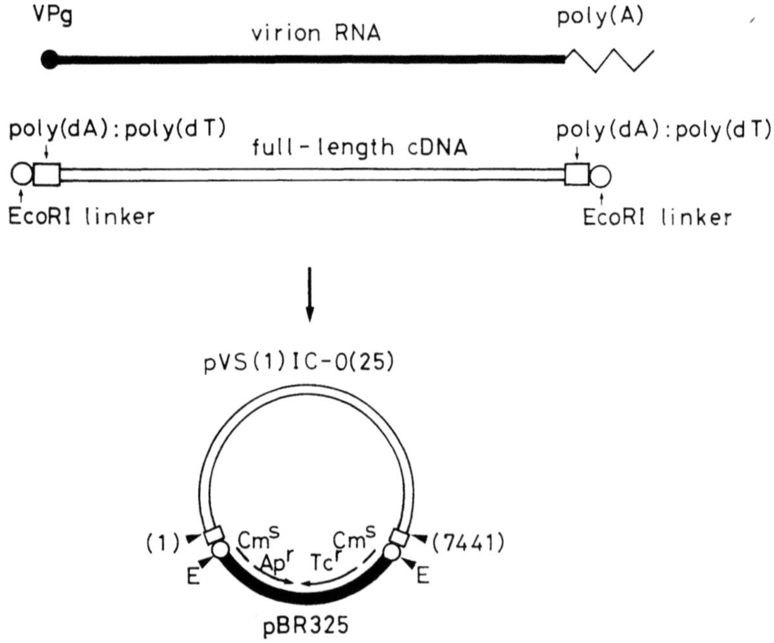

Figure 3. Infectious cDNA clone of the Sabin 1 strain. The virion RNA of the Sabin 1 strain is shown at the top. The full-length cDNA of the Sabin 1 genome is shown below the corresponding RNA. Open square and open circle represent poly(dA) : poly(dT) and EcoR1 linker nucleotide, respectively. In the plasmid pVS(1)IC–0(25), the filled bar indicates the sequence of bacterial plasmid pBR325. Tc^r and Ap^r represent tetracycline- and ampicillin-resistant genes, respectively. The gene that has lost the function of chloramphenicol resistance is indicated by Cm^s. Numbers in parentheses represent nucleotide numbers on the corresponding Sabin 1 genome.

Table 3. Infectivity of Sabin 1 Infectious Clones Measured in a 6-cm Plastic Dish[a]

DNA	μg[b]	Average number of plaques[c]	
		AGMK cells	HeLa S3 cells
pVS(1)IC–0(25)	10	2	38
	2	ND[d]	5
pVS(1)IC–O(T)	10	104	ND
	2	23	—[e]
	0.4	6	ND
	0.016	ND	28

[a] Transfection period was 4 hr. Results obtained from experiments using AGMK cells are from Kohara *et al.* (1986).
[b] Total amount of DNA was adjusted to 10 μg per dish by adding salmon sperm DNA as carrier. No plaques were observed by the transfection with 10 μg salmon sperm DNA per dish.
[c] Average plaque number of three to four experiments.
[d] No experiment was done.
[e] Impossible to count because of the enormous number of plaques.

cultured mammalian cells. Semler *et al.* (1984) constructed a recombinant plasmid containing an independently derived full-length cDNA clone of type 1 Mahoney RNA and transcription and replication reguratory sequences from simian virus 40 (SV40). The above recombinant plasmid produced high levels of infectious poliovirus when transfected into monkey cells.

Full-length cDNA of the poliovirus vaccine Sabin 1 strain has been inserted into the EcoR1 site of bacterial plasmid pBR325 using EcoR1 linkers (Omata *et al.*, 1984) (Fig. 3). This cloned DNA is infectious in mammalian cells and designated as pVS(1)IC–0(25). Recently, Semler and Johnson (manuscript in preparation) constructed a highly infectious cDNA clone of the Mahoney strain of type 1 poliovirus. In addition to a full-length cDNA copy of the type 1 Mahoney genome, this plasmid (pPVA55) contains the SV40 origin of replication as well as the promoter and coding region for the SV40 large T antigen. To obtain a highly infectious Sabin 1 clone, the Mahoney sequence in the plasmid was replaced by the Sabin 1 sequence. The cDNA clone of the Sabin 1 strain thus obtained was designated pVS(1)IC–0(T) (Kohara *et al.*, 1986) and was tested for infectivity on AGMK cells. For transfection experiments in which calcium phosphate–DNA precipitates were left on the cells for 4 hr, we observed a specific infectivity 50-fold higher than that obtained with pVS(1)IC–0(25) (Table 3). A similar transfection experiment was also carried out using HeLa S3 monolayer cells. The results indicated more than 500-fold higher specific infectivity for pVS(1)IC–0(T) than that of pVS(1)IC–0(25) (Table 3).

The virus recovered from AGMK cells transfected with 10 μg pVS(1)IC–0(25) or pVS(1)IC–0(T) was designated PV1(Sab)IC–0(A) or PV1(Sab)IC–0(TA), respectively. PV1(Sab)IC–0(A) and PV1(Sab)IC–0(TA) were further passaged once in AGMK cells, and the passaged viruses were used for biologic tests.

B. Characteristics of the Recovered Viruses

Virus recovered from HeLa S3 cells transfected with the plasmid pVS(1)IC–0(25) had been shown by physical methods and *in vitro* marker tests (*rct*, *d*, and plaque size marker tests) to be indistinguishable from the Sabin 1 vaccine reference virus (Omata *et al.*, 1984; Kohara *et al.*, 1985). The test using reproductive capacity at different temperatures (*rct*) test is based on the temperature sensitivity of multiplication of poliovirus vaccine strains. The delayed growth (*d*) test is based on the phenotype that the vaccine strains replicate poorly at low bicarbonate concentrations under agar overlays. Plaque size is also one of the marker tests used as an indication of the rate of viral multiplication, and the plaque size of vaccine strains is smaller than that of virulent strains.

These *in vitro* marker tests were performed to determine the characteristics of PV1(Sab)IC–0(A); the results are shown in Table 4. Virus WR1S–77ABF had been obtained by one passage of LSc,2ab/KP_3 (SOM) in AGMK cell culture, now being used as reference virus in *in vitro* marker tests on newly manufactured type 1 polio live vaccines in the Japan Poliomyelitis Research Institute, Tokyo. Similarly, Mahoney Y-3 is the reference of the virulent Mahoney strain of type 1 poliovirus. The results did not indicate any differences in these *in vitro* phenotypes between virus recovered by DNA transfection and the reference vaccine virus.

The tests were carried out on viruses of three different isolates produced by DNA transfections into HeLa S3 cells and of four different isolates produced by the transfections into AGMK cells. No differences among these isolates were observed (data not shown; see Omata *et al.*, 1984; Kohara *et al.*, 1986). These results indicate that viruses of indistinguishable properties were obtained from all transfections independently carried out with the same infectious cDNA clone of the Sabin 1 genome. The results of *in vitro* marker tests of PV1(Sab)IC–0(TA) showed that the virus also had *in vitro* phenotypes indistinguishable from the reference vaccine virus (Kohara *et al.*, 1986).

Neurovirulence tests of viruses PV1(Sab)IC–0(A) and PV1(Sab)IC–0(TA) were carried out using cynomolgus monkeys with the intraspinal route of infection. The results

Table 4. *In Vitro* Marker Tests of Recovered Virus and Reference Virus[a]

Virus	*rct*[b] Log difference between temperatures			*d*[c] Log difference between concn. (%)	Plaque size[d] 5 days postinoculation (mm diameter)
	36/39	36/39.5	36/40	0.225/0.03	
WR1S-77ABF	5.20	>7.05	—	4.21	2–5
PV1(Sab)IC-O(A)	5.73	>7.58	—	5.33	3–5
Mahoney Y-3	—	—	0.44	0.41	16

[a] From Kohara *et al.* (1986).
[b] Values shown are the logarithmic differences of virus titers obtained at temperatures indicated.
[c] Values shown are the logarithmic differences of virus titers obtained at two different sodium bicarbonate concentrations.
[d] Diameters of approximately 100 plaques for each virus were measured.

Table 5. Monkey Neurovirulence Tests of Recovered Viruses and Reference Virus[a]

Virus	Number of monkeys	Average lesion score
NIH Ref.	10	1.07
PV1(Sab)IC–O(A)	8	0.76
PV1(Sab)IC–O(TA)	9	0.82

[a] From Kohara *et al.* (1986).

are shown in Table 5 (Kohara *et al.*, 1986). The average lesion scores of PV1(Sab)IC–0(A) and PV1(Sab)IC–0(TA) were similar to that of the vaccine reference, suggesting that these viruses are attenuated enough to be considered as candidate viruses of the oral polio live vaccine.

V. IN VITRO CONSTRUCTION OF A CANDIDATE VACCINE STRAIN OF TYPE 3 POLIOVIRUS

Knowledge of the total nucleotide sequences of the genomes of the virulent Mahoney and the attenuated Sabin 1 strains of poliovirus and the availability of infectious cDNA clones of both strains suggested the use of a molecular genetic approach to identify mutations that influence the biologic differences between the two strains. In recent studies (Omata *et al.*, 1986), biologic tests including monkey neurovirulence tests were carried out on recombinant viruses constructed *in vitro* by allele replacement of the genome segments between both strains.

The results of monkey neurovirulence tests on the recombinant viruses revealed that many of 55 mutations discovered in the genome of the Sabin 1 strain influence the phenotype of viral attenuation. The loci influencing attenuation are spread over wide areas of the viral genome, including the 5′-noncoding region (Omata *et al.*, 1986). These conditions are compatible with the observation that the Sabin 1 virus is a relatively stable vaccine, as multiple mutations would be required to regain the neurovirulent phenotype. Whereas sequence studies of the genomes of the neurovirulent Leon strain of type 3 poliovirus and its attenuated derivative, the Sabin 3 strain, revealed only 10 nucleotide differences, of these mutations, only three led to amino acid replacements (Stanway *et al.*, 1984).

More stable (hence safer) vaccine strains of type 2 and type 3 polioviruses might therefore be constructed *in vitro* by the replacement of only the sequence encoding the antigenic determinants of the Sabin 1 genome by the corresponding sequences of the type 2 and type 3 genomes, respectively. Using this strategy, we replaced the Sabin 1 sequence of pVS(1)IC–0(T) encoding the whole-coat (P1) proteins by the corresponding sequence of a Sabin 3 cDNA clone, pVS(3)2603 (Toyoda *et al.*, 1984). The recombinant cDNA clone thus obtained was designated pVSS(1/3)IC-BN; the virus recovered from AGMK cells transfected with pVSS(1/3)IC-BN was designated PV1/3(SS)BN. The genome structure of the recovered virus is shown in Fig. 4.

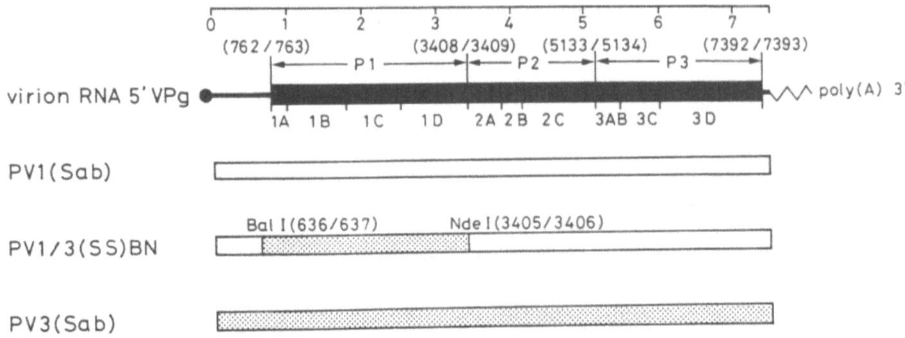

Figure 4. Genome structure of a recombinant virus between the Sabin 1 and the Sabin 3. The virion RNA and the gene organization of poliovirus are shown at the top. The length of nucleotides is indicated in kilobases above the RNA. Numbers in parentheses indicate the consensus nucleotide numbers of all three poliovirus serotypes (Toyoda *et al.*, 1984). Open and shadowed bars represent sequences of the Sabin 1 and Sabin 3 genomes, respectively.

PV1/3(SS)BN was recognized and neutralized by both monoclonal and polyclonal anti-type 3 poliovirus antibodies (data not shown). The *rct* marker test on the recombinant virus was carried out; the results are shown in Table 6. Virus F-310 is the Sabin 3 vaccine virus practically being used in Japan. PV1(M)pDS306 is the virus recovered from AGMK cells transfected with an infectious cDNA clone of the virulent Mahoney strain of type 1 poliovirus (Kohara *et al.*, 1985; Omata *et al.*, 1986). The result did not indicate any difference in temperature sensitivity between viruses PV1/3(SS)BN and PV1(Sab)IC–O(TA), and the former virus was even more sensitive to temperature than the vaccine virus F-310 (Table 6). Preliminary biologic tests including monkey neurovirulence tests were performed on the recombinant virus PV1/3(SS)BN; the results suggested that the recombinant virus might be used as a vaccine candidate of type 3 poliovirus (data not shown).

VI. OTHER PROSPECTS FOR CONSTRUCTION OF NEW POLIO VACCINE STRAINS

Recently, many polioviruses carrying modified genome sequences have been constructed using the infectious cDNA clones, and the biologic properties of these viruses

Table 6. Reproductive Capacity at Different Temperatures of Recovered Viruses and Reference Viruses ($\log_{10}$PFU/ml)

Virus	Log difference between temperatures		
	36/39	36/39.5	36/40
PV1(Sab)IC–O(A)	5.38	>8.46	—
PV1/3(SS)BN	5.57	>7.92	—
F-310	4.41	5.64	—
PV1(M)pDS306	—	0.04	0.17

have been analyzed (Kohara *et al.*, 1985, 1986; Omata *et al.*, 1986; LaMonica *et al.*, 1986; Semler *et al.*, 1986; Sarnow *et al.*, 1986). These studies have provided important examples of how recombinant DNA technology can be used to investigate the relationship of gene structure and functions in poliovirus.

Omata *et al.* (1986) suggested by allele-replacement experiments that mutations in the 5′-noncoding region of the Sabin 1 strain significantly influence the attenuation phenotype. The mechanism for this phenomenon is unknown but could be a modulation in one or all of the following steps in viral proliferation: initiation of protein synthesis, RNA replication, or morphogenesis. Indeed, the genomes of the Mahoney and the Sabin 1 strains differ in their efficiencies in *in vitro* translation (Svitkin *et al.*, 1985). Moreover, a single base change in the 5′-untranslated region of type 3 poliovirus has been correlated with the attenuation phenotype of this virus (Evans *et al.*, 1985).

It might therefore be possible to construct new live polio vaccines with a novel conception that viral attenuation is attributed to lowered efficiency of certain replication steps essential for viral multiplication. The attenuation phenotype must be very stable if (1) the efficiency of the certain stage of viral replication is reduced by introducing deletion of nucleotides into the viral genome, or (2) all the revertants that regain the normal efficiency of the replication step are designed to be selectively eliminated because of more serious deficiency in other replication steps. As examples for the former strategy, we recently succeeded in constructing deletion mutants that have a stable low rate of viral multiplication, although the viral replication steps modulated by the mutations are now obscure. The construction of viruses according to the latter strategy is not easy but seems possible, since different and possibly distant genome locus of single-stranded RNA genomes (e.g., the genome of poliovirus) may have specific and complex interactions with each other, resulting in the formation of multifunctional structures essential for viral replication. In any event, molecular genetic and biochemical studies designed to identify the functional structures of the viral RNA and proteins must be required to establish the construction strategies of new live polio vaccines, which might be extrapolated to construction of new live vaccines of other picornaviruses.

REFERENCES

Chin, J. (1984). *Rev. Infect. Dis. (Suppl.)* **6,** 581–585.

Diamond, D. C., Jameson, B. A., Bonin, J., Kohara, M., Abe, S., Itoh, H., Komatsu, T., Arita, M., Kuge, S., Nomoto, A., Osterhaus, A. D. M. E., Crainic, R., and Wimmer, E. (1985). *Science* **229,** 1090–1093.

Emini, E. A., Jameson, B. A., Lewis, A. J., Larsen, G. R., and Wimmer, E. (1982). *J. Virol.* **43,** 997–1005.

Emini, E. A., Jameson, B. A., and Wimmer, E. (1983). *Nature (Lond.)* **304,** 699–703.

Evans, D. M. A., Dunn, G., Minor, P. D., Schild, G. C., Cann, A. J., Stanway, G., Almond, J. W., Currey, K., and Maizel, J. V. (1985). *Nature (Lond.)* **314,** 548–550.

Gregg, M. B. (1984). *Rev. Infect. Dis. (Suppl.)* **6,** 577–580.

Hanecak, R., Semler, B. L., Ariga, H., Anderson, C. W., and Wimmer, E. (1984). *Cell* **37,** 1063–1073.

Hogle, J. M., Chow, M., and Filman, D. J. (1985). *Science* **229,** 1358–1363.

Holland, J., Spindler, K., Horodyski, F., Grabau, E., Nichol, S., and VandePol, S. (1982). *Science* **215,** 1577–1585.

Kew, O. M., Nottay, B. K., Hatch, M. H., Nakano, J. H., and Obijeski, J. F. (1981). *J. Gen. Virol.* **56,** 337–347.

Kitamura, N., Semler, B. L., Rothberg, P. G., Larsen, G. R., Adler, C. J., Dorner, A. J., Emini, E. A.,

Hanecak, R., Lee, J. J., van der Werf, S., Anderson, C. W., and Wimmer, E. (1981). *Nature (Lond.)* **291,** 547–553.

Kohara, M., Omata, T., Kameda, A., Semler, B. L., Itoh, H., Wimmer, E., and Nomoto, A. (1985). *J. Virol.* **53,** 786–792.

Kohara, M., Abe, S., Kuge, S., Semler, B. L., Komatsu, T., Arita, M., Itoh, H., and Nomoto, A. (1986). *Virology* **151,** 21–30.

Kuge, S., Chisaka, O., Kohara, M., Imura, N., Matsubara, K., and Nomoto, A. (1984). *J. Biochem.* **96,** 305–312.

LaMonica, N., Meriam, C., and Racaniello, V. R. (1986). *J. Virol.* **57,** 515–525.

Melnick, J. L. (1984). *Rev. Infect. Dis. (Suppl.)* **6,** 323–327.

Minor, P. D. (1982). *J. Gen. Virol.* **59,** 307–317.

Nomoto, A., Kitamura, N., Golini, F., and Wimmer, E. (1977). *Proc. Natl. Acad. Sci. USA* **74,** 5345–5349.

Nomoto, A., Omata, T., Toyoda, H., Kuge, S., Horie, H., Kataoka, Y., Genba, Y., Nakano, Y., and Imura, N. (1982). *Proc. Natl. Acad. Sci. USA* **79,** 5793–5797.

Omata, T., Kohara, M., Sakai, Y., Kameda, A., Imura, N., and Nomoto, A. (1984). *Gene* **32,** 1–10.

Omata, T., Kohara, M., Kuge, S., Komatsu, T., Abe, S., Semler, B. L., Kameda, A., Itoh, H., Arita, M., Wimmer, E., and Nomoto, A. (1986). *J. Virol.* **58,** 348–358.

Paul, J. R. (1971). *A History of Poliomyelitis.* Yale University Press, New Haven, Connecticut.

Racaniello, V. R., and Baltimore, D. (1981*a*). *Proc. Natl. Acad. Sci. USA* **78,** 4887–4891.

Racaniello, V. R., and Baltimore, D. (1981*b*). *Science* **214,** 916–919.

Rossman, M. G., Arnold, E., Erickson, J. W., Frankenberger, E. A., Griffith, J. P., Hecht, H.-J., Johnson, J., Kamer, G., Luo, M., Mosser, A. G., Rueckert, R. R., Sherry, B., and Vriend, G. (1985). *Nature (Lond.)* **317,** 145–153.

Rueckert, R. R., and Wimmer, E. (1984). *J. Virol.* **50,** 957–959.

Sabin, A. B., and Boulger, L. R. (1973). *J. Biol. Stand.* **1,** 115–118.

Salk, J. E. (1960). *Lancet* **2,** 715–723.

Sarnow, P., Bernstein, H. D., and Baltimore, D. (1986). *Proc. Natl. Acad. Sci. USA* **83,** 571–575.

Semler, B. L., Dorner, A. J., and Wimmer, E. (1984). *Nucleic Acids Res.* **12,** 5123–5141.

Semler, B. L., Johnson, V. H., and Tracy, S. (1986). *Proc. Natl. Acad. Sci. USA* **83,** 1777–1781.

Stanway, G., Çann, A. J., Hauptmann, R., Hughes, P., Clarke, L. D., Mountford, R. C., Minor, P. D., Schild, G. C., and Almond, J. W. (1983). *Nucleic Acids Res.* **11,** 5629–5643.

Stanway, G., Hughes, P. J., Mountford, R. C., Reeve, P., Minor, P. D., Schild, G. C., and Almond, J. W. (1984). *Proc. Natl. Acad. Sci. USA* **81,** 1539–1543.

Svitkin, Y. V., Maslova, S. V., and Agol, V. I. (1985). *Virology* **147,** 243–252.

Toyoda, H., Kohara, M., Kataoka, Y., Suganuma, T., Omata, T., Imura, N., and Nomoto, A. (1984). *J. Mol. Biol.* **174,** 561–585.

Toyoda, H., Nicklin, M. J. H., Murray, M. G., Anderson, C. W., Dunn, J. J., Studier, F. W., and Wimmer, E. (1986). *Cell* **45,** 761–770.

van der Werf, S., Dreano, M., Bruneau, P., Kopecka, H., and Girard, M. (1983). *Gene* **23,** 85–93.

Wetz, K., and Habermehl, K. O. (1979). *J. Gen. Virol.* **44,** 525–534.

WHO Consultative Group on Oral Poliomyelitis Vaccine (Sabin Strains) (1977). WHO, BLG/POLIO/77.1, Annex 5.

Wimmer, E. (1982). *Cell* **28,** 199–201.

Yogo, Y., and Wimmer, E. (1972). *Proc. Natl. Acad. Sci. USA* **69,** 1877–1882.

5

Rotavirus

New Vaccine and Vaccination

T. Vesikari, E. Isolauri, T. Ruuska, A. Delem, and F. E. André

I. INTRODUCTION

Rotavirus gastroenteritis is one of the major infectious diseases in the world today, as judged by mortality. Precise figures are not available, but it has been estimated that rotavirus diarrhea may take the lives of 500,000 or more children annually (Butler, 1984). Rotavirus diarrhea frequently results in dehydration and therefore runs a more severe course than do most other forms of infectious diarrhea in young children. While the share of rotavirus in all episodes of diarrhea in the developing countries is small, rotaviruses are much more frequently found among severe cases. A 2-year survey in Bangladesh showed that rotavirus to be responsible for 46% of the diarrheal episodes leading to dehydration (Black *et al.*, 1981). In other etiologic studies of childhood diarrhea requiring hospital admission, rotavirus was found in 45% of cases in Costa Rica (Mata *et al.*, 1983) and 49% in Ethiopia (Thorén *et al.*, 1982).

In developed countries such as Finland, where acute gastroenteritis is no longer responsible for significant mortality, rotavirus is associated with more than 50% of cases of childhood diarrhea seen in pediatric wards (Vesikari *et al.*, 1981). In Finland severe rotavirus diarrhea typically occurs in children aged 6–18 months (Mäki, 1981), i.e., in the first winter season after the infant has reached the age of 6 months. Before that age, rotavirus gastroenteritis usually runs a milder course, and in neonates the infection is most often asymptomatic (Chrystie *et al.*, 1978). It is probable that maternal antibodies and breast-feeding give partial protection against rotavirus diarrhea during the first 6 months of life, although the protective mechanisms in humans are unknown.

Infection as a neonate by a nursery strain of rotavirus could be seen as a kind of natural vaccination, as neonatal infection may permit the development of active immunity against rotavirus in the protection of maternally derived antibodies. Conversely, it is possible that the severe cases of rotavirus diarrhea at 6–18 months of age occur in those

T. Vesikari, E. Isolauri, and T. Ruuska • Department of Clinical Sciences, University of Tampere, SF-33520 Tampere, Finland. *A. Delem and F. E. André* • Biological Division, Smith Kline RIT, B-1330 Rixensart, Belgium.

children who have escaped rotavirus infection at an earlier age. Neonatal rotavirus infection has been shown to provide protection against severe rotavirus diarrhea, but not against rotavirus infection, up to 3 years of age (Bishop *et al.*, 1983). This observation clearly has implications for rotavirus vaccines and vaccination in the future. It would be reasonable to assume that a similar subclinical intestinal infection by a live attenuated rotavirus vaccine strain might also induce protection; in fact, this seems to be the case. On the other hand, if immunity derived from a natural human rotavirus infection during the neonatal period results in only partial protection against diarrhea upon subsequent exposure to rotavirus, it would be unreasonable to expect that a rotavirus vaccine could do much better.

II. GOAL OF ROTAVIRUS VACCINATION

A realistic goal for a rotavirus vaccination could be set as prevention of severe rotavirus diarrhea up to the age of 3 years. In developed countries this approach would eliminate the winter epidemics of rotavirus diarrhea and cut down the hospital admissions for all gastroenteritis in childhood by 50–60%. As diarrhea is due to other causes, but rotavirus tends to be milder than rotavirus diarrhea (Mäki, 1981), the actual impact of a successful rotavirus vaccination on pediatric diarrhea would be even greater.

It is more difficult to estimate the health impact of a rotavirus vaccination in developed countries. If a rotavirus vaccine could modify the course of rotavirus diarrhea from severe to mild in children living in less privileged conditions, it would possibly prevent the majority of the estimated 500,000 diarrheal deaths. Such a vaccine should have a high priority for use in developing countries.

III. ROTAVIRUSES

Rotaviruses are common in animal kingdom, and human and many animal rotaviruses, like bovine and monkey rotaviruses, share a common group antigen with human rotaviruses (Kapikian *et al.*, 1976). The group antigen is located in the inner capsid of the virus particles, and complement-fixing antibodies against the group antigen develop in human rotavirus infection. It is not known, however, what the role of immunity (antibodies) is to the group antigen in protection against rotavirus infection and disease.

Human rotaviruses can be further divided to two subgroups and four serotypes (WHO, 1984). Each of the known serotypes of human rotaviruses causes clinical diarrhea, but it is apparent that serotype 1 (subgroup 2) is associated with the winter outbreaks of rotavirus gastroenteritis more often than the other serotypes (Brandt *et al.*, 1979). Therefore, while any future rotavirus vaccine should preferably give protection against all four serotypes, protection against serotype 1 might be crucial. The neutralizing antigen specificities (VP7) that mainly determine the four serotypes are encoded by gene g of the rotavirus sequenced genome (Kapikian *et al.*, 1986). Human and animal rotaviruses do not generally show cross-neutralization with each other, but there are exceptions. Rhesus monkey rotavirus strain MU 18006, which has been the parent strain for the current rhesus rotavirus candidate vaccine RRV-1, shares neutralizing antigen with human rotavirus serotype 3 (Kapikian *et al.*, 1986). Bovine rotavirus vaccine strain RIT 4237 does not usually elicit cross-neutralizing antibodies to human rotaviruses.

IV. ROTAVIRUS CANDIDATE VACCINES

For the moment, the strongest candidate vaccines for human rotavirus vaccination are live attenuated animal rotaviruses of bovine (RIT 4237) and simian (RRV-1) origin. Initially, the bovine strain RIT 4237, derived from Nebraska Calf Diarrhea Virus, was chosen for development as vaccine strain because it grows to a high titer (10^8/ml) in tissue culture and because there was evidence of cross-protection with human rotaviruses in experimental animals (Zissis *et al.*, 1983; Delem *et al.*, 1984). The RIT 4237 virus was attenuated by passaging 147 times in primary fetal bovine kidney cells and is produced at 154th passage level in primary monkey kidney cells (Delem *et al.*, 1984). The high passage level is apparently unnecessary for use as a human vaccine but gives a good safety margin for revertance to virulent bovine rotavirus in case of accidental spread to cattle.

The rhesus rotavirus vaccine, RRV-1, has a much shorter passage history of a total of 14 tissue culture passages. The vaccine bulk for human studies was produced in primary monkey kidney cells, but the virus can also be propagated in simian diploid fibroblast cells (Kapikian *et al.*, 1985).

Other approaches for the development of human rotavirus vaccines include human and animal rotavirus reassortants and direct use of live attenuated human rotaviruses (Kapikian *et al.*, 1980, 1983). Further exploration of these possibilities will depend to some extent on the success of the heterologous rotavirus vaccines, which at the moment are more readily available for large-scale production. The following discussion summarizes the clinical experience of RIT 4237 rotavirus vaccine trials in Finland.

V. CLINICAL TRIALS OF RIT 4237 VACCINE

The cross-protection between RIT 4237 and human rotaviruses was first demonstrated in a piglet model (Zissis *et al.*, 1983). Thereafter, the RIT 4237 rotavirus vaccine was tested in Tampere, Finland, in successive trials for safety, immunogenicity, and clinical protection in different age groups, starting from adult volunteers and ending in neonates. As of today, more than 1000 children have received the RIT 4237 vaccine (Table 1).

A. Safety and Immunogenicity

Oral administration of the RIT 4237 vaccine virus leads to subclinical intestinal infection, which can be indirectly demonstrated by homologous antibody response. Most vaccinees do not excrete detectable amounts of the vaccine virus. In one study 21% of all vaccinees and 28% of those with serologic response shed the RIT 4237 virus, as shown by virus isolation (Vesikari *et al.*, 1986*b*). The RIT 4237 vaccine does not seem to cause any appreciable clinical symptoms in any target population, including neonates and rotavirus seronegative young infants. Apart from being safe for the vaccinee, the RIT 4237 rotavirus vaccine also appears safe for the environment. It is unlikely that the vaccine virus would spread to contact humans, although this has not been specifically studied. Accidental transmission to cattle also seems a remote possibility and, even if it occurred, the virus would be highly attenuated for cattle.

Evidence of the "take" of the RIT 4237 vaccine, as determined by rotavirus en-

Table 1. Sequence of Clinical Trials of RIT 4237 Rotavirus Vaccine in Tampere, Finland

Study code	Year(s)	Description[a]	Reference
01	1982	Immunogenicity and safety in adults ($N=20$)	Vesikari *et al.* (1983)
03	1982	Immunogenicity and safety in seropositive and seronegative 2-year-old and 6-month-old children ($N=25$)	Vesikari *et al.* (1983)
05	1983–1984	Clinical protection in 8–11-month-old children ($N=86$)	Vesikari *et al.* (1984*a*)
09	1983–1985	Clinical protection in 6–12-month-old children ($N=160$)	Vesikari *et al.* (1985*a*) Vesikari *et al.* (1986*a*)
22 35	1983–1985	Immunogenicity in breast and bottle-fed infants; dose-response studies ($N=357$)	Vesikari *et al.* (1985*b*) Vesikari *et al.* (1986*c*)
34	1984	Longitudinal protection study in children immunized as newborns ($N=347$)	Unpublished data

[a] N = number of children vaccinated with RIT 4237.

zyme-linked immunosorbent assay (ELISA) antibody response, is presented in Table 2. The "take" of the vaccine is clearly dependent on the dose of vaccine, prevaccination immunity status of the subject, and feeding before vaccination. The standard "full" dose of the vaccine is about 10^8 tissue culture infective doses per milliliter, which is the amount of virus that can easily be produced in primary monkey kidney cells. A 10-fold reduction in the vaccine dose leads to a significantly lower antibody response rate (Vesikari *et al.*, 1985*b*, 1986*c*). The vaccine virus is acid labile and may therefore be inactivated in the stomach before infecting the small intestine (Vesikari *et al.*, 1984*b*). Milk-feeding before vaccination may neutralize gastric acid and therefore increase the take of the vaccine. This

Table 2. Serologic Response to RIT 4237 Rotavirus Vaccine as Detected by Rotavirus ELISA Antibody Test Using Homologous Antigen[a]

Study code	Age group (months)	Prevaccine immunity	Feeding prevaccination	No. with response/ No. studied
05	8–11	Negative	Free	28/56 (50%)
		Positive	Free	12/30 (40%)
09	6–12	Negative	Free	77/146 (53%)
		Positive	Free	5/22 (23%)
17	5–6	Negative	Infant formula	14/16 (88%)
		Negative	None	9/20 (45%)
22	4–6	Negative	Breast milk	13/17 (76%)
		Negative	Formula	6/7 (86%)
		Positive	Breast milk	7/16 (44%)
		Positive	Formula	4/7 (57%)
35	6–12	Negative	Breast milk	26/32 (81%)
		Negative	Formula	32/37 (86%)
		Positive	Breast milk	11/16 (69%)
		Positive	Formula	9/14 (64%)
34	Newborn	Positive	Breast milk	41/116 (34%)

[a] Studies 05, 09, 17, 22 ELISA IgG responses only; studies 34 and 35 combined ELISA IgM and IgG responses.

Table 3. Summary of Clinical Protection against Rotavirus Diarrhea by RIT 4237 Rotavirus Vaccine in Two Trials of Children Aged 6–12 Months at Vaccination

Study code	Follow-up season[b]	Cases of rotavirus diarrhea		Protection rate (%)
		RIT 4237	Placebo	
05[a]	1st season	2/86	18/92	88
	2nd season	0/86	2/92	100
09[a]	1st season	5/160	26/168	82
	2nd season	1/160	6/168	82
	Total for 2 years	8/246	52/260	84

[a] Both studies were randomized double-blind placebo controlled trials.
[b] Winter and spring (January to May)

was not known at the time of the first clinical protection trials, and therefore the serologic response rate in those studies was lower than in subsequent studies in which the vaccinees were systematically fed before the administration of the vaccine (Table 2). It was also feared that breast-feeding might be inhibitory for the RIT 4237 vaccine, but recent studies of breast- and formula-feeding before vaccination have failed to show any significant difference in the vaccine virus infection rate between breast- and bottle-fed children (Vesikari *et al.*, 1986*c*) (Table 3).

In the most recent study, when the RIT 4237 vaccine was given under standard conditions to 6–12-month-old children following a meal of breast or bottle milk, the serologic response in initially seronegative children by rotavirus ELISA IgG or IgM antibody was 84% and in seropositive children 68% (Vesikari *et al.*, 1986*c*). In neonates, all of whom possess transplacentally acquired rotavirus antibody, the serologic response rate was only 34%, as demonstrated by ELISA IgM antibodies (Table 2). It should be emphasized that all the tests for serologic response presented in Table 2 indicate only "take" of the vaccine. There may be additional cases with successful vaccination but no demonstrable ELISA antibody response; conversely, the results of ELISA tests do not necessarily indicate clinical protection against diarrhea. However, it is known that those with a demonstrable antibody response after vaccination show better protection against rotavirus diarrhea than those who failed to respond by ELISA antibodies (Vesikari *et al.*, 1984*a,b*).

B. Clinical Protection

It is questionable whether the RIT 4237 vaccine induces any protection against human rotavirus infection as such. In the largest clinical protection study so far (Vesikari *et al.*, 1985*a*), 60 of the 134 (44%) of the control children showed serologic evidence of rotavirus infection during one epidemic season from December to May. In the same study, 59 children remained seronegative by rotavirus ELISA after vaccination and of those 37 (63%) had a significant increase over the season (Vesikari *et al.*, 1986*a*). It is not clear why the seroconversion rate was actually higher in the vaccine-treated group, but this could be due to late seroconversions by the vaccine or to priming after vaccination with booster response upon exposure to natural rotavirus. In any case, the serologic data do not imply protection against primary rotavirus infection by the RIT 4237 vaccine.

Rather than protecting against rotavirus infection at the first line of defense, the RIT 4237 vaccine seems to reduce the clinical severity of symptoms associated with human rotavirus infection. Vaccination of 6–12-month-old children before their first rotavirus epidemic season gave highly significant protection against severe rotavirus diarrhea during the season in two separate placebo-control double-blind studies in Finland (Table 3), (Vesikari *et al.*, 1984*a*, 1985*a*). The protection rate against severe rotavirus diarrhea was more than 80% but was only about 50% if all cases of mild gastrointestinal (GI) upset with rotavirus in the stools were included. Such cases would not ordinarily have required medical attention but were recorded during the close follow-up in the epidemic season. The follow-up was continued for the second year (rotavirus season only) in both study groups. Mild cases were not reported, possibly because they no longer alerted the parents, but the distribution of clinical rotavirus gastroenteritis between the vaccinees and the placebo recipients was similar as in the first season (Table 3) (Vesikari *et al.*, 1986*a*). It is therefore apparent that the RIT 4237 vaccine gives protection against rotavirus diarrhea for 2 years, covering the critical period of greatest susceptibility for rotavirus gastroenteritis.

There is only preliminary evidence of protection against the various serotypes of rotavirus. In the first protection study, the rotavirus isolates were not serotyped according to the present classification, but were subgrouped by G. Zissis (Free University of Brussels); 19 of the 20 subgroupable rotaviruses were of subgroup 2, hence possibly serotype 1. In the second study, 14 rotavirus isolates were found to be of serotype 1 (T. H. Flewett, Birmingham), and of those three were found in the vaccine group and 11 in the placebo-treated group. Thus the evidence accumulated thus far suggests protection against the most common rotavirus serotype encountered in the typical winter epidemics in temperate countries.

C. Neonatal Vaccination

An extensive clinical trial of neonatal rotavirus vaccination with the RIT 4237 vaccine is currently under way in Tampere, Finland (Table 1). Vaccination in the neonatal period appears an alternative for the administration of an oral rotavirus vaccine, particularly in developing countries, where many children may not be available for regular health care services at any other time. So far it seems that the serologic response to RIT 4237 vaccine is lower than in older children (Table 2). Preliminary data indicate that only 34% of the infants responded to the vaccination by ELISA-IgM antibody (Table 2), but 40% by neutralizing antibody giving a cumulative response rate of 47% (Vesikari, Ruuska, and Delem, unpublished data). Clinical protection data following neonatal vaccination is not yet available.

VI. RHESUS ROTAVIRUS VACCINE RRV-1

The RRV-1 vaccine also holds promise for a human rotavirus vaccine but so far the clinical efficacy data is lacking. It is clear that the RRV-1 vaccine is less attenuated for humans than the RIT 4237 vaccine. In a trial in Finland the RRV-1 vaccine caused reactions in 60% of the in 6–8 month-old children who usually had fever and general irritability 3–4 days after oral vaccination (Vesikari *et al.*, 1986*b*). On the other hand the RRV-1 vaccine is clearly more immunogenic than the RIT 4237. In the above trial, the RRV-1 vaccine induced an 88% serologic response rate and considerably higher rotavirus antibody titers than did the RIT 4237 vaccine (Vesikari *et al.*, 1985*b*).

It appeared that the RRV-1 vaccine in the dose used (1 : 10 of the bulk) was too reactogenic for the target population of 6–12-month-old children in Finland. However, the good immunogenicity of the RRV-1 vaccine might form an advantage if the reaction rate could be decreased by further attentuation, by administration of a smaller dose, or by vaccination in the protection of passively acquired maternal antibodies. Recent data from Venezuela indicates that most of the immunogenicity of the RRV-1 vaccine is retained even if only a 1 : 100 dilution of the vaccine bulk is given (A. Z. Kapikian, personal communication).

VII. CONCLUSION

Oral vaccination with live attenuated heterologous rotavirus appears to be a feasible approach for the prevention or amelioration of human rotavirus diarrhea in young infants. Bovine rotavirus strain RIT 4237 has been studied most extensively; its properties as a human vaccine are characterized as follows. The RIT 4237 vaccine is essentially nonreactogenic in infants. The overall ''take'' rate in the primary target group, children aged 6–12 months of age, is around 80%; thus, more than one vaccination will be needed for full coverage. In well-nourished children, the vaccine gives more than 80% protection against epidemic rotavirus diarrhea in winter, and the protection lasts up to 3 years of age or beyond the critical period when severe rotavirus diarrhea typically occurs. There is insufficient information of protection against various rotavirus serotypes. There is no evidence of efficacy in malnourished children living in less privileged conditions.

The rhesus rotavirus vaccine RRV-1 is less attenuated and more immunogenic in humans than is the RIT 4237 vaccine. The protective efficacy of RRV-1 vaccine in developed countries is currently being investigated. The next priority will be the extension of clinical trials of both candidate rotavirus vaccines to developing countries.

REFERENCES

Bishop, R. F., Barnes, G. L., Cipriani, E., and Lund, J. S. (1983). *N. Engl. J. Med.* **309,** 72–76.
Black, R. E., Merson, M. H., Huq, I., Alim, A. R. M. A., and Yunus, M. D. (1981). *Lancet* **1,** 141–143.
Brandt, C. D., Kim, H. W., Yolken, R. H., Kapikian, A. Z., Arrobio, J. O., Rodriguez, W. J., Wyatt, R. G., Chanock, R. M., Parrott, R. H. (1979). *Am. J. Epidemiol.* **110,** 243–254.
Butler, T. C. (1984). *J. Diarr. Dis. Res.* **2,** 137–141.
Chrystie, I. L., Totterdell, B. M., and Banatvala, J. E. (1978). *Lancet* **1,** 1176–1178.
Delem, A., Lobmann, M., and Zygraich, N. (1984). *J. Biol. Stand.* **12,** 443–445.
Kapikian, A. Z., Cline, W. L., Kim, H. W., Kalica, A. R., Wyatt, R. G., VanKirk, D. H., Chanock, R. M., James, H. D., Jr., and Vaughn, A. L. (1976). *Proc. Soc. Exp. Biol. Med.* **152,** 535–539.
Kapikian, A. Z., Wyatt, R. G., Greenberg, H. B., Kalica, A. R., Kim, H. W., Brandt, C. D., Rodriguez, W. J., Parrot, R. H., and Chanock, R. M. (1980). *Rev. Infect. Dis.* **2,** 459–469.

Kapikian, A. Z., Wyatt, R. G., Levine, M. M., Yolken, R. H., VanKirk, D. H., Dolin, R., Greenberg, H. B., and Chanock, R. M. (1983). *J. Infect. Dis.* **147,** 95–106.

Kapikian, A. Z., Midthun, K., Hoshino, Y., Flores, J., Wyatt, R. G., Glass, R. I., Askaa, J., Nakagomi, O., Nakagomi, T., Chanock, R. M., Levine, M. M., Clements, M. L., Dolin, R., Wright, P. F., Belshe, R. B., Anderson, E. L., and Potash, L. (1985). In *Vaccines 85* (R. A. Lerner, R. M. Chanock, and F. Brown, eds.) pp. 357–367, Cold Spring Harbor Laboratory, Cold Spring Harbor, New York.

Kapikian, A. Z., Hoshino, Y., Flores, J., Midthun, K., Glas, R. I., Nakagomi, O., Chanock, R. M., Potash, L., Levine, M. M., Dolin, R., Wright, P. F., Belshe, R. E., Anderson, E. L., Vesikari, T., Gothefors, L., Wadell, G., and Perez-Schael, I. (1986). In *Developments of Vaccines and Drugs against Diarrhea: Eleventh Nobel Conference (Stockholm)* (J. Holmgren, A. Lindberg, R. Mölby, eds.), pp. 192–214, Student Litteratur, Lund.

Mäki, M. (1981). *Acta Paediatr. Scand.* **70,** 107–113.

Mata, L., Simhon A., Padilla, R., Gamboa, M., Vargas, G., Hernandez, F., Mohs, E., and Lizano, C. (1983). *J. Trop. Med. Hyg.* **32,** 146–153.

Thorén, A., Stintzing, G., Tufvesson, B., Walder, M., and Habte, D. (1982). *J. Trop. Pediatr.* **28,** 127–131.

Vesikari, T., Mäki, M., Sarkkinen, H. K., Arstila, P. P., and Halonen, P. E. (1981). *Arch. Dis. Child.* **56,** 264–270.

Vesikari, T., Isolauri, E., Delem, A., d'Hondt, E., André, F. E., and Zissis, G. (1983). *Lancet* **2,** 807–811.

Vesikari, T., Isolauri, E., d'Hondt, E., Delem, A., and Zissis, G. (1984*a*). *Lancet* **1,** 977–981.

Vesikari, T., Isolauri, E., d'Hondt, E., Delem, A., and André, F. E. (1984*b*). *Lancet* **2,** 700.

Vesikari, T., Isolauri, E., Delem, A., d'Hondt, E., André, F. E., Beards, G. M., and Flewett, T. H. (1985*a*). *J. Pediatr.* **107,** 189–194.

Vesikari, T., Ruuska, T., Bogaerts, H., Delem, A., and André, F. (1985*b*). *Pediatr. Infect. Dis.* **4,** 622–625.

Vesikari, T., Isolauri, E., André, F. E. (1986*a*). In *Developments of Vaccines and Drugs against Diarrhea: Eleventh Nobel Conference (Stockholm)* (J. Holmgren, A. Lindberg, R. Mölby, eds.), pp. 185–191, Student Litteratur, Lund.

Vesikari, T., Kapikian, A. Z., Delem, A., and Zissis, G. (1986*b*). *J. Infect. Dis.* **153,** 832–859.

Vesikari, T., Ruuska, T., Delem, A., and André, F. E. (1986*c*). *Acta Paediatr. Scand.* **75,** 573–578.

WHO (1984). *Bull. WHO* **62,** 501–503.

Zissis, G., Lambert, J. P., Marbehant, P., Marissens, D., Lobmann, M., Charlier, P., Delem, A., and Zygraich, N. (1983). *J. Infect. Dis.* **148,** 1061–1068.

6

Expression of Hepatitis B Virus Surface Antigens Containing Pre-s Regions by Recombinant Vaccinia Viruses

Bernard Moss, Kuo-Chi Cheng, and Geoffrey L. Smith

I. INTRODUCTION

Hepatitis B is a major problem throughout the world. Approximately 200 million people are chronically infected with hepatitis B virus (HBV), and numerous deaths result from fulminant hepatitis, cirrhosis, and primary hepatocellular carcinoma. The 42-nm blood-borne infectious particle is composed of a nucleocapsid and a lipoprotein envelope. During acute infections as well as in chronic carrier states, excess envelope protein, hepatitis B virus surface antigen (HB_sAg), is secreted into the blood as 22-nm empty particles. The latter have been purified and used as a subunit vaccine. Despite the efficacy and safety of the plasma-derived product, the high cost associated with purification and testing and the limited quantities available have prevented extensive use of the vaccine in areas of the world where hepatitis B is highly endemic. For these reasons, alternative methods of producing HB_sAg are being developed. Genetic engineering has been used to produce HB_sAg in yeast and mammalian cells (reviewed by Tiollais *et al.*, 1985), offering the possibility of a more widely available recombinant vaccine. Nevertheless, the requirement for extensive protein purification, refrigerated storage of the vaccine, and a schedule of repeated inoculations may still impose limits on the global use of HBV vaccine for the immediate future.

The eradication of variola virus, the cause of smallpox, was the most effective mass vaccination campaign ever carried out. The success of this public health measure was attributable to both the method of immunization and the nature of the disease itself. Smallpox vaccine was composed of a dried preparation of live vaccinia virus, a member of the poxvirus family closely related to variola virus. The vaccine was economically

Bernard Moss • Laboratory of Viral Diseases, National Institute of Allergy and Infectious Diseases, National Institutes of Health, Bethesda, Maryland 20892. *Kuo-Chi Cheng* • Department of Pediatrics, Cornell University Medical College, New York, New York 10021. *Geoffrey L. Smith* • Department of Pathology, University of Cambridge, Cambridge CB2 1QP, England.

produced, transported without refrigeration, and easily administered by a single skin scratch.

Advances in genetic engineering have now made it possible to insert and express genes from other sources in vaccinia virus, raising the possibility of using live recombinant viruses as new vaccines. This chapter summarizes the progress made in expressing HB_sAg in vaccinia virus.

II. FORMATION OF RECOMBINANT VACCINIA VIRUS

Poxviruses, including vaccinia virus, are large-complex double-stranded DNA viruses that replicate in the cytoplasm of infected cells (reviewed by Moss, 1985). The genome of vaccinia virus is approximately 185,000 base pairs (bp) in length and encodes about 200 genes, not all of which are essential for replication. The DNA is packaged within the virus core, which also contains a complete transcription system, including a virus-coded RNA polymerase. DNA sequences upstream of individual early and late genes contain the regulatory signals for early or late expression. Recent studies indicate that regulatory signals of eukaryotic cells or of unrelated viruses are not recognized by the vaccinia transcription system.

Foreign genes can be inserted into vaccinia virus by homologous recombination (Panicali and Paoletti, 1982; Mackett *et al.*, 1982). Expression of these genes may occur either because of vaccinia transcriptional regulatory elements at or near the site of insertion or because of more precise genetic engineering. Plasmid vectors that facilitate insertion and expression of foreign genes have been developed (Mackett *et al.*, 1984). These vectors contain a vaccinia promoter, at least one restriction endonuclease site for insertion of the foreign gene, and flanking DNA from a nonessential region of the vaccinia genome. The choice of promoter determines both the time and level of expression, whereas the flanking DNA sequence specifies the site of homologous recombination. The latter is carried out by infecting cells with standard vaccinia virus and transfecting with the plasmid vector.

Since the vast majority of virus particles produced are of the parental type, selective procedures have been developed to pick out the recombinants. By using segments of the nonessential vaccinia virus thymidine kinase (TK) gene to flank the foreign DNA, the latter will be inserted into the TK locus (Mackett *et al.*, 1984). Consequently, the vaccinia TK gene will be inactivated. Conveniently, it is possible to select TK^- virus by plaque assay on TK^- cells with 5-bromodeoxyuridine in the agar overlay. In cells infected with the parental TK^+ vaccinia virus, the nucleoside analogue would be kinased and lethally incorporated into the virus genome. A second procedure that has recently been employed involves the coexpression of *Escherichia coli* β-galactosidase and staining of recombinant plaques with a color indicator (Chakrabarti *et al.*, 1985).

III. EXPRESSION OF HBV SURFACE PROTEINS

The region of the HBV genome that encodes surface proteins contains a long open-reading frame that is divided into pre-s1, pre-s2 and s regions (reviewed by Tiollais *et al.*, 1985). The major surface protein, designated S, is encoded entirely within the s region. A disulfide-bonded dimer of S forms the structural component of the HBV envelope. The next larger size protein, MS, is encoded by pre-s2 and s, whereas the largest protein, LS, is encoded by the entire open-reading frame. MS and LS are minor constituents of the HBV envelope and may be involved in binding to hepatocytes. Expression of S alone or a combination of S and MS has been achieved in mammalian cells by recombinant DNA techniques. Only in yeast has selective expression of MS been achieved (Valenzuela *et al.*, 1986).

We investigated the use of vaccinia virus as a vector to express individual HBV

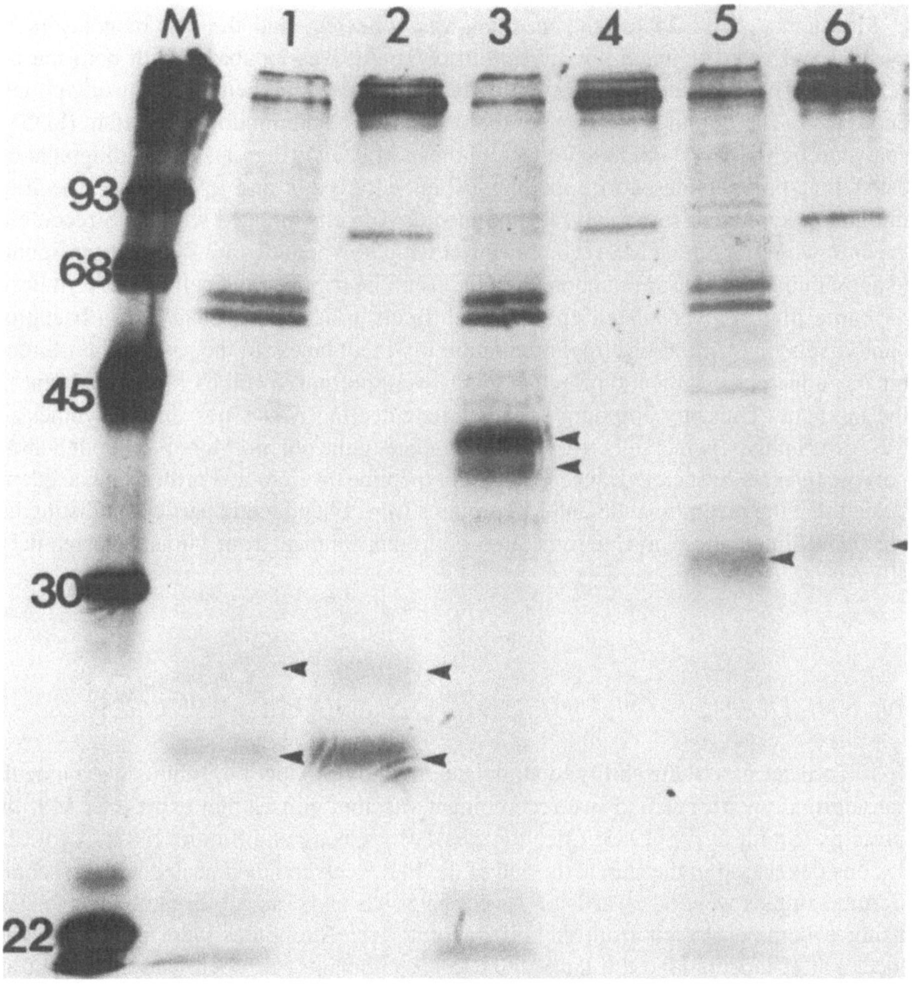

Figure 1. Autoradiograph showing expression of HBV surface proteins by recombinant vaccinia viruses. CV-1 cells were infected with recombinant vaccinia viruses containing HBV s (lanes 1 and 2), pre-s1 + pre-s2 + s (lanes 3 and 4), and pre-s2 + s (lanes 5 and 6), coding regions and metabolically labeled with [^{35}S]methionine and [^{35}S]cysteine for 24 hr. Cell-associated (odd-number) and secreted (even-number) proteins were immunoprecipitated with anti-HB_sAg guinea pig serum and analyzed by polyacrylamide gel electrophoresis. M, marker proteins with molecular weights indicated as kilodaltons. Arrowheads indicate position of expressed HBV proteins.

surface proteins selectively. DNA segments containing either the HBV s region, the pre-s2 and s, or pre-s1, pre-s2, and s were inserted into a plasmid vector just downstream of a vaccinia promoter (P7.5) with early and late regulatory signals (Smith *et al.*, 1983; Cheng *et al.*, 1986; Cheng and Moss, 1987). In each case, the first potential translation initiation codon following the vaccinia P7.5 promoter was the one predicted to start the corresponding S, MS, or LS HBV protein. Homologous recombination was carried out and TK^- virus plaques were isolated. The presence of the appropriate HBV DNA was checked by restriction endonuclease analysis and DNA hybridization.

Evidence for HBV protein synthesis was obtained by infecting tissue culture cells with the recombinant viruses and metabolically labeling them with [^{35}S]methionine and [^{35}S]cysteine. After 24 hr, the medium was collected, and the cell monolayers were washed and lysed. Guinea pig antiserum to HB_sAg was incubated with both the intracellular and extracellular proteins; antibody complexes were then isolated using protein A Sepharose. The bound proteins were then eluted with sodium dodecyl sulfate (SDS) and analyzed by polyacrylamide gel electrophoresis (PAGE) (Fig. 1). Autoradiographs indicated that polypeptides corresponding to authentic p24s and gp27s were specifically immunoprecipitated from the lysate and medium of cells infected with the S recombinant vaccinia virus. With the MS recombinant virus, a polypeptide of 3,300 M_r corresponding to authentic gp33s was immunoprecipitated from the medium and cell lysate. Synthesis of the more highly glycosylated gp36s was difficult to ascertain because of a background band displaying similar electrophoretic mobility in all lanes. In the case of the LS recombinant, apparently authentic p39s and gp42s were present in cell extracts but absent from the medium. Thus, by appropriate genetic engineering it was possible to synthesize S, MS, or LS polypeptides selectively in the absence of the others. Moreover, both S and MS polypeptides were secreted, whereas LS polypeptides were not. Further analysis demonstrated that the recombinant S and MS proteins were assembled in particles indistinguishable by sedimentation or electron microscopic appearance from human plasma-derived HB_sAg.

IV. STIMULATION OF ANTIBODY TO S AND PRE-S EPITOPES

To determine their ability to stimulate antibody production, rabbits were inoculated intradermally with purified live recombinant vaccinia viruses that express S, MS, or LS proteins (Smith *et al.*, 1983; Cheng *et al.*, 1986; Cheng and Smith, 1987). Typical pox lesions developed in the skin at the end of the first week and then healed without sequelae. Serum samples were taken at 1- or 2-week intervals and assayed for antibodies to HB_sAg using a commercial kit from Abbott Laboratories. Antibodies were detected within 2 weeks after inoculation with the S and MS recombinants, but there was a delay of a few additional weeks before antibody was detected in rabbits receiving LS recombinant. Because of the small number and outbred nature of the animals, the significance of this difference is not clear.

Antibodies to pre-s epitopes were detected in the sera from rabbits inoculated with MS or LS recombinants. These were detected by radioimmunoassay (RIA) using synthetic peptides corresponding to the pre-s1 and pre-s2 N-terminal regions. The kinetics of ppearance of pre-s and s antibodies were similar.

V. PROTECTION OF CHIMPANZEES

Except for humans, chimpanzees are the only animals susceptible to HBV. To test the potential of vaccinia recombinants as a hepatitis B vaccine, two chimpanzees received intradermal inoculations of the S recombinant vaccinia virus and one control received the parental vaccinia virus (Moss *et al.*, 1984). All three animals developed typical pox lesions as well as antibodies that neutralized vaccinia virus, although the lesion size was smaller and the antibody titers were somewhat lower in the animals receiving recombinant virus. Surprisingly, neither of the chimpanzees vaccinated with recombinant virus had significant anti-HB_sAg titers. Upon intravenous HBV challenge, however, they exhibited a rapid rise in anti-Hb_sAg, indicating that they had been primed. Moreover, the animals vaccinated with recombinant virus had no detectable circulating HB_sAg and developed no signs of hepatitis. By contrast, within a few weeks the control chimpanzee had high levels of circulating HB_sAg and developed other signs of hepatitis but did not have measurable antibodies to HB_sAg for 6 months.

VI. CONCLUSIONS

Recombinant vaccinia viruses were shown to synthesize HBV S, MS, or LS proteins selectively in mammalian cells. Using other recombinant DNA systems, only S or a mixture of S and MS were made. Thus, vaccinia virus provides a unique system for analyzing the synthesis, assembly, and secretion of these HBV proteins. Unexpectedly, we found that LS does not by itself form secreted particles. Rabbits vaccinated with the recombinant vaccinia viruses make antibodies to s and pre-s epitopes. Thus far, only the S recombinant vaccinia virus has been tested in chimpanzees. The results were a qualified success, since the animals did not have measurable anti-HB_sAg titers prior to challenge with HBV but nevertheless were protected against hepatitis. Most likely, as a result of priming, the chimpanzees developed a rapid anamnestic response to the challenge virus and at most had only a mild inapparent infection. Further work is needed to improve the immune response. It will be of particular interest to examine chimpanzees that receive the pre-s recombinants, since the pre-s epitopes are highly immunogenic and in some mouse strains enhance the response to s epitopes (Milich *et al.*, 1985).

REFERENCES

Chakrabarti, S., Brechling, K., and Moss, B. (1985). *Mol. Cell. Biol.* **5,** 3403–3409.

Cheng, K.-C., and Moss, B. (1987). *J. Virol.* **61,** 1286–1290.

Cheng, K.-C., Smith, G. L., and Moss, B. (1986). *J. Virol.* **60,** 337–344.

Mackett, M., Smith, G. L., and Moss, B. (1982). *Proc. Natl. Acad. Sci. USA* **79,** 7415–7419.

Mackett, M., Smith, G. L., and Moss, B. (1984). *J. Virol.* **49,** 857–864.

Milich, D. R., McNamara, M. K., McLachlan, A., Thornton, G. B., and Chisari, F. V. (1985). *Proc. Natl. Acad. Sci. USA* **82,** 8168–8172.

Moss, B., Smith, G. L., Gerin, J. L., and Purcell, R. H. (1984). *Nature (Lond.)* **311,** 67–69.

Moss, B. (1985). In *Virology* (B. N. Fields, D. M. Knipe, R. M. Chanock, J. Melnick, B. Roizman, and R. Shope, eds.), pp. 685–703, Raven Press, New York.

Panicali, D., and Paoletti, E. (1982). *Proc. Natl. Acad. Sci. USA* **79,** 4927–4931.

Smith, G. L., Mackett, M., and Moss, B. (1983). *Nature (Lond.)* **302,** 490–495.

Tiollais, P. C., Pourcell, C., and Dejean, A. (1985). *Nature (Lond.)* **317,** 489–495.

Valenzuela, P., Coit, D., and Kuo, C. H. (1985). *Biotechnology* **3,** 317–320.

7

New Generation of Rabies Vaccine

Vaccinia–Rabies Glycoprotein Recombinant Virus

T. J. Wiktor, M. P. Kieny, and R. Lathe

I. INTRODUCTION

Rabies is a disease of major significance to human and veterinary medicine. Derivatives of the vaccine developed by Pasteur and associates 100 years ago (1885) are still in use in many parts of the world. During the past two decades considerable progress has been made in improving the efficiency and safety of rabies vaccines through the use of tissue-culture techniques. The Wistar Institute human diploid rabies vaccine, introduced in 1970, is now in general use in the United States and Western Europe. In spite of having a nearly perfect record of safety and efficacy, this vaccine is too costly for extensive use in many of the developing countries, particularly those where rabies vaccine is most needed. Furthermore, rabies is unique in that vaccine treatment is generally applied, in humans, after exposure to the virus: although postexposure vaccination is effective, the precise mechanism of its action is not yet clearly understood.

The research effort of several laboratories is now directed toward the development of new low-cost and effective rabies vaccines. These include recombinant viruses expressing the rabies virus glycoprotein, preparations of anti-idiotypic antibodies, and synthetic peptides. Of these approaches, a vaccinia–rabies glycoprotein recombinant virus (V–RG) given to experimental animals by different inoculation routes (including oral administration) has proved to be an extremely effective anti-rabies vaccine, and promising results have been obtained in preventing rabies in wildlife by administration of baits containing V–RG virus.

II. RABIES VIRUS STRUCTURE

Rabies virus is an enveloped rhabdovirus with a single, nonsegmented negative-strand RNA. Virions contain multiple copies of five proteins: the virion transcriptase (L),

T. J. Wiktor† • The Wistar Institute, Philadelphia, Pennsylvania 19104. *M. P. Kieny* • Transgene S. A., Strasbourg 67000, France *R. Lathe* • LGME–CNRS and U184–INSERM, Strasbourg 67085, France.

†Deceased. Readers will be sad to learn of the tragic death of Dr. Tadeusz Wiktor after completing this manuscript. His deep understanding of the biology of viral disease and his abundant enthusiasm for new solutions to old problems will be sorely missed by all of us: MPK, RL.

glycoprotein (G), nucleoprotein (N), the nucleocapsid-associated protein (NS) and the matrix (M) protein. The L, N and NS proteins are noncovalently bound to the viral RNA. The resulting nucleocapsid (NC) complex is coiled within the virion into a helical structure. The NC is surrounded by a lipoprotein envelope containing M protein and through which the surface projections (spikes) of G protein extend to the exterior of the virus.

Rabies virus G protein is the major antigen responsible for the induction of virus-neutralizing (VN) antibodies and for conferring immunity against lethal infection with rabies virus (Wiktor *et al.*, 1973). The G coding sequence was the first rabies virus gene to be cloned and analyzed (Anilionis *et al.*, 1981). To define the antigenic structure of rabies virus G protein, the nucleotide sequence of the virus-specific G-mRNA of rabies virus (ERA strain) was determined from the cloned cDNA copy. From the nucleotide sequence, a polypeptide 505 amino acids long was deduced, preceded by a signal sequence of 19 amino acids.

The immunogenic activity of purified native G protein and of smaller fragments of G protein, both occurring naturally or derived by chemical cleavage of the rabies virus G protein, have been compared in a variety of ways to determine the structural basis of VN antibody production following immunization (Dietzschold *et al.*, 1982). Rabies virus-infected cells shed a soluble glycoprotein (G_s protein) which can be purified from virion-depleted tissue culture fluid by immunoadsorption chromatography (Dietzschold *et al.*, 1983). No antigenic differences between G and G_s proteins could be detected by binding to monoclonal VN antibodies that recognized four distinct antigenic sites of ERA G protein, but only the G protein was able to protect against lethal rabies challenge. The difference in the molecular weights of G_s (61,000 M_r) and G proteins (67,000 M_r) may be the important factor accounting for their disparate immunogenicities as this difference results from the loss of 58 amino acid residues from the carboxy terminus of Gs (Dietzschold *et al.*, 1983).

III. VACCINIA VIRUS

Vaccinia virus, a member of the *Orthopoxvirus* genus, has been used successfully as a live vaccine for the prevention and elimination of the serious human disease, smallpox. The eradication of smallpox rendered the vaccine obselete and was considered a strong contraindication to any further immunization with vaccinia virus. Other members of *Orthopoxvirus* genus include cowpox, horsepox, monkeypox, camelpox, and mousepox (ectromelia). The origins of vaccinia virus are not known, but since the virus is not found naturally it is presumed to have evolved from one of the mammalian poxviruses. As the genomes of all vaccinia virus strains are very similar to each other but different from those of smallpox and cowpox viruses, it is unlikely that the vaccinia virus evolved directly from either smallpox or cowpox since this would require considerable changes in the genome (Mackett and Archard, 1979). During the nineteenth century, smallpox vaccines were developed from horsepox, a virus now extinct in nature. It is possible that the clinical suitability of horsepox vaccines led to their retention, and to rejection of cowpox vaccines. This would explain the survival of a closely related collection of vaccine strains, not found naturally, which are not obviously derived from cowpox or smallpox (Baxby, 1981).

Vaccinia is a dermotropic virus which usually requires inoculation into superficial layers of the skin in order to infect. Infection produces a lesion caused by epidermal hyperplasia and proliferation and inflammatory infiltration that progresses from a papule through a vesicle and pustule to a crust. Transient viremia probably occurs. Generalized lesions are rare in the immunocompetent person, but serious complications can occur in the immunodeficient and eczematous individual. Vaccination induces an adequate humoral and cellular immune response.

Vaccinia has a wide host range, infecting man and most domestic and laboratory animals. There is nonetheless no evidence that vaccinia virus can become established in animal populations. This is despite the fact that smallpox vaccination of humans has been conducted on a massive scale in both developed and developing countries where smallpox vaccines were commonly produced in cows, buffalos, or sheep.

Poxviruses are distinguished by their large size and complex morphology, a high-molecular-weight DNA genome encoding vaccinia-specific enzymes for DNA and RNA synthesis, and a cytoplasmic site of replication. Vaccinia virus is easily grown and has been widely studied as a typical poxvirus (Dales and Pogo, 1981).

The genome of vaccinia virus consists of a linear double-stranded DNA molecule of approximately 185,000 base pairs (bp) located within the core structure and closed with a hairpin loop at either end (Geshelin and Berns, 1974). The size of the genome and its mode of replication make vaccinia an excellent candidate for the expression of foreign genes for vaccine purposes.

IV. CONSTRUCTION OF A VACCINIA–RABIES G RECOMBINANT VIRUS

Recent technical advances have permitted the development of vaccinia virus as a cloning and expression vector (Panicali and Paoletti, 1982; Smith *et al.*, 1982) and vaccinia recombinants bearing foreign antigen coding sequences have been used in experiments to immunize against the cognate diseases (reviewed by Smith *et al.*, 1984; Smith and Moss, 1984).

Expression of exogenous protein-coding sequences in vaccinia virus involves essentially two steps. First, the exogenous coding sequence if aligned with a vaccinia promoter and inserted *in vitro* at a site within a (nonessential) segment of vaccinia DNA cloned into a suitable bacterial plasmid replicon. Second, the flanking vaccinia sequences now permit homologous recombination *in vivo* between the plasmid and the viral genome. Double reciprocal recombination results in transfer of the DNA insert from the plasmid to the viral genome, wherein it is propagated and expressed (Fig. 1).

A double-stranded cDNA copy of the ERA strain rabies glycoprotein (G) mRNA has been isolated and cloned, and the complete nucleotide sequence has been determined (Anilionis *et al.*, 1981). However, expression of this sequence in a bacterial host or in tissue culture cells has failed to yield material capable of immunizing animals against rabies (Lathe *et al.*, 1984; Kieny *et al.*, 1986). A comparison of the predicted amino acid sequence determined from the cDNA nucleotide sequence with that found by direct amino acid sequencing of G protein revealed a discrepancy at amino acid position 8 of the mature protein, where the cDNA coded for leucine while authentic viral G protein carried a proline residue at this position. Assuming that this change near the amino acid terminus

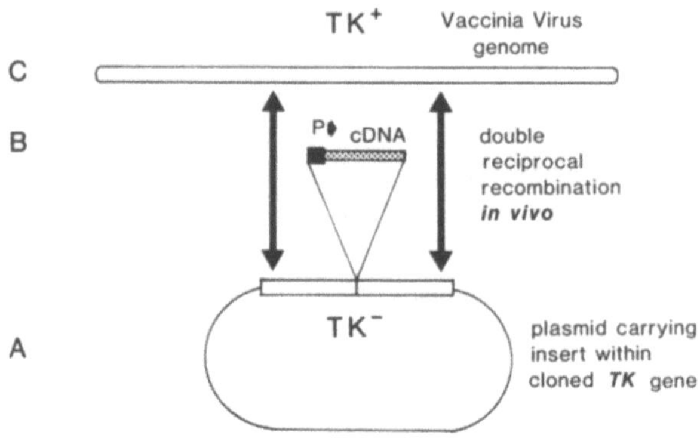

Figure 1. Cloning in vaccinia virus. In A the cloned TK gene stationed in a bacterial plasmid is interrupted by the insertion of the expression block, B, comprising a vaccinia promoter sequence (solid) and a cDNA coding sequence (hatched). Reciprocal recombination *in vivo* transfers the expression block to the vaccinia genome, C, creating a selectable *TK*-deficient virus expressing the integrated coding sequence.

might have a considerable impact upon the folding and assembly of G protein, the cDNA clone was modified *in vitro* to correct the eighth codon of the cDNA from leucine to proline (Kieny *et al.*, 1984).

A. DNA Constructions

The upstream G/C tail of the rabies G coding sequence (a by-product of the cDNA cloning procedure) was first removed, since such tracts can interfere with expression of the encoded protein. A unique *Mst*II site overlapping codons for amino acids 2–4 of the primary translation product was linked, through a synthetic double-stranded adaptor oligonucleotide, to an upstream *Bgl*II site located within the plasmid vector (Fig. 2a). In the resulting plasmid, codon 8 is bounded by an upstream *Eco*RI site and a *Hind*III site overlapping the codons for amino acids 10–11 mature G. This short segment was cloned into a single-stranded bacteriophage M13 vector and localized mutagenesis using a synthetic 19-mer oligonucleotide was used to correct the triplet CTA (leucine) to CCA (proline) (Fig. 2b). An adjacent A to C alteration created a recognition site for *Sau*3A (GATC), facilitating the selection and recognition of correctly altered clones. This resulted in a silent change in an adjacent isoleucine codon (from ATA to ATC), an acceptable alteration in terms of codon utilization for rabies G. Following mutagenesis, the correctly altered segment was restored to the parental plasmid by exchange of *Eco*RI–*Hind*III fragments (ptg155–PRO).

The next step was to isolate the fragments of vaccinia virus DNA necessary for the expression and recombinational transfer of the rabies G coding sequence. Mutants of vaccinia virus lacking thymidine kinase *(TK)* are able to be propagated both *in vitro* and *in vivo*, demonstrating that the *TK* gene is a nonessential function. Furthermore, transfer of mutations or inserts disrupting the *TK* gene to the vaccinia genome may be recognized by a simple chemical selection procedure. The 4.6-kilobase (kb) *Hind*III fragment (Hin-J) of the vaccinia genome containing the complete *TK* gene (Drillien and Spehner, 1983) was transferred to an *E. coli* miniplasmid (ptg1H, 2049 bp) bearing a β-lactamase gene (ampicillin resistance), an origin of replication, and a unique *Hind*III site. This construct, ptg1H-TK, was used as the carrier plasmid in subsequent experiments.

A strong promoter sequence (to drive the expression of rabies G) was next isolated from the vaccinia genome. The promoter of an early vaccinia gene encoding a 7.5 kilodalton (7.5K) protein has been used to direct the expression of foreign coding sequences (Smith *et al.*, 1983). This promoter lies on the Sal-S fragment of the vaccinia genome, one of the smallest fragments generated upon digestion of vaccinia DNA by SalI. As small DNA fragments are preferentially cloned, the Sal-S segment was rapidly isolated by shotgun ligation of SalI fragments of vaccinia DNA into a suitably cleaved bacteriophage M13 vector. A *Bgl*II linker was next introduced immediately downstream of the

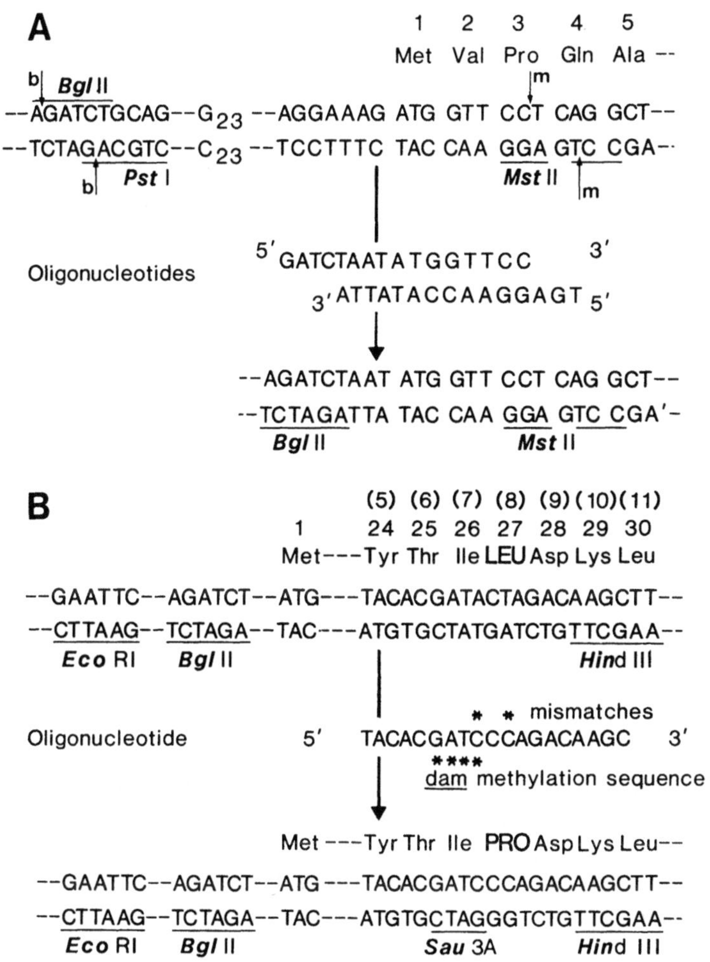

Figure 2. Tailoring of the rabies glycoprotein cDNA prior to transfer to vaccinia virus. In A the upstream G/C tail is eliminated by introducing a synthetic double-stranded adaptor oligonucleotide between an upstream *Bgl*II site and a unique *Mst*II site overlapping codons 3–5 of the rabies glycoprotein cDNA. Exact cleavage sites are indicated *b* and *m* respectively. In B a mutant leucine codon CTA is replaced by the codon CCA, restoring a proline residue at position 8 of the mature glycoprotein. An adjacent A-C transversion introduces a new GATC sequence (*dam* methylation; *Sau*3A) to aid selection of correctly altered clones. Figures give amino acid positions of the rabies glycoprotein precursor; figures in parentheses are the corresponding positions in the mature glycoprotein after removal of the N-terminal signal sequence.

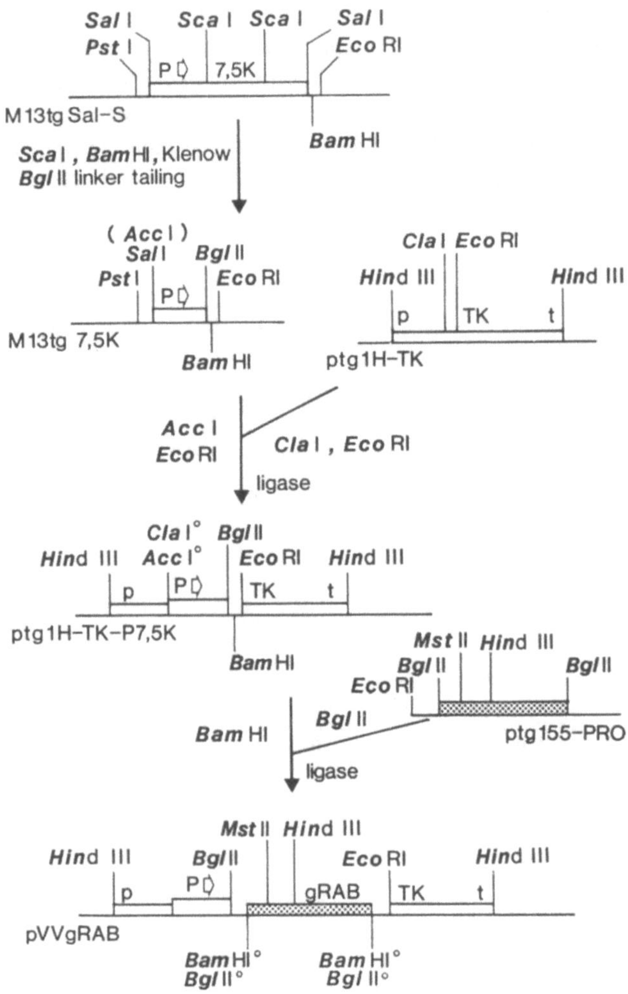

Figure 3. Construction of the expression plasmid for exchange into the vaccinia genome. The Sal-S fragment of the vaccinia genome containing the promoter of the early 7.5K gene has been inserted into an M13 bacteriophage vector to create M13tgSal-S. The upstream *Sca*I site lies immediately downstream of the start site(s) for transcription of the E7.5K gene (Cochran *et al.*, 1985) and a *Bgl*II linker was introduced between this site and the *Bam*HI site of the M13 vector, recreating the *Bam*HI site. In ptg1H-TK the Hin-F fragment of vaccinia DNA has been stationed at the unique *Hind*III site of plasmid ptg1H. The *TK* coding sequence contains unique recognition sequences for *Cla*I and *Eco*RI (Weir and Moss, 1984) and the 270 nucleotide, promoter fragment present in M13tg7.5K, was next excised with *Acc*I (cleaving at the *Sal*I site) and *Eco*RI, and ligated between the compatible termini generated by *Cla*I and *Eco*RI digestion of ptg1H-TK, deleting 33 nucleotides and generating ptg1H-TK-P7.5K. In a final step the restructured rabies glycoprotein cDNA from ptg155-PRO was excised with *Bgl*II and ligated into the unique *Bam*HI site to generate pVVgRAB.

7.5K gene promoter to generate M13tg7.5K and the resulting promoter segment (270 bp) was transferred to the TK gene-coding sequence carried by plasmid ptg1H-TK (Fig. 3).

In a final step the corrected rabies glycoprotein coding sequence from ptg155-PRO was excised with *Bgl*II and introduced into a *Bam*HI site immediately downstream of the 7.5K gene promoter, within the *TK* gene, to generate pVVgRAB.

B. Transfer to the Vaccinia Genome

The strategy devised by Smith *et al.* (1983) relies on exchange *in vivo* between a plasmid bearing an insert within the vaccinia *TK* gene and the wild-type viral genome to inactivate the virus-borne *TK* gene. TK-negative viruses may be selected by plating upon a *TK*-negative cell line in the presence of 5-bromodeoxyuridine (5BUdR). TK phosphorylates 5BUdR to the 5′-monophosphate, which is subsequently converted to the triphosphate. This compound is an analogue of dTTP and its incorporation into DNA prevents the correct development of the virus. A *TK*-negative virus can, however, replicate its DNA normally and give rise to visible plaques on a monolayer of an appropriate TK^- cell line.

Vaccinia propagates in the cytoplasm of infected cells rather than in the nucleus. It is unable to take advantage of host DNA replication and transcription machineries and instead carries in the virion those components necessary for the onset of viral gene expression. Purified vaccinia DNA is noninfective. To generate recombinants it is thus necessary to infect cells simultaneously with vaccinia virus and introduce the DNA construct of interest by calcium-mediated transfection.

The generation of recombinants is nevertheless limited to that fraction of the cell population competent for DNA transfection, whereas all infected cells yield progeny. The virus stock obtained after one cycle of growth and lysis thus contains a majority of wild-type TK^+ helper virus together with a few rare *TK*-negative recombinants. These rare recombinants cannot easily be selected by plating in the presence of 5BUdR: cells infected by the recombinant TK^- virus will invariably be infected by a TK^+ parental virus and cell death will result. This effect may be overcome by employing high dilution factors, however the number of plates required to detect the few rare recombinants may often be restrictive.

We instead employed an indirect "congruence" protocol to reduce the background of nonrecombinant parental viruses (Fig. 4). This was achieved by using as helper virus a temperature-sensitive (*ts*) mutant unable to propagate at the nonpermissive temperature (Drillien and Spehner, 1983). When cells are infected with a *ts* mutant under nonper-

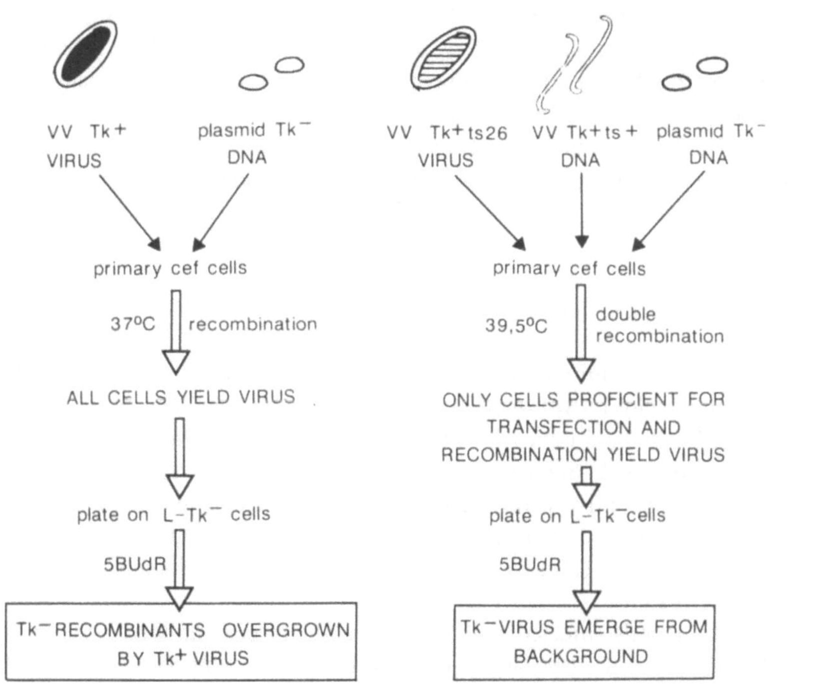

Figure 4. Congruence strategy employed to detect vaccinia recombinants.

missive conditions and simultaneously transfected with wild-type virus DNA, viral multiplication occurs only in those cells that are competent for transfection; no yield is obtained from cells infected with the helper virus alone. When a recombinant plasmid (such as pVVgRAB) carrying a fragment of vaccinia DNA is included with the wild-type DNA in the transfection mix, it will also be taken up by competent cells and may thus participate in homologous recombination with vaccinia DNA. This procedure results in a considerable improvement in the ratio of recombinant TK^- to parental TK^+ virus, and the detection of rare recombinants is considerably enhanced.

Monolayers of primary chick embryo fibroblasts were infected with the *ts26* mutant of the vaccinia copenhagen strain. After 2 hr at the permissive temperature (33°C), the cells were transfected with a calcium phosphate coprecipitate of vaccinia wild-type DNA and the recombinant plasmid pVVgRAB. Two hours later, the cells were rinsed and incubated for 48 hr at the non-permissive temperature (39.5°C). Dilutions of virus emerging were reinfected onto a monolayer of mouse L-TK^- cells at 37°C in the presence of 5BUdR. Several TK^- plaques were obtained from cells that had received the recombinant plasmid, whereas in control cultures without the plasmid no plaques were observed.

Double-reciprocal recombination between the hybrid vaccinia–rabies G plasmid and the vaccinia genome is expected to exchange the viral *TK* gene for the insert-bearing *TK* gene present on the plasmid. In the vaccinia genome the *TK* gene is present on a single *Hind*III fragment, Hin-J, whereas in recombinants having acquired the rabies G coding sequence and accompanying promoter the Hin-J fragment should carry an additional *Hind*III site overlapping codons 10 and 11 of the rabies G cDNA. DNA purified from *TK*-negative viruses was accordingly cleaved with *Hind*III. Fragments were resolved by agarose gel electrophoresis, immobilized by transfer to nitrocellulose, and probed for hybridization with a radiolabeled DNA prepared from the isolated *TK* gene fragment. As expected, the 4.6 kilobase Hin-J fragment was absent from the recombinant virus and two new fragments of approximately 1.1 and 5.5 kb revealed the presence of an insert bearing an internal *Hind*III site. Further analysis revealed that the insert hybridized to the rabies-G and one such recombinant, V-RG was retained. [The recombinant virus V-RG used in this work bears the technical designation VVTGgRAB-26D3 (Kieny *et al.*, 1984).]

V. EXPRESSION OF RABIES G IN CELLS INFECTED WITH V–RG VIRUS

Monolayers of semi-confluent mouse L-TK^- cells were infected with the V–RG virus and labeled with [^{35}S]-L-methionine for 4 hr. After brief sonication, the cell extract was incubated with anti-rabies-G antibodies and cross-reacting proteins recovered by adsorption to protein A–Sepharose and displayed by gel electrophoresis and fluorography. The polyclonal anti-G serum as well as two different virus-neutralizing monoclonal antibodies react with a protein migrating as a diffuse band with an apparent molecular weight of about 66,000 (Fig. 5). This molecular weight corresponds to the size, 67,000 M_r, of ERA virus G.

In BHK-21 cells infected with V–RG, a fluorescent antibody-staining technique revealed that the recombinant rabies G was present in the cytoplasm of acetone-fixed cells but predominantly at the cell surface (Fig. 6).

The rabies antigen expressed by this recombinant was tested by radioimmunoassay

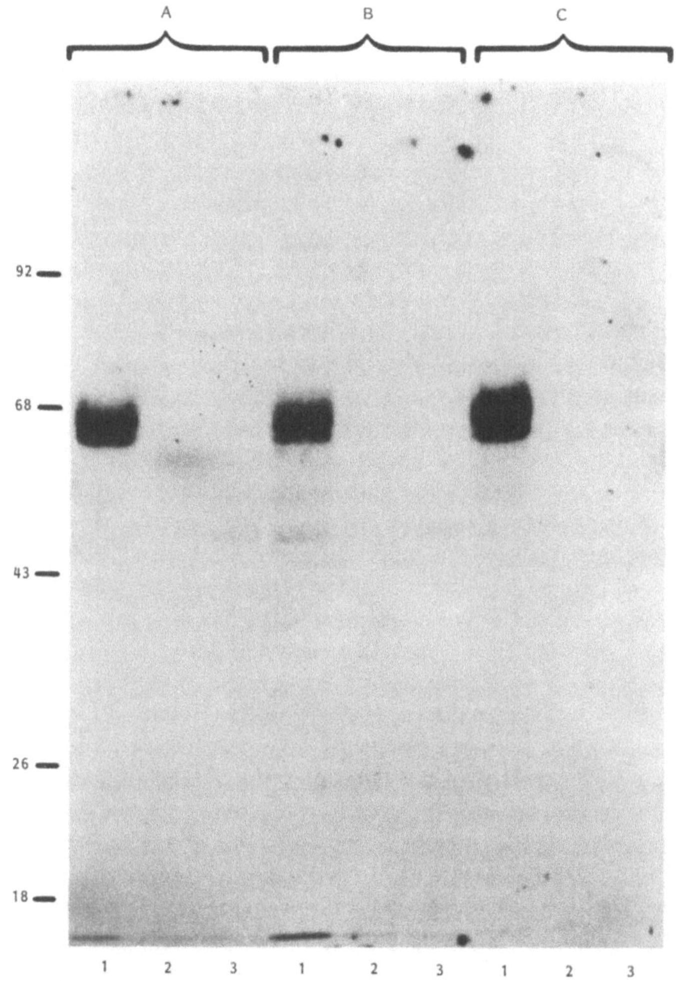

Figure 5. Rabies antigen synthesized by V-RG. Mouse LTK$^-$ cells were infected with V-RG and newly-synthesized proteins labeled by the addition of ^{35}S-methionine to the medium. Cells were disrupted after 4 hr and incubated with anti-rabies sera; bound proteins were collected by adsorption to resin-bound protein A and resolved by SDS gel electrophoresis and autoradiography. Cells were infected with (1) V-RG (2) wild-type vaccinia (3) no virus, and treated with (A) polyclonal antiserum R215 or (B), (C) monoclonal antibodies 509-6, 101-1 which recognize different epitopes on the glycoprotein. All three antisera detect a major protein species at 67 kilodaltons which corresponds in molecular weight to that of authentic rabies glycoprotein.

for reaction with a panel of monoclonal antibodies directed against rabies glycoprotein and other viral proteins (N, NS, and M). The binding activity of the V–RG protein with 44 anti-G monoclonal antibodies was almost identical to that observed with purified ERA rabies virus, whereas only the ERA virus reacted with anti-N, -NS, and -M antibodies (Fig. 7). This demonstrates that the rabies glycoprotein produced by V–RG-infected cells is qualitatively indistinguishable from the native glycoprotein of rabies virus.

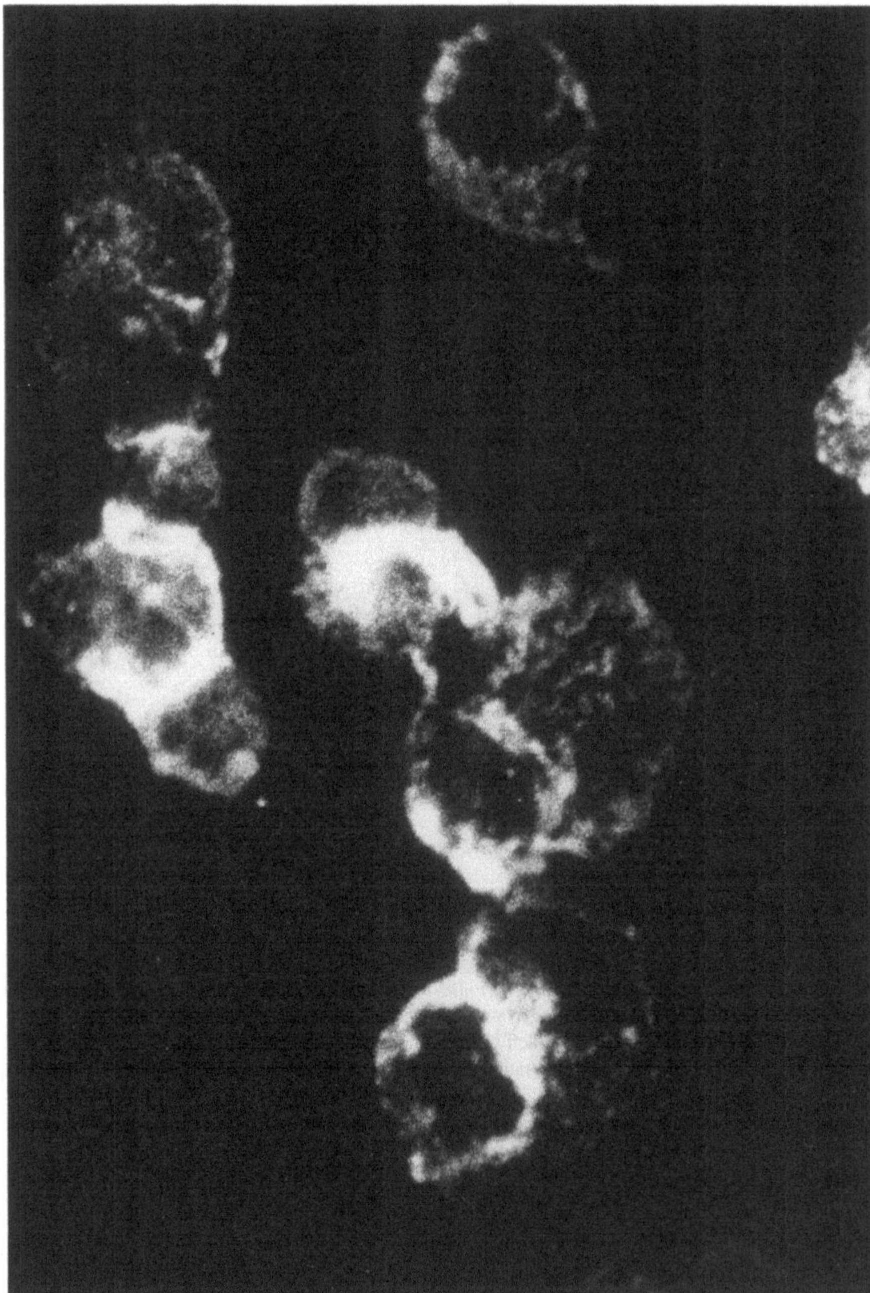

Figure 6. Immunofluorescence of rodent cells infected with V-RG.

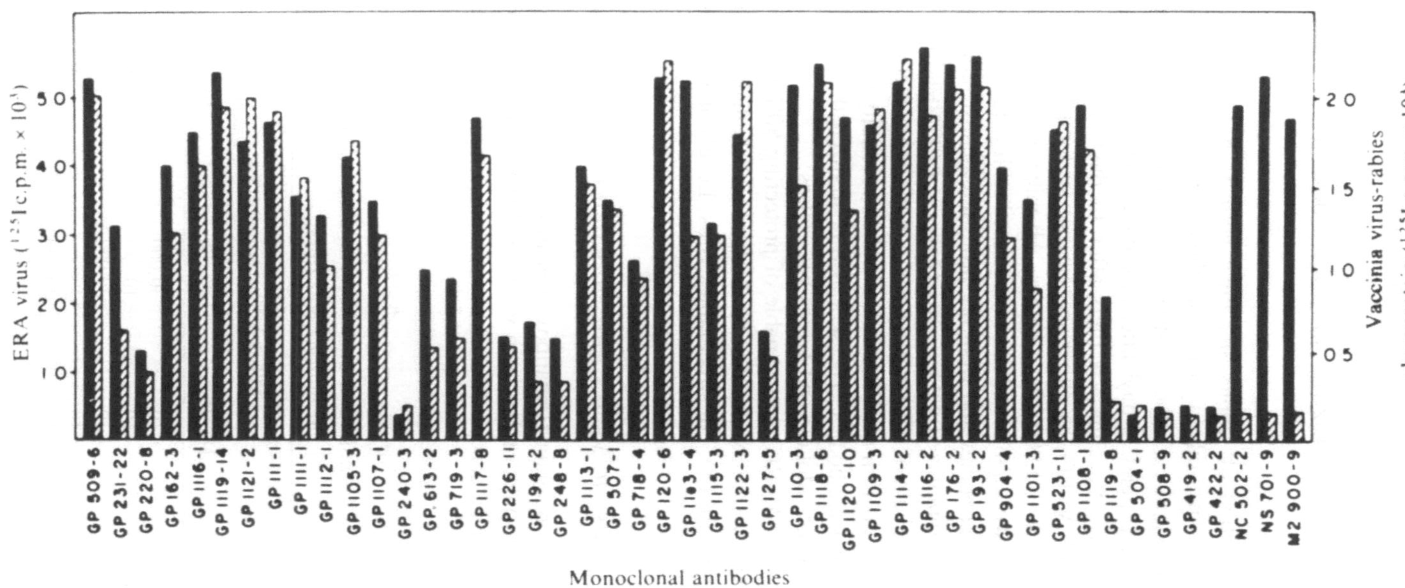

Figure 7. Comparison of V-RG and rabies ERA strain antigens for reaction with a panel of monoclonal antibodies. Lysates of infected cells were adsorbed to microtitre plates and incubated with 1/1000 dilutions of ascites fluid. Bound antibody was detected by the addition of ^{125}I-labeled goat anti-mouse immunoglobulin, washing, and liquid scintillation counting. Solid bars: ERA strain rabies virus, hatched bars: V-RG virus.

VI. IMMUNOGENIC PROPERTIES OF VACCINIA RECOMBINANT VIRUS EXPRESSING THE RABIES GLYCOPROTEIN

A. Induction of Virus-Neutralizing Antibodies and Protection against Rabies

Intradermal inoculation of rabbits and mice with V–RG virus resulted in rapid induction of rabies virus-neutralizing (VN) antibodies and in protection from severe intracerebral challenge with large doses of street rabies virus (Table 1). In rabbits, rabies VN antibody titers at 5, 11, and 14 days after inoculation were 800, 10,000, and greater than 30,000, respectively. Vaccinia VN titers after 14 days were substantially lower. Inoculation of mice with V–RG, by either scarification or footpad injection, resulted in rabies–VN titers of at least 30,000 after 14 days. All mice were protected against challenge (Wiktor *et al.*, 1984). Rabbits and mice inoculated with wild-type vaccinia virus did not develop rabies VN antibodies and were not protected against rabies virus challenge.

Sera from rabbits immunized with V–RG (day 14) neutralized between $10^{5.3}$ and $10^{6.6}$ tissue culture LD_{50} both of ERA rabies virus and of three street rabies virus isolates previously shown to differ from the ERA strain in their reactivity with the panel of anti-glycoprotein monoclonal antibodies. The sera also neutralized the Duvenhage strain of rabies-related virus but not Mokola and Lagos bat viruses.

A minimum dose of V–RG virus capable of protecting 50% of recipient mice was 10^4 PFU (Fig. 8). Mice were inoculated in the footpad and challenged intracerebrally 15 days later with 2400-mouse LD_{50} of street rabies virus. Levels of VNA were determined at 7 and 14 days.

B. Secondary Response to V–RG Virus Inoculation

Although vaccination against smallpox was discontinued some 5 years ago, the majority of the human population has been vaccinated at least once in their lifetimes. If V–RG virus should ever be accepted for human immunization, it is important to know

Table 1. Induction of VN Antibodies and Protection from Rabies by Live V-RG Virus in Rabbits and Mice[a]

Animals	Vaccine	Rabies VNA Day 0	Rabies VNA Day 14	Vaccinia VNA Day 14	Protection
Rabbits	V-RG	<10	>30,000	250	15/16
	None	<10	<10	<10	0/6
Mice I.D.	V-RG	<10	>30,000	250	12/12
	Vaccinia	<10	<10	250	0/12
Mice F.P.	V-RG	<10	>30,000	1,250	12/12
	None	<10	<10	<10	0/12

[a] Results are the combined data of two experiments.

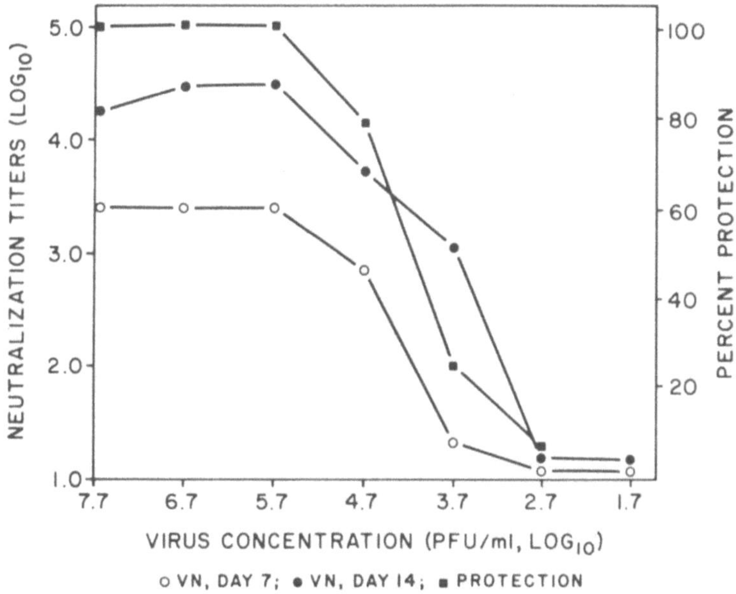

Figure 8. Minimum protective dose of V-RG. Mice were inoculated (footpad) with dilutions of live V-RG and levels of rabies VNA determined after 7 (○) and 14 (●) days. Mice were challenged (day 15) by intracerebral inoculation of 2400 LD_{50} units of street rabies virus (■).

whether previous vaccination interferes with the immune response to a second inoculation with V–RG virus.

Three rabbits immunized intradermally with $10^{7.6}$ PFU of V–RG virus and showing a VN titer >30,000 15 days after vaccination were inoculated intradermally 6 months later with the same dose of virus. Animals were bled before the booster inoculation and 3, 5, 7, and 15 days later. Three weeks after the booster, they received 24,000 mouse LD_{50} of street rabies virus strain MD5951 intracerebrally. As shown in Table 2, the level of rabies VN antibody on the day of the booster inoculation was 8000 for two animals and 12,000 for the third rabbit (significantly lower than that observed 15 days after the primary vaccination). Levels of VN antibodies in all three animals increased drastically starting on day 3 (titers 24,000) and reached titers of 70,000 or higher by day 15. As

Table 2. Booster Response to V–RG Virus in Rabbits

Rabbit	Virus-neutralizing antibody (day) 0	3	5	15	Protection (Day 21)
A	8,000	24,000	70,000	>70,000	Yes
B	8,000	24,000	70,000	>70,000	Yes
C	12,000	24,000	>70,000	>70,000	Yes

Table 3. Immune Response of Rabbits to V–RG Virus Administered by Different Routes

Route of inoculation	Virus-neutralizing antibody titers (day) 0	3	5	7	15	21	Protection day 21
ID	<3	6[a]	3,500	7,200	8,500	24,000	Yes
IM	<3	45	7,200	48,000	48,000	135,000	Yes
SC	<3	45	7,200	70,000	70,000	100,000	Yes
Oral	<3	<3	2,400	24,000	70,000	115,000	Yes
None	<3	—	—	—	—	—	No

[a] Mean value for two rabbits.

expected, all animals resisted challenge with street rabies virus. These results indicate that the immunity induced by V–RG virus did not interfere with the response to the same immunogen given 6 months later.

C. Immune Response to V–RG Virus Administered by Different Routes

Since intradermal inoculation may not be a practical route of administration for a rabies vaccine, we compared the immune response to V–RG virus administered by the intradermal (ID), intramuscular (IM), subcutaneous (SC), and oral (O) routes.

Pairs of rabbits were injected with, or induced to swallow, 0.2 ml V–RG virus ($10^{7.8}$ PFU). Levels of VN antibodies were evaluated at 0, 3, 5, 7, and 15 days, and on day 21 all rabbits and two unimmunized controls were challenged with street rabies virus as above.

VN antibodies could be detected on day 3 (Table 3) in rabbits injected by the ID, IM, or SC route. By day 5 all animals showed high levels of antibodies raching titers of 100,000–120,000 on day 21. All resisted challenge with street rabies virus.

D. Immune Response to V–RG Virus in Raccoons

The mid-Atlantic region of North America is experiencing a raccoon rabies outbreak that may be one of the most intense wildlife epizootics on record. To assess the possibility that the V–RG virus vaccine might be applied in controlling wildlife rabies, we inoculated raccoons with $10^{7.8}$ PFU of V–RG virus by several routes (Table 4). In two animals, a 1.0-ml volume of V–RG was administered intramuscularly, resulting in the rapid induction of rabies VN antibodies, with detectable levels still apparent several months later. Three other raccoons received 1.0 ml V–RG virus on the oral mucosa. They responded less well via the oral route as compared with the parenteral inoculation. However, within 3 weeks of an oral booster inoculation (with or without buccal mucosa scarification), all animals responded with higher levels of VN antibodies than after the primary vaccination.

A group of 10 captive raccoons was given 1.0 ml (10^8 PFU) of V–RG injected into an experimental sponge (currently under consideration as a vaccine field distribution bait).

Table 4. Induction of VN Antibodies in Raccoons by V–RG Virus

		Virus-neutralizing antibody titer (day)				
Animal	Route of inoculation	0	10	28	42 (Booster)	63
A	I.M.	<10	1,215	1,215	None	400
B	I.M.	<10	405	1,215	None	400
C	Oral	<10	135	450	Oral + scar.	8,000
D	Oral	<10	135	450	Oral + scar.	11,000
E	Oral	<10	135	450	Oral	2,400
F	None	<10	—	<10	—	—

Two animals that had not ingested the bait within 48 hr received 1.0 ml of either undiluted stock of V–RG virus or of the V–RG virus that had been placed in sponge baits and maintained for 2 weeks at room temperature (residual infectivity titer of 10^6 PFU/m) directly into the oral cavity.

No VNA could be detected prior to vaccination (titers <1 : 15) in any animal in the study; 4 weeks after bait consumption, all vaccinated raccoons had demonstrable VN antibodies (titers 1 : 30–1 : 2700) and 9 out of 10 were protected from intramuscular challenge with a virulent rabies virus (Fig. 9). The only animal that was not protected

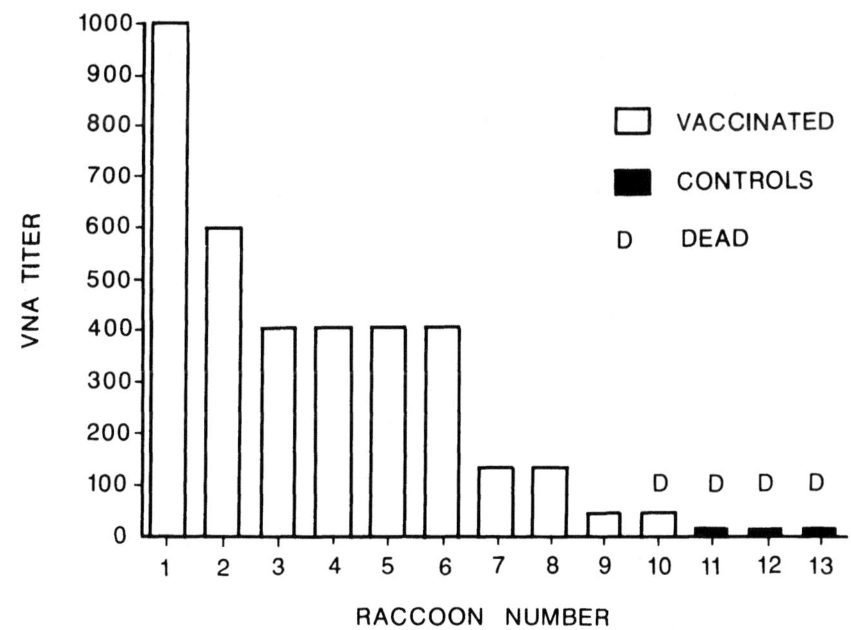

Figure 9. Antibody titers and resistance to challenge in raccoons receiving oral V-RG. (See text for details.)

received the low-titer V–RG virus. Six nonvaccinated control raccoons did not present VNA and were not protected.

A similar level of protection has been obtained by oral or intramuscular vaccination of European foxes with V–RG (J. Blancou and M. P. Kieny, unpublished observations); these results offer great promise that recombinant virus research can lead to an effective oral wildlife rabies vaccine.

E. Immunogenicity of Inactivated V–RG Virus

Live vaccinia virus has a long history of safe use as a vaccine for humans, despite a low incidence of serious complications. Nevertheless, the reintroduction of vaccinia virus-based vaccines may be controversial. We therefore evaluated the immunogenicity of vaccinia and V–RG virus-infected cell extracts, purified inactivated V–RG virus, and the rabies glycoprotein isolated from V–RG virus-infected cell extracts using an affinity column prepared with anti-rabies virus glycoprotein antibody. Mice were vaccinated twice at weekly intervals and challenged intracerebrally with 240 LD_{50} of MD5951 street rabies virus 2 weeks after the first vaccine dose. All preparations, with the exception of the inactivated vaccinia virus-infected cell extract, induced high levels of rabies VN antibodies and protected against rabies virus challenge (Table 5).

F. Use of Live and Inactivated V–RG Virus in Postexposure Protection

The principal application of rabies vaccine for humans is the postexposure treatment of persons bitten by rabid or potentially rabid animals. An experimental vaccine was prepared from sonicated V–RG virus-infected BHK cells by centrifugation through a 36% sucrose cushion. A portion of the resuspended virus (infectivity titer of $10^{8.7}$ PFU/m) was inactivated by β-propiolactone treatment. The antigenic value (AV) of this V–RG vaccine preparation was 17.2 IU, as determined by the standard NIH potency test for inactivated rabies vaccines.

Groups of twelve hamsters were inoculated intramuscularly with 0.1 ml of live or inactivated V–RG vaccine, diluted 1 : 3, at either 24 hr before, or 1 hr or 3 days after challenge with 0.1 ml of MD5951 street rabies virus (10–20 hamster LD_{50}). An additional group of twelve hamsters was treated with an experimental PM-Vero inactivated vaccine (AV 10.0), also diluted 1 : 3.

Table 5. Induction of VN Antibodies and Protection from Rabies by Inactivated V–RG Virus

Vaccine	Rabies VNA titers Day 0	Rabies VNA titers Day 14	Protection
Vaccinia cell lysate	<10	<10	0/12
V–RG cell lysate	<10	8,000	12/12
V–RG virus (purified)	<10	4,000	12/12
V–RG G protein	<10	15,000	12/12

Table 6. Postexposure Protection of Hamsters with V–RG Vaccine

Vaccine type	Virus PFU/hamster ($\log_{10}$)	Antigenic value (NIH/ml)	Percentage protection	
			Vaccination −1 day	Vaccination + 1 hr and 3 days
V–RG live	7.7	—	50	83
V–RG BPL	—	5.7	42	50
PM–Vero	—	3.3	33	17
None	—	—	0	—

As shown in Table 6, the live and inactivated V–RG preparations protected 50–80% of hamsters treated either 24 hr before or at 1 hr or 3 days after challenge. All the unvaccinated control hamsters died. Protection by the PM-Vero vaccine was less evident.

VII. IMMUNOLOGIC PROPERTIES OF V–RG BEARING LEU INSTEAD OF PRO AT POSITION 8

The rabies G cDNA isolated by Anilionis *et al.* (1981) bears a codon for leucine at position 8 rather than for proline. Although this was successfully corrected prior to the construction of V–RG, it was of evident interest to investigate the immunologic properties of the leucine variant. The rabies glycoprotein cDNA (leucine variant) was transferred to the vaccinia genome as before to create recombinant V–RG–leu.

Mice were inoculated via the intradermal or footpad routes with 2×10^8 PFU (intradermal) or 5×10^7 PFU (footpad) of the recombinant viruses V–RG or V–RG–leu. Titres

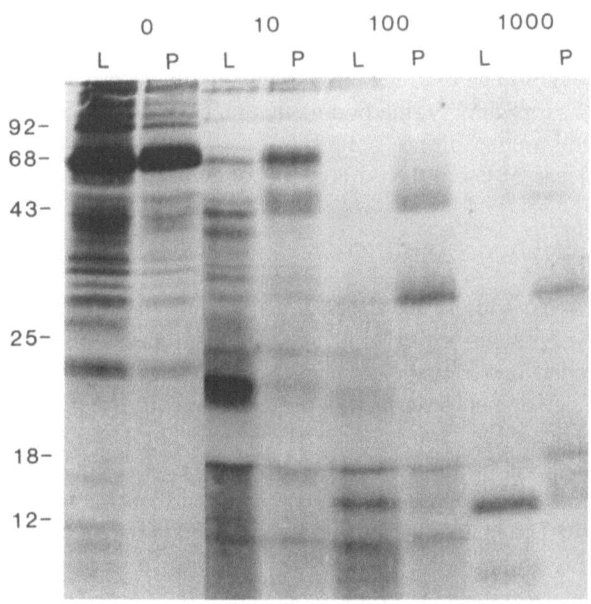

Figure 10. Altered conformation of G-leu probed by trypsin digestion. Rabies G antigen synthesized upon infection of rodent cells with either V-RG-pro (P) or V-RG-leu (L) was recovered by immunoadsorption to anti-virion serum R215 and protein A sepharose. Immune complexes were treated with trypsin (1 hr, 37°C) and digests resolved by SDS gel electrophoresis and autoradiography. Trypsin concentrations are given in µg/ml (above); molecular weight standards are in kilodaltons (left).

of rabies-neutralizing antisera were determined at 14 days, and animals were challenged with 2400 LD_{50} MD5951 rabies virus on day 14. Although high titers (>30,000) of VNA appeared in sera of animals vaccinated with V–RG, VNA were not detectable in sera from animals vaccinated with V–RG–leu.

Examination of tissue-culture cells infected with V–RG–leu revealed the presence of material capable of reacting with a polyclonal anti-rabies serum but reacting only weakly with monoclonal virus-neutralizing antibodies. These results suggest that the folding and/or processing of the leucine-G are abnormal. To confirm this, labeled cell extracts infected with V–RG or V–RG–leu viruses were subjected to mild treatment with trypsin after immunoadsorption to polyclonal antiserum and protein-A Sepharose. Proteins were resolved by SDS gel electrophoresis and autoradiography. As expected, the profiles displayed by the G-leucine and G-proline were quite different (Fig. 10) demonstrating that the pro-leu alteration has a dramatic effect on the processing of the glycoprotein. Further experiments are required in order to pinpoint the precise defect in G maturation.

VIII. A RECOMBINANT VACCINIA EXPRESSING THE SOLUBLE RABIES GLYCOPROTEIN

Cells infected with rabies virus shed a soluble glycoprotein species into the extracellular medium. Amino acid sequence analysis has revealed that the soluble G protein, Gs, lacks the hydrophobic transmembrane anchor zone present at the C-terminal of the glycoprotein. Gs is presumed to result from proteolytic cleavage of rabies G *in vivo*. Purified Gs fails to aggregate in the presence of low concentrations of detergent and, although the conformation and glycosylation of the Gs species appear to be identical to G (Dietzschold *et al.*, 1983), the protective activity of Gs in mice is drastically reduced compared with full-length G. Aggregation and/or membrane binding may thus be important parameters in induction of an anti-rabies immune response. Indeed, copresentation of antigen at the cell surface in juxtaposition with histocompatibility determinants is now thought to be of crucial importance in the generation of an appropriate immune response.

Intriguingly, the finding that inoculation of mice with rabies G may possibly provoke autoimmune disease (Burrage *et al.*, 1985) might suggest a further biologic role for release of Gs in the etiology of rabies.

A. Hybrid Construction

To address these questions we introduced a translation stop-codon into the rabies glycoprotein coding sequence. The appropriate restriction fragment of the G cDNA was cloned into a single-stranded M13 bacteriophage vector, and localized oligonucleotide-directed mutagenesis was used to introduce a translation stop codon at the proposed cleavage site, Ser-Ala-Gly-Ala- ↓ -Leu-Thr, at the beginning of the transmembrane zone. This was accomplished by deleting one nucleotide; the codons CTG(Leu)–ACT(Thr) generate a stop codon (TGA) when the C of the leucine codon is removed.

The mutated fragment was restored to the vaccinia TK plasmid by procedures analogous to those described earlier, and *in vivo* recombination was employed to transfer the mutated rabies G coding sequence to the vaccinia genome, generating V–RG–S.

B. Immunologic Properties of V–RG–S

BHK-21 cell monolayers were infected with V–RG–S or the V–RG control virus and adsorbed with fluorescent-labeled antibodies directed against rabies G. Discrete fluorescent staining was observed in the cytoplasm of acetone-fixed cells infected with V–RG–S virus following treatment with anti-virion or anti-G sera. Little or no fluorescence was observed upon staining with monoclonal anti-G antibodies, suggesting that the cytoplasmic material observed represents incomplete or partially processed forms of rabies G. Surprisingly, Gs present in the medium only reacted poorly with anti-rabies serum.

Mice were injected by footpad inoculation with 10^7 PFU V–RG–S or control viruses and bled 10 days later. Serum was tested for the presence of virus-neutralizing and binding antibodies by radioimmunoassay and VN tests. Control serum from mice immunized with V–RG gave high levels of binding antibody. In the same test, titers of binding antibody in mice immunized with V–RG–S were approximately 30-fold less.

Virus-neutralizing antibody levels in mice injected with V–RG–S were low (1 : 200) compared with the titers obtained in sera from mice injected with V–RG ($\geq$1 : 30,000). As expected, mice immunized with V–RG–S virus succumbed to challenge with street rabies virus, whereas mice vaccinated with V–RG resisted severe challenge infection.

These results underline the importance of cell-surface presentation of antigens to the immune system. In addition, the theoretical possibility that rabies-associated immunosuppression (Wiktor *et al.*, 1977) operating through Gs release might contribute to the lack of Gs immunogenicity cannot be ruled out.

The low level of antigen detected in extracts of cells infected with V–RG–S virus is not understood. Apart from the chain-terminating signal introduced by localized mutagenesis the elements of the construction are essentially the same as those used in the assembly of V–RG. These data indicate that the transmembrane zone of rabies glycoprotein may not only tether the glycoprotein in the viral envelope but also play a role in the processing, assembly, or stability of the viral glycoprotein spikes.

IX. INDUCTION OF CELL-MEDIATED IMMUNITY BY V–RG VIRUS

Cytotoxic T-lymphocytes (CTL) have been postulated to play a major role in the postexposure protection against rabies virus infection (Wiktor, 1978). Therefore, the ability to induce CTL may be a highly desirable property of any contemplated rabies vaccine.

A/J mice immunized with V–RG virus generated a substantial secondary CTL response *in vitro* after re-exposure of lymphocytes to PM or ERA rabies viruses (Fig. 11) or *in vivo* after inoculation of V–RG virus-immunized mice with ERA rabies virus. Figure 11 also shows that inactivated V–RG virus failed to stimulate rabies-virus-specific CTL; however in other experiments live V–RG virus was found to be highly effective. Despite the ability of V–RG virus to prime for CTL memory, primary rabies virus specific CTL responses were weak. Since V–RG virus induced a strong vaccinia virus-specific response, this finding may reflect some form of immunodominance; however, the mechanisms involved are not clear.

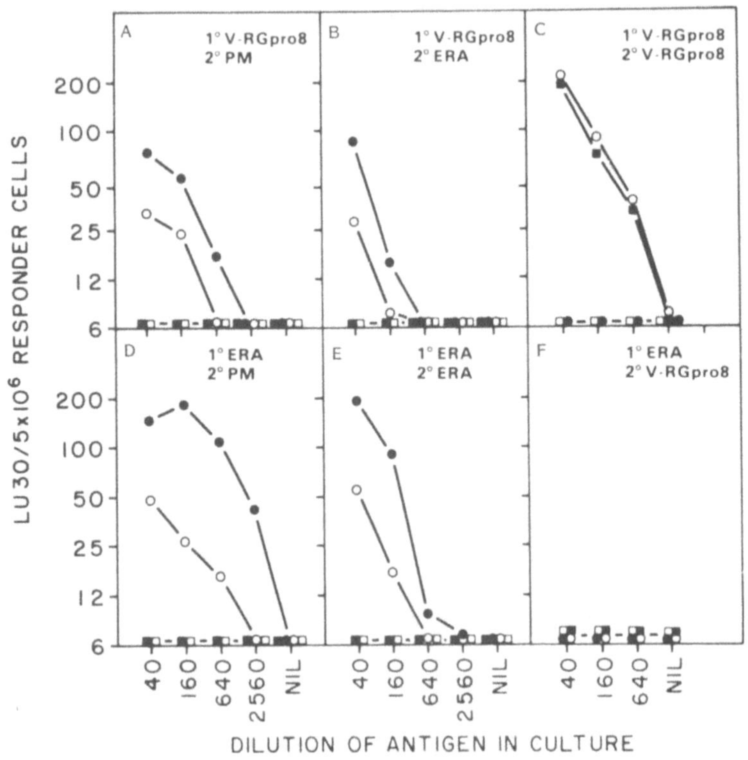

Figure 11. Secondary CTL response stimulated by V-RG and rabies viruses *in vitro*. A/J mice were inoculated intravenously with V-RG virus (A, B and C) or intraperitoneally with ERA rabies virus (D, E and F). Four weeks later, spleen cells were cultured with dilutions of inactivated ERA (B and E) or PM (A and D) rabies viruses, or with inactivated V-RG virus (C and F). After 5 days, each culture was titrated for CTL activity which is expressed as lytic units per 5×10^6 responder spleen cells. Target cells: uninfected (□), ERA rabies virus-infected (●), vaccinia virus-infected (■), or V-RG virus-infected (○) ^{51}Cr-labeled NA cells.

The derivation of the V–RG virus afforded the opportunity to analyze the target antigens recognized by rabies virus-specific CTL. First, it is clear that at least some CTL recognize the rabies glycoprotein since infection of syngeneic target cells by V–RG virus rendered them susceptible to lysis (Fig. 11). However, there was always a clear difference in susceptibility between rabies virus and V–RG virus-infected cells. This was at least partially explained by a series of experiments in which the lytic specificities of a panel of cloned rabies virus-specific CTL were examined. Seven cell lines were derived by limiting-dilution cloning of A/J mouse spleen cells stimulated *in vitro* with inactivated ERA rabies virus. All the cell lines were restricted by H-2K^k, and all demonstrated high lytic activity against rabies virus-infected target cells yet failed to lyse V–RG virus-infected cells. These results suggest that the CTL may recognize another rabies virus structural protein at the surface of infected cells, or that rabies-G-specific epitopes present at the surface of rabies virus are absent from V–RG-infected cells. Perhaps significantly, three of the seven cloned CTL lines also recognized Mokola virus; there is very little serologic cross-reactivity between Mokola and rabies virus glycoproteins.

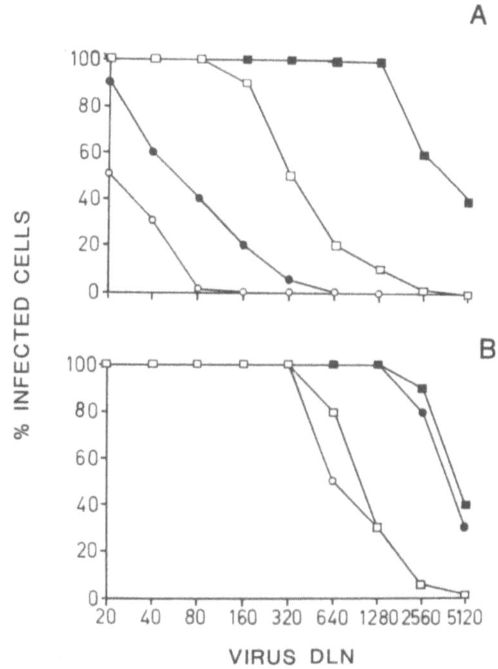

Figure 12. Anti-viral activity of cloned CTL. Two thousand cloned CTL (cell line 284.3D5.F5) (●, ○), or the same volume of medium (□, ■), were incubated together with 5,000 syngeneic NA (panel A) or allogeneic MC57G (panel B) cells that had been infected with ERA rabies virus (1 hr, 37°C). Cultures were in 15 μl in Terasaki tissue culture plates, in the presence (open symbols) or absence (closed symbols) of rabies virus-neutralizing antibody. Human interleukin 2 of recombinant origin was also present. After 48 hr, the plates were fixed in 80% acetone, and the percentage of infected cells determined by immunofluorescence.

Evidence for a role of CTL in the pathogenesis of rabies virus infection is mainly indirect; we therefore attempted to demonstrate an antiviral effect of cloned CTL *in vivo* or *in vitro*. Adoptive transfer experiments were inconclusive, since CTL lines display abnormal migration properties. However, in another experiment, 2000 cloned CTL (cell line 204.3D5.F5) were incubated together with 5000 allogenic MC57G or syngeneic NA indicator cells and dilutions of ERA rabies virus. After 48 hr, the percentage of infected cells was determined by immunofluorescence. Figure 12 shows that this cloned CTL line was even more effective than virus-neutralizing antibody in limiting the spread of rabies virus in syngeneic NA cell cultures. No effect against MC57G ($H\text{-}2^b$) cells was observed. This antiviral mechanism, possibly dependent on lymphokines (IFNγ in particular) rather than on lysis, is not specific for the rabies virus glycoprotein. This indicates that further work in rabies vaccine development should take into account the possibility that other proteins contribute to immunity.

X. CONCLUSION

The ability to construct recombinant viruses expressing foreign genes clearly has great potential for vaccine development. In the case of rabies, the V–RG virus has proved to be highly immunogenic and has outperformed conventional rabies vaccines in every comparative protection test. The question remains, however, whether recombinant vaccines based on vaccinia viruses will ever be accepted for human or veterinary use. In this regard, it should be noted that the lessons learned in the construction and evaluation of V–

RG virus can be rapidly applied for the development of other recombinant viruses (e.g., based on adenoviruses or herpes viruses).

Since completion of this manuscript reports have appeared attesting to the efficacy of the V-RG vaccine in foxes (Blancou *et al.*, 1986) and in raccoons (Rupprecht *et al.*, 1986).

ACKNOWLEDGMENTS

We would like to thank our many co-workers for their contributions to this project and for permission to cite unpublished data. We thank Dr. H. Koprowski and Dr. J. P. Lecocq in whose laboratories much of the work described was carried out.

REFERENCES

Anilionis, A., Wunner, W. H., and Curtis, P. J. (1981). *Nature (London)* **294,** 275–278.
Baxby, D. (1981). *Jenner's Smallpox Vaccine*. Heinemann Educational Books, London.
Blancou, J., Kieny, M. P., Lathe, R., Lecocq, J. P., Pastoret, P. P., Soulebot, J. P., and Desmettre, P. (1986). *Nature* **322,** 373–375.
Burrage, T. G., Tignor, G. H., and Smith, A. L. (1985). *Virus Res.* **2,** 273–279.
Cochran, M. A., Puckett, C., and Moss, B. (1985). *J. Virol.* **54,** 30–37.
Dales, S., and Pogo, B. G. T. (1981). *Biology of Poxviruses, Virology Monograph* **18** (D. W. Kingsbury and H. Zwihausen, eds.), Springer-Verlag, New York.
Dietzschold, B., Wiktor, T. J., MacFarlan, R., and Varrichio, A. (1982). *J. Virol.* **44,** 595–602.
Dietzschold, B., Wiktor, T. J., Wunner, W. H., and Varrichio, A. (1983). *Virology* **24,** 330–337.
Drillin, R., and Spehner, D. (1983). *Virology* **131,** 385–393.
Geshelin, P., and Berns, K. I. (1974). *J. Mol. Biol.* **88,** 785–796.
Lathe, R., Kieny, M. P., Schmitt, D., Curtis, P., and Lecocq, J. P. (1984). *J. Mol. Appl. Genet.* **2,** 331–342.
Kieny, M. P., Lathe, R., Drillien, R., Spehner, D., Skory, S., Schmitt, D., Wiktor, T., Koprowski, H., and Lecocq, J. P. (1984). *Nature (Lond.)* **312,** 163–166.
Kieny, M. P., Desmettre, P., Soulebot, J. P., and Lathe, R. (1987). *Prog. Vet. Microbiol. Immunol.* **3,** 73–111.
Mackett, M., and Achard, L. G. (1979). *J. Gen. Virol.* **45,** 689–701.
Panicali, D., and Paoletti, E. (1982). *Proc. Natl. Acad. Sci. USA* **79,** 4927–4931.
Pasteur, L. (1885). *C.R. Acad. Sci.* **101,** 765–772.
Rupprecht, C. E., Wiktor, T. J., Johnston, D. H., Hamir, A. N., Dietzschold, B., Wunner, W. H., Glickman, L. T., and Koprowski, H. (1986). *Proc. Natl. Acad. Sci. USA* **83,** 7947–7950.
Smith, G. L., and Moss, B. (1984). *Bioessays* **1,** 120–124.
Smith, G. L., Mackett, M., and Moss, B. (1983). *Nature (Lond.)* **302,** 490–495.
Smith, G. L., Mackett, M., and Moss, B. (1984). *Biotech. Genet. Eng. Rev.* **2,** 383–407.
Weir, J. P., and Moss, B. (1983). *J. Virol.* **46,** 530–537.
Wiktor, T. J. (1978). *Dev. Biol. Stand.* **40,** 255–264.
Wiktor, T. J., Gyorgy, E., Schlumberger, H. D., Sokol, F., and Koprowski, H. (1973). *J. Immunol.* **110,** 269–276.
Wiktor, T. J., Doherty, P. C., and Koprowski, H. (1977). *J. Exp. Med.* **145,** 1600–1617.
Wiktor, T. J., MacFarlan, R. I., Reagan, K. J., Dietzschold, B., Curtis, P. J., Wunner, W. H., Kieny, M. P., Lathe, R., Lecocq, J. P., Mackett, M., Moss, B., and Koprowski, H. (1984). *Proc. Natl. Acad. Sci. USA* **81,** 7194–7198.

II

Vaccination with Synthetic Peptides

8

Synthetic Peptides as Immunogens

F. Brown

I. INTRODUCTION

The concept of using short fragments of proteins for immunization is a logical extension of the demonstration, more than 20 years ago, that for most viruses the protective activity is normally associated with only one of the constituent proteins. During the 1960s, it was shown that the immunizing activity of measles, influenza, rabies, vesicular stomatitis, and other lipid-containing viruses is associated with the surface projections clearly observable by electron microscopy. With the exception of those from vesicular stomatitis virus, the activity of the isolated surface projections in eliciting neutralizing protective antibody was lower than when they were attached to virus particles. It soon became apparent, however, that the activity of the isolated projections could be increased by restoring the configuration they possessed when forming part of the virus particle.

Also in the 1960s, Anderer (1963*a,b*) made the very important observation that a short fragment of the protein of tobacco mosaic virus, obtained by tryptic cleavage, would elicit antibody that neutralized the infectivity of the virus. This was followed by his demonstration that a synthetic peptide corresponding in sequence to the fragment would also elicit neutralizing antibody.

The work by Anderer was possible only because of the availability of the amino acid sequence of the virus protein (Tsugita *et al.*, 1960). However, very few sequences were available at that time, and it was not until Sela and colleagues, working with the RNA containing MS_2 bacteriophage, demonstrated that a fragment of the coat protein and the synthetic peptide corresponding to it would elicit antibodies that reacted with the virus particle that the concept of synthetic peptides as immunogens started to receive the attention it deserved (Langebeheim *et al.*, 1976).

The publication in 1977 of the two methods for the sequencing of DNA provided the means for deriving the amino acid sequences of many biologically active proteins and allowed the concept of Anderer and Sela and colleagues to be explored with other proteins. Viruses have been recognized as excellent models for testing the concept because of their relatively simple structure as compared with bacteria and protozoa. This chapter concentrates on one example, foot-and-mouth disease virus, which provides evidence that the peptide approach has great potential for future vaccine products.

F. Brown • Wellcome Biotechnology Ltd., Beckenham, Kent BR3 3BS, England.

II. POTENTIAL ADVANTAGES OF PEPTIDE IMMUNOGENS

One of the great success stories in veterinary and human medicine is the use of prophylactic vaccination against a wide range of diseases. In the western world the number of cases of many infectious diseases, which only a few years ago were counted in the tens of thousands, has now decreased to vanishing point. Indeed, the greatest achievement of all in this field, the eradication of smallpox, depended on the provision of a potent and readily available vaccine. These great achievements relied on vaccines that in principle were no different from those used by Jenner for smallpox, by Pasteur for rabies, and by Salk and Sabin for poliomyelitis.

However, the production of most vaccines presents the manufacturer with problems. If the vaccine is attenuated, the following questions need to be answered:

1. Are the cell substrate and growth medium free from adventitious infectious agents?
2. Is the vaccine at the correct level of attenuation?
3. Is the vaccine stable at the temperatures likely to be encountered in the field?
4. What is the shelf life of the product?

Even with the best vaccines, such as the Sabin poliovaccine and the venerable vaccinia virus vaccine, the side reactions, sometimes leading to death, are sufficient to stimulate vaccine manufacturers to search for safer products. The physical stability of aqueous products is always likely to cause problems, and the wonder is that they are so stable.

With killed vaccines, several problems face the manufacturer:

1. The innocuity of the product
2. Its potency
3. The stability of the vaccine at the temperatures that will be encountered in the field
4. The shelf life of the product
5. The anaphylactic reactions that sometimes occur

Questions 3 and 4 have already been considered with attenuated vaccines. The innocuity of the product is a question of statistics, and even the best-controlled manufacturing process must be carefully monitored to ensure that the virus is completely inactivated. Potency can often be a problem, relying as it does on the adequate growth of the virus. An added problem is the need to provide disease-secure premises when virulent agents are being used. Examples of this problem are provided by rabies virus, which is a hazard to the staff, and foot-and-mouth disease virus, which is a hazard to the environment. An additional problem is the rare but sufficiently frequent occurrence of anaphylactic shock following the injection of vaccines.

With these problems, it is hardly surprising that manufacturers and researchers alike have looked to alternative methods of producing vaccines. The concept of a peptide vaccine envisaged many years ago by Anderer and Sela together with the encouraging early results using peptides as antigens for the stimulation of an immune response against foot-and-mouth disease (Bittle *et al.*, 1982; Pfaff *et al.*, 1982) and hepatitis B (Lerner *et al.*, 1981) make it worthwhile to summarize the advantages of a peptide vaccine over the conventional vaccines. Provided that the correct region(s) of the immunogenic protein can be identified, there are several advantages of a peptide vaccine:

1. The product would be chemically defined and not subject to the variable yields frequently encountered when viruses are grown in culture, so there would be no problems with the consistency of its quality.
2. There would be no question of the presence of any infectious agent.
3. The product would be stable. Peptides can be stored for long periods and almost certainly indefinitely without loss of activity. This means that a cold chain would not be necessary for either storage or transportation. The transportation of a dry product would also be much cheaper than one consisting mainly of water which requires to be kept cold.
4. The need for downstream processing would be eliminated.
5. The product could be designed so that the immune responses could be varied by the insertion of appropriate sequences or amino acid residues.
6. The stability of the product, opening up the possibility of different vaccine formulations and regimens (e.g., pulse-release mechanisms), wherein the peptide is encapuslated in soluble glass and released at predetermined intervals to provide a booster injection.
7. Cost is clearly an important consideration, but improving methods for synthesis and the fact that a large production plant would not be required mean that large-scale production of peptides would compare favorably with the existing tissue culture technology.

III. IDENTIFICATION OF POSSIBLE IMMUNOGENIC SITES

Attempts to develop synthetic peptides depend on the ability to predict the location of epitopes on the virus particle. Epitopes are generally defined as sequential and conformational. Sequential epitopes comprise peptides that are specifically recognized in their unfolded random coil form, so that a particular sequence provides sufficient information for recognition by the antibody against the protein. Discontinuous epitopes are made up of residues that are not contiguous in sequence but that are brought together at the surface of the protein or virus particle either by the folding of the polypeptide chain or by the constraints imposed by the architecture of the virus. The discontinuous nature of an epitope could probably best be demonstrated by a crystallographic study of the virus–antibody complex, but the specificity of monoclonal antibody molecules can provide useful information.

Several methods have been used to locate immunogenic sites of virus proteins. These are now considered.

A. Biologic Activity of Fragments

Measurement of the biologic activity of fragments of the protein followed by aligning the active sequences obtained by different methods of cleavage is the most direct approach. It suffers from the possible drawback that the agents cleaving the protein may destroy the immunogenic site. For this reason, there are advantages in using more than one reagent to cleave the protein. This approach has been used by Strohmaier *et al.* (1982) with foot-and-mouth disease virus; its application to that virus is described fully in Section IV.

B. Solvent-Accessible Regions

The aim of the use of solvent-accessible regions is to be able to predict whether a given amino acid sequence is on the surface and therefore able to interact with the immune system. Hopp and Woods (1981) investigated the possibility that antigenic determinants might be associated with amino acid sequences containing a large number of charged and polar residues (i.e., hydrophilic residues). They assigned a hydrophilicity value to each amino acid based on the solvent parameter values assigned by Levitt (1976). These values were then respectively averaged down the protein chain, thereby generating a series of local hydrophilicity values. By choosing sequences of six amino acids, generally accepted as the average size of an antigenic determinant, and plotting the hydrophilicity value versus the sequence position, Hopp and Woods were able to predict correctly the antigenic sites of 12 proteins for which extensive immunochemical information was already available. However, as will be shown for foot-and-mouth disease virus, several of the maxima correspond to regions with no demonstrable immunogenic activity.

In an extension of this approach, solvent-accessible regions were recently explored by Connolly (1983). These regions were originally defined by Lee and Richards (1971) as the areas traced out by the center of a spherical probe, representing a solvent molecule, as it rolled over the surface of the molecule under study. By the use of computer graphics, a graphic display of the solvent-accessible regions can be obtained, providing greater visual realism and bringing a better appreciation of the folding of the polypeptide chain. By relating this information to that derived from studies of the immunogenicity of segments of proteins, more reliable predictions regarding the location of immunogenic sites may be possible.

C. Interpreting the Secondary Structure of Proteins

The secondary structure of proteins can also give clues in identifying immunogenic sites. Thus, amphipathic α-helical regions have been proposed as important sites, and β-turns frequently occur in proteins as elbows jutting out from the main body. The amino acids most frequently occurring in turns are asparagine, aspartic acid, proline, and glycine. At this stage, there is no rule that enables immunogenic sites to be predicted from secondary structure. However, the observation that the N and C terminals of proteins possess higher than average antigenicity has been attributed to their being less constrained than other parts of the proteins; two groups of workers recently reported that segmental mobility along the polypeptide chain is correlated with the lcation of continuous epitopes. With the recent report by Karplus and Schulz (1985) that it may be possible to predict flexibility from the primary structure of proteins, mobility could be used to predict antigenic sites in the same way as hydrophilicity. The potential of this method can be assessed from the observations of Van Regenmortel and his colleagues with tobacco mosaic virus (Westhof *et al.*, 1984). Six of seven known continuous epitopes in the molecule were found to correspond to peaks of the mobility plot. It is also significant that three hexapeptides corresponding to highly accessible regions of the virus protein situated outside mobile regions exhibited no antigenic activity. This approach has been reviewed by Van Regenmortel (1986).

D. Predicting from Antigenic Variation

The amino acid sequences of antigenic variants can be compared on the assumption that the sequences will differ at the antigenic sites. The value of this method is best illustrated by the example provided by foot-and-mouth disease virus (Bittle *et al.*, 1982). Essentially, the method relies on the interpretation of sequence data from several serotypes of the virus. These showed that most of the sequence of the immunogenic protein VP1 was highly conserved, but there was considerable variation in three regions. Two of these corresponded to hydrophilic regions, and of these one elicited high levels of neutralizing antibody, whereas the other gave lower but still significant levels.

E. Reaction with Neutralizing Monoclonal Antibody

Antigenic variants can be created in the laboratory by growing the virus in the presence of neutralizing monoclonal antibody. The sequences of the parent virus and escape mutant are then compared. This method is based on the assumption that the difference in sequence identifies the sites involved in neutralization. It was applied by Minor *et al.* (1983) to a study of poliovirus type 3. These workers showed that the parent strain and escape mutant differed at positions 277–300 on the part of the RNA genome coding for VP1, corresponding to a sequence of eight amino acids. Subsequent work with synthetic peptides corresponding to this region indicate that it contains an antigenic site (Ferguson *et al.*, 1985).

F. Screening the Proteins Expressed from Fragments of the Coding Gene

Screening the proteins expressed from fragments of the coding gene has been used by Nunberg *et al.* (1984) with feline leukemia virus. The complementary DNA (cDNA) corresponding to the gene coding for the immunogenic protein of the virus was fragmented with DNAse; these randomly generated fragments were expressed separately in λ-phage. The phage library was then screened with neutralizing monoclonal antibody in order to identify those expressing the specific antigenic determinant. Sequencing of the DNA fragment of the immunoreactive phage then permitted mapping of the antigenic determinant reacting with the monoclonal antibody. Antibody binding was mapped to a region comprising 14 amino acids; a synthetic peptide corresponding to this region elicited virus neutralizing antibody in guinea pigs.

G. Direct Mapping of the Epitope with Neutralizing Antibody

Direct mapping of the epitope with neutralizing antibody has been used by Geysen and colleagues (1984) to map the VP1 protein of foot-and-mouth disease virus. Hexapeptides corresponding to amino acids 1–6, 2–7, 3–8 . . . 208–213 were synthesized on plastic sticks, which were then allowed to react with neutralizing antibody. The reactive regions were then detected with an antispecies antiserum. This method has the advantage of rapidity and of being done with very small amounts of peptide.

IV. APPLICATION OF THE PREDICTIVE METHODS TO THE IDENTIFICATION OF THE IMMUNOGENIC SITES OF FOOT-AND-MOUTH DISEASE VIRUS

Experience has shown that one predictive method alone is usually insufficient to provide the information required. The value of using a combination of predictive methods is demonstrated by the application of methods A, C, D, and F to the identification of immunogenic sites to the foot-and-mouth disease virus particle. This virus, which belongs to the family *Picornaviridae,* comprises one molecule of ssRNA, molecular weight 2.6×10^6 and 60 copies of four proteins VP1–VP3 (molecular weight 24×10^3) and VP4 (molecular weight 10×10^3). The protein VP1 was identified as carrying the immunogenic activity of the virus by the observation that incubation with trypsin led to the loss of most of the activity of a virus belonging to serotype 0 (Wild and Brown, 1967; Wild *et al.*, 1969). Polyacrylamide gel electrophoretic (PAGE) analysis of the native and enzyme-treated particles showed that only VP1 appeared to be altered, being cleaved into two fragments having a combined molecular weight similar to that of the native protein. The inference that VP1 was the only protein affected by the treatment led to the view that this protein alone carried immunogenic activity and the subsequent observation, first made by Laporte *et al.* (1973) that VP1 isolated from the virus particle elicited low but significant levels of neutralizing antibody in several species, whereas the other proteins did not, appeared to confirm this supposition.

Most preparations of VP1 possess very low immunogenic activity compared with the intact virus particle. Whereas 5 μg of virus particles as one injection generally gives protection, two suitably spaced injections of c 250 μg of VP1 are necessary to give the same level of immunity (Bachrach *et al.*, 1975). Nevertheless, the activity of the isolated protein has acted as a stimulus to several groups to identify the immunogenic sites. The first experiments were described by Strohmaier *et al.* (1982) using the approach of Anderer and Sela. The protein was cleaved by two methods. In the first, the isolated protein was cleaved with cyanogen bromide and the separated fragments tested for their ability to elicit neutralizing antibody. In the second, the protein was cleaved *in situ* by several different proteolytic enzymes, and the fragments separated following disruption of the treated virus. These fragments were also tested for immunogenic activity. The results of these experiments, which are summarized in Fig. 1, led the German workers to predict that sequences 146–154 and 200–213 of the 213-amino acid protein would form part of immunogenic sites.

Hydrophilic plots gave seven peaks, two of which were at the same positions as those predicted by Strohmaier and colleagues. Further predictions from secondary structure were made by Pfaff *et al.* (1982). This group predicted, on the basis of the presence of a strong helical region that displayed hydrophilic and hydrophobic sequences on opposite sides of the helix, that residues 144–159 would form an immunogenic site.

On the assumption that in a highly variable virus the amino acid sequence(s) corre-

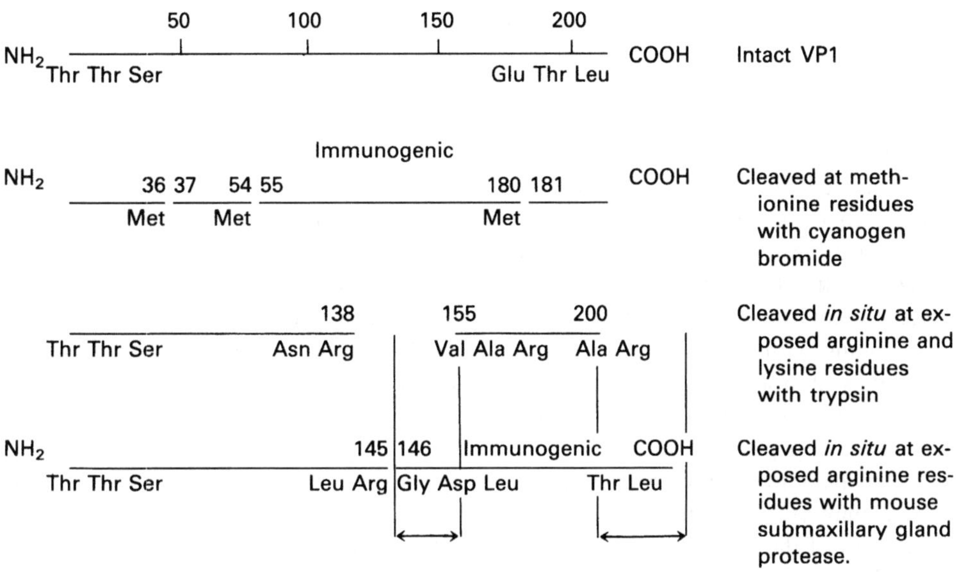

Figure 1. Location of immunogenic sites on VP1 of foot-and-mouth disease virus. (Adapted from Strohmaier *et al.*, 1982.)

sponding to the antigenic sites would also vary, Bittle *et al.* (1982) compared the sequences of four viruses, two belonging to serotype A and the other two to serotypes 0 and C. Most of the sequences were found to be conserved, but three regions, corresponding to amino acids 42–61, 138–160, and 193–204, are highly variable. Taken together with the hydrophilic plots, it seemed unlikely that 42–61 would be immunogenic because it is hydrophobic. By contrast, the other two sequences were in accord with the predictions of Strohmaier's group.

Finally, the immunoactivity of the consecutive overlapping hexapeptides has been measured for three viruses belonging to the three European serotypes A, 0, and C. These so-called peptide scans have highlighted sequences starting at positions 144, 146, and 143 respectively, for the three serotypes (Meloen and Barteling, 1986) and have led the authors to propound the merits of a sequence comprising amino acids 146–152.

V. IMMUNOGENIC PEPTIDES OF FOOT-AND-MOUTH DISEASE VIRUS

The consensus from the predictions described in Section IV indicated that a peptide encompassing amino acids 138–160 would be antigenically active. This consensus was found to be essentially correct when synthetic peptides containing 20 amino acids (e.g., 1–20, 21–40) corresponding to the entire sequence of VP1 were tested for their immunogenic activity. The sequence 141–160 was by far the most active, all the other peptides exhibiting 1% or less of its activity. Nevertheless, the neutralizing antibody elicited by other regions indicated that those too may have a role in the overall immunogenicity of the virus particle. Direct evidence to support this supposition has come from studies with neutralizing monoclonal antibodies (Parry *et al.*, 1985). This work is described later in this Section.

In the early experiments it was shown that one injection of 200 μg of peptide 141–160, linked to keyhole limpet hemocyanin via an added cysteine residue at the C terminal and mixed with an adjuvant such as aluminium hydroxide or incomplete Freund's adjuvant, elicited high levels of neutralizing antibody in guinea pigs and protected them against challenge infection with a large dose of virus. Subsequent experiments have shown that the method of coupling to the carrier protein is not critical in eliciting neutralizing antibody. Good responses have also been obtained when the peptide is linked to tetanus toxoid or bovine serum albumin (BSA), either via a C-terminal cysteine or by crosslinking with glutaraldehyde (Bittle *et al.*, 1984). Since the use of carrier proteins could present problems, particularly if repeated injections were made, the recent observations by Francis *et al.* (1985) that good neutralizing antibody responses can be obtained with the free peptide are important.

To obtain the optimal neutralizing antibody response, the presentation of the peptide to the immune system in a conformation that corresponds to its conformation when it forms part of the virus particle would seem to be essential. Walter and Doolittle (1983) have shown, for example, that antisera prepared against fragments of ribonuclease react well with the fragments but poorly with the native enzyme. Conversely, antibodies against the native protein react weakly with the fragments. These results are best explained by assuming that the fragments exist in a range of conformations, only a small proportion of which resembles the corresponding region on the native protein.

In our work with foot-and-mouth disease virus peptides, we have found that the neutralizing activity of serum from guinea pigs and rabbits that had received one or more injections of the 141–160 peptide was only reduced about 30% by absorption with excess virus particles, although the activity of the same sera could be completely absorbed with the peptide (N. R. Parry, unpublished observations). This type of experiment emphasizes the need to consider the manner in which peptides are presented to the host. It seems reasonable to expect that if the peptide corresponding to the immunogenic site could be fixed in the conformation it assumes on the virus particle, the neutralizing antibody response would be enhanced. This would represent a major step forward in the potential value of peptides as immunogens.

The observation by Bittle *et al.* (1982) that there are two sequences, 141–160 and 200–213, of the VP1 protein that elicit neutralizing antibody raised the question of the relative contributions of the two sites to the total immunogenicity of the protein. Although the two antipeptide sera had similar enzyme-linked immunosorbent assay (ELISA) titers against intact virus particles, the neutralizing activity of the anti-141–160 serum was about 100 times greater than that of the anti-200–213 serum. It seemed necessary to define the precise contribution of each sequence to the immunogenic activity of the protein. The activity of each antipeptide antiserum in blocking the reaction of neutralizing monoclonal antibody with the virus particle was compared with the blocking activity of anti-146S virus particle and anti-VP1 antisera (Parry *et al.*, 1985). Although the latter two sera were able to block completely the attachment of the monoclonal antibody to the virus particles, neither of the anti-peptide antisera competed by more than 60%. However, 100% blocking was achieved with a mixture of the two antipeptide antisera, suggesting that both sequences contribute to a single site. The peptides have also been used to block the attachment of monoclonal antibody to virus particles. A peptide consisting of both sequences was found to inhibit the binding of the antibody much more efficiently than the individual peptides. These results show that this fine analysis with monoclonal antibodies will be valuable in elucidating the composition of the immunogenic site. Nevertheless, it seems that the 141–160 sequence contains the immunodominant site of the virus protein.

VI. MOLECULAR BASIS FOR ANTIGENIC VARIATION

Although antigenic variation in foot-and-mouth disease is a vaccine manufacturer's nightmare, examination of the amino acid sequences of the VP1 proteins of several serotypes and subtypes has provided interesting information on the molecular basis for antigenic variation. Most of the VP1 sequences are highly conserved, even between serotypes. The major variation in sequence occurs between amino acids 138–160 and 41–60, but since this latter sequence is hydrophobic and antibody against it does not react with virus particles, it is unlikely to be on the surface and hence less likely to play any part in antigenic variation.

Sera against viruses belonging to two different subtypes, A10 and A12, of serotype A differentiated clearly between them. The sera prepared against the 141–160 peptides corresponding to the same viruses were equally discriminating, indicating the dominant nature of this amino acid sequence in neutralization tests (Table 1).

The amino acids involved in the neutralization reaction have been identified even more precisely with the A12 virus (Rowlands *et al.*, 1983). An isolate of this virus was found to contain at least three antigenic variants, clearly differentiated in neutralization tests (Table 2). The amino acid sequences of the viruses differ in two positions only, at residues 148 and 153 of VP1. Antisera to these peptides also differentiated between the isolates. Further evidence for the importance of the residues was obtained from another virus of the same subtype. This virus had been derived from the mixed virus population referred to earlier, by passaging it many times in a variety of tissue culture cells. The derived virus differed from virus B, one of the antigenic variants, only in having a Phe

Table 1. Serologic Relationships between Two Subtypes of Foot-and-Mouth Disease Virus Demonstrated by Antivirus Particles and Antipeptide Sera

Serum	Log_{10} virus neutralized by 0.015 ml serum	
	A10	A12
Antivirus particle	2.5	0.9
Antipeptide	2.7	0.5
Antivirus particle	0.7	3.5
Antipeptide	0.5	2.5

Table 2. Effect of Changes at Residues 148 and 153 of VP1 on the Serologic Relationships between Three Naturally Occurring Variants of Viruses Belonging to the A12 Subtype

Serum	Log_{10} virus neutralizing by 0.015 ml serum			
	A Ser 148 Leu 153	B Leu 148 Pro 153	C Ser 148 Ser 153	D Phe 148 Pro 153
A				
Antivirus particle	3.3	1.3	1.7	0.7
Antipeptide	2.5	1.1	1.5	0.9
B				
Antivirus particle	0.9	3.7	0.9	2.1
Antipeptide	0.0	3.2	0.5	2.2
C				
Antivirus particle	1.9	0.5	2.5	0.5
Antipeptide	1.5	0.1	2.6	1.5
D				
Antivirus particle	1.3	2.5	1.1	3.7
Antipeptide	0.0	2.0	0.0	2.7
Antipeptide (148 Leu, 153 Ser)	0.3	0.9	0.5	0.7
Antipeptide (148 Phe, 153 Leu)	0.0	0.4	0.4	0.3

residue at position 148 instead of Leu, but it could be differentiated from it using either antivirus particle or antipeptide sera in neutralization tests (Table 2).

Nonsense peptides that differ from those of the four variants in having different residues at positions 148 and 153 elicited antibody that failed to neutralize any of the viruses (Table 2). These results reinforced the previous evidence that residues 148 and 153 are important in determining specificity.

Recent evidence has also pinpointed residue 148 as being important in the serotype 0 viruses, although the antigenic site in this serotype appears to be more complex than that of serotype A (Ouldridge *et al.*, 1986). Thus, although the 141–160 sequences of viruses classified as belonging to subtype 1 on the basis of serum neutralization tests are identical, the viruses do show small but significant differences in the neutralization test. To account for these differences, it seems that part of the immunogenic site must lie outside the 141–160 sequence or the structure of this sequence is influenced by distal regions of the VP1 protein or the other structural proteins.

We have found, contrary to expectation, that antipeptide sera are more cross-reactive in neutralization tests with other viruses of the same serotype than are the corresponding antivirion antisera (Table 3). Sequencing studies have shown that even when the amino acids in positions 140–145 differed, the antipeptide serum to the 141–160 sequence of subtype 1 would neutralize the virus. However, in those examples in which the antipeptide serum did not neutralize the virus, the amino acid at position 148 also differed. This is demonstrated clearly in the sequences for the virus of subtype 6 and the isolate from Thailand (TAI 1/80) shown in Table 4. With the virus of subtype 6 the change is from Leu to Thr and in the Thailand isolate it is from Leu to Arg.

The cross-reactivity of the antiserum against the 141–160 peptide with heterologous

Table 3. Cross-Neutralization Tests on Isolates of Viruses Belonging to Foot-and-Mouth Disease Virus, Serotype 0[a]

	Virus							
Serum	Kauf B64	Kauf B7	BFS 1848	BFS 1860	OV1 subtype 6	Hong Kong	Indonesia 7/83	Thailand 1/80
Antivirus particle (BFS 1860)	0.2	0.2	0.2	1.0	<0.1	0.5	0.2	0.1
Antipeptide (141–160)	0.6	0.6	1.0	1.0	<0.1	1.0	0.9	0.1

[a] Values are expressed as the ratio $\frac{\text{neutralization of heterologous virus}}{\text{neutralization of homologous virus}}$

strains of virus may have practical implications for vaccine design. Sequence 145–151 is highly conserved in the viruses of serotype 0 and, suitably presented, may be the basis for a universal serotype 0 vaccine.

The Arg–Gly–Asp sequence at positions 145, 146, and 147 may also have some significance in the furnishing of a universal vaccine. This sequence occurs in all the viruses so far examined, with the exception of serotype A10. With this virus, the sequence is Ser–Gly–Asp at the same positions. By inserting an Arg residue between the Ser and

Table 4. Amino Acid Sequences of the 141–160 Region of VP1 of Several Isolates of Foot-and-Mouth Disease Virus, Serotype 0[a]

	Amino acid sequence									
Virus	141	142	143	144	145	146	147	148	149	150
Kaufbeuren B64	Val	Pro	Asn	Leu	Arg	Gly	Asp	Leu	Gln	Val
Kaufbeuren B7	•	•	•	•	•	•	•	•	•	•
BFS 1848	•	•	•	•	•	•	•	•	•	•
BFS 1860	•	•	•	•	•	•	•	•	•	•
0-V1 (Subtype 6)	•	•	•	Val	•	•	•	Thr	•	•
Hong Kong	Met	Ser	•	Val	•	•	•	•	•	•
Indonesia 7/83	Thr	Thr	•	Val	•	•	•	•	•	•
Thailand 1/80	Leu	Thr	•	Val	•	•	•	Arg	•	•

	Amino acid sequence									
	151	152	153	154	155	156	157	158	159	160
Kaufbeuren B64	Leu	Ala	Gln	Lys	Val	Ala	Arg	Thr	Leu	Pro
Kaufbeuren B7	•	•	•	•	•	•	•	•	•	•
BFS 1848	•	•	•	•	•	•	•	•	•	•
BFS 1860	•	•	•	•	•	•	•	•	•	•
0-V1 (Subtype 6)	•	Asp	•	•	•	Ser	•	Ala	•	•
Hong Kong	•	Thr	•	•	Ala	Ser	•	Ala	•	•
Indonesia 7/83	•	•	•	•	Ala	Ala	•	•	•	•
Thailand 1/80	•	•	•	•	Ala	•	•	Pro	•	•

[a] •, No change from Kaufbeuren B64 sequence.

Gly residues in the A10 peptide, the antiserum that it then elicited cross-neutralized some of the A12 viruses, whereas the "natural" A10 antipeptide antiserum does not neutralize the A12 viruses.

These observations provide firm leads for the further exploration of the molecular basis for antigenic variation. Clearly, a universal vaccine against all serotypes is a worthy target, but a vaccine that provided immunity against the variants within a serotype would also have enormous practical advantages.

VII. PRACTICAL CONSIDERATIONS

Most killed vaccines are given as multiple injections, suitably spaced to obtain an effective booster response. The prospect of a vaccine composed of a stable product opens up another method of giving multiple doses at intervals. Guinea pigs that have received a dose of inactivated vaccine or peptide insufficient to elicit a neutralizing antibody response produce high levels of neutralizing antibody when they receive a similar subimmunizing dose of peptide (Francis *et al.*, 1985). The stability of the peptide in the dry state permits it to be incorporated in a delayed-release mechanism from which it can be released at a preselected time. Such an arrangement would have considerable advantages over the method currently used, in which multiple injections are given. The attendance of people at clinics for a second or third injection of a vaccine is difficult to arrange. The mustering of animals in difficult terrain such as that found in Africa or South America is even more difficult, so that it is usually only feasible to inoculate them at infrequent intervals. A system in which two or more doses could be delivered without the need to muster the animals would be a considerable boon.

Synthesis of peptides on a large scale should not present undue problems. Indeed, if the amount of peptide required for a single injection is about 1 mg, the amount required for one million people would be 1 kg, which is small scale in chemical terms.

VIII. PROSPECTS FOR PEPTIDE VACCINES

The advantages of using a stable solid as a vaccine were described in Section II. For these practical reasons, research on the use of peptides as vaccines is likely to continue at an increasing pace. The prospects of achieving the ultimate goal depend heavily on obtaining a better neutralizing antibody response. The key consideration appears to be the presentation of the peptide in a configuration similar to that it assumes on the virus particle. Our evidence with the foot-and-mouth disease virus peptide indicates that only about 30% of the antipeptide antibody that is induced is useful in neutralizing virus particles (N. R. Parry, unpublished observations). By altering the configuration of the peptide, it may be possible to increase its specificity with regard to neutralization. There is some evidence from work with a hepatitis B surface antigen (HB_5Ag) peptide that this can be achieved by allowing cyclization to take place via two cysteine residues that form part of the sequence (Dreesman *et al.*, 1984). Similar observations have been made with the 141–160 peptide of foot-and-mouth disease virus, whereby cyclization and/or polymerization was made possible by the addition of a cysteine residue to each end of the peptide followed by air oxidation (Bittle *et al.*, 1984).

Furthermore, little attention has been paid to the way in which peptides are attached to the carrier proteins that have been used. A variety of coupling agents has been used, often taking no account of the blocking of potentially important groups such as the $-NH_2$ of lysine residues, which occurs when glutaraldehyde is used or the free COOH groups of aspartic and glutamic acid residues when carbodiimides are used. Residues suitable for attachment to proteins can be built into the synthetic peptide, even though they do not form part of the natural sequence. Cysteine residues are particularly useful for this purpose.

The loss of mobility of the peptide that occurs when disulfide bridges are formed may mean that the range of antibody molecules would be reduced; thus, when antigenic variants occur as in foot-and-mouth disease virus, the cross-neutralizing activity may be diminished. The effect of such attempts to restrict the mobility of peptides on the antibody produced needs study at the molecular level.

The response to peptides of the other arm of the immune system (i.e., the T-cell repertoire) has not been studied in the same detail. It is now emerging, however, from Berzofsky's work with amphipathic structures (i.e., those with separated hydrophilic and hydrophobic surfaces) that the stimulation of T lymphocytes is dependent on such a structure (Delisi and Berzofsky 1986). The 151–160 sequence of the foot-and-mouth disease virus peptide appears to be amphipathic. The unpublished work of Francis *et al.* (1985) with a nested set of peptides embracing the 141–160 sequence suggests that only the 141–150 sequence is related to the evocation of neutralizing antibody. The fine analysis of the immune response should permit the design of more immunogenic peptides.

The considerable increase in immune response obtained when the poorlv immunogenic surface projections of influenza and measles virus are incorporated into immunostimulating complexes (Morein *et al.*, 1984) lends support to the idea that peptides incorporated similarly may display greater activity. If this step were accompanied by the increase in immunogenic activity obtained with surface projections, the future for peptide vaccines would be very bright.

Exploration of these avenues will provide information that may enable us to take peptide vaccines out of the laboratory and into the field. The advantages of vaccines that are defined in precise chemical terms are sufficient goals to stimulate intensive work in this area over the next few years.

REFERENCES

Anderer, F. A. (1963*a*). *Z. Naturforsch.* **18,** 1010–1014.

Anderer, F. A. (1963*b*). *Biochim. Biophys. Acta* **71,** 246–248.

Bachrach, H. L., Moore, D. M., McKercher, P. D., and Polatnick, J. (1975). *J. Immunol.* **115,** 1636–1641.

Bittle, J. L., Houghten, R. A., Alexander, H., Shinnick, T. M., Sutcliffe, J. G., Lerner, R. A., Rowlands, D. J., and Brown, F. (1982*a*). *Nature (Lond.)* **298,** 30–33.

Bittle, J. L., Worrell, P., Houghten, R. A., Lerner, R. A., Rowlands, D. J., and Brown, F., (1984). In *Modern Approaches to Vaccines* (R. M. Channock and R. A. Lerner, eds.), pp. 103–107, Cold Spring Harbor Laboratory, Cold Spring Harbor, New York.

Connolly, M. L. (1983). *Science* **221,** 709–713.

Delisi, C., and Berzofsky, J. A. (1986). *Proc. Natl. Acad. Sci. USA* **82,** 7048–7052.

Dreesman, G. R., Sparrow, J. T., Kennedy, R. C., and Melnick, J. L. (1984). In *Modern Approaches to Vaccines* (R. M. Chanock and R. A. Lerner, eds.), pp. 115–119, Cold Spring Harbor Laboratory, Cold Springs Harbor, New York.

Ferguson, M., Evans, D. M. A., Magrath, D. I., Minor, P. D., Almond, J. W., and Schild, G. C. (1985). *Virology* **143,** 505–515.

Francis, J. J., Fry, C. M., Rowlands, D. J., Brown, F., Bittle, J. L., Houghten, R. A., and Lerner, R. A. (1985). *J. Gen. Virol.* **66,** 2347–2354.

Geysen, H. M., Meloen, R. H., and Barteling, S. J. (1984). *Proc. Natl. Acad. Sci. USA* **81,** 3998–4002.

Hopp, T. P., and Woods, K. R. (1981). *Proc. Natl. Acad. Sci. USA* **78,** 3824–3828.

Karplus, P. A., and Schulz, G. E. (1985). *Naturwissenschaften* **72,** 212–213.

Langbeheim, H., Arnon, R., and Sela, M. (1976). *Proc. Natl. Acad. Sci. USA* **73,** 4636–4640.

Laporte, J., Grosclaude, J., Wantyghem, J., Bernard, S., and Rouze, P. (1973). *C. R. Acad. Sci. Paris* **276,** 3399–3401.

Lee, B., and Richards, F. M. (1971). *J. Mol. Biol.* **55,** 379–400.

Lerner, R. A., Green, N., Alexander, H., Liu, F. T., and Sutcliffe, J. G. (1981). *Proc. Natl. Acad. Sci. USA* **78,** 3403–3407.

Levitt, M. (1976). *J. Mol. Biol.* **104,** 59–107.

Meloen, R. H., and Barteling, S. J. (1986). *Virology* **149,** 55–63.

Minor, P. D., Schild, G. C., Bootman, J., Evans, D. M. A., Ferguson, M., Reeve, P., Spitz, M., Stanway, G., Cann, A. J., Hauptman, R., Clarke, L. D., Mountford, R. C., and Almond, J. W. (1983). *Nature (Lond.)* **301,** 674–679.

Morein, B., Sundquist, B., Hoglund, S., Dalsgaard, K., and Osterhaus, A. (1984). In *Modern Approaches to Vaccines* (R. M. Chanock and R. A. Lerner, eds.), pp. 363–367. Cold Spring Harbor Laboratory, Cold Spring Harbor, New York.

Nunberg, J. H., Rodgers, G., Gilbert, J. H., and Snead, R. M. (1984). *Proc. Natl. Acad. Sci. USA* **81,** 3675–3679.

Ouldridge, E. J., Parry, N. R., Barnett, P. V., Bolwell, C., Rowlands, D. J., Brown, F., Bittle, J. L., Houghten, R. A., and Lerner, R. A. (1986). In *Vaccines 86* (F. Brown, R. M. Chanock, and R. A. Lerner, eds.), pp. 45–49, Cold Spring Harbor Laboratory, Cold Spring Harbor, New York.

Parry, N. R., Ouldridge, E. J., Barnett, P. V., Rowlands, D. J., Brown, F., Bittle, J. L., Houghten, R. A., and Lerner, R. A. (1985). In *Vaccines 85* (R. M. Channock, R. A. Lerner, and F. Brown, eds.), pp. 211–216, Cold Spring Harbor Laboratory, Cold Spring Harbor, New York.

Pfaff, E., Mussgay, M., Bohm, H. O., Schulz, G. E., and Schaller, H. (1982). *EMBO J.* **1,** 869–874.

Rowlands, D. J., Clarke, B. E., Carroll, A. R., Brown, F., Nicholson, B. H., Bittle, J. L., Houghten, R. A., and Lerner, R. A. (1983). *Nature (Lond)* **306,** 694–697.

Strohmaier, K., Franze, R., and Adam, K.-H. (1982). *J. Gen. Virol.* **59,** 295–306.

Tainer, J. A., Gretzoff, E. D., Alexander, H., Houghten, R. A., Olson, A. J., Hendrickson, W. A., and Lerner, R. A. (1984). *Nature (Lond.)* **312,** 127–133.

Tsugita, A., Gish, D., Young, G. J., Fraenkel-Conrat, H., Knight, C. A., and Stanley, W. M. (1960). *Proc. Natl. Acad. Sci. USA* **46,** 1463–1469.

Van Regenmortel, M. H. V. (1986). *Trends Biochem. Sci.* **11,** 36–39.

Walter, G., and Doolittle, R. F. (1983). In *Genetic Engineering: Principles and Methods* (J. K. Satlow and A. Hollander, eds.), pp. 61–69, Academic, Orlando, Florida.

Westhof, E., Altschuh, D., Moras, D., Bloomer, A. C., Mondragon, A., Klug, A., and Van Regenmortel, M. H. V. (1984). *Nature (Lond.)* **311,** 123–126.

Wild, T. F., and Brown, F. (1967). *J. Gen. Virol.* **1,** 247–250.

Wild, T. F., Burroughs, J. N., and Brown, F. (1969). *J. Gen. Virol.* **4,** 313–320.

9

Vaccination with Synthetic Hepatitis B Virus Peptides

A. R. Neurath, S. B. H. Kent, N. Strick, and K. Parker

I. INTRODUCTION

Until recently, it was generally accepted that the protein moiety of the hepatitis B virus (HBV) envelope (env) and of subviral hepatitis B surface antigen particles (HB_sAg) consists of a single 25,000-M_r species existing in either nonglycosylated (P25) or glycosylated forms (GP29). Higher-molecular-weight components detected by polyacrylamide gel electrophoresis (PAGE) of HBV (HB_sAg) were considered to represent artifacts of the technique or contaminants (Koistinen, 1980; Peterson *et al.*, 1984). This conclusion was not compatible with results indicating the presence on HBV and on tubular forms of HB_sAg of antigenic determinants not present on P25/GP29 (Alberti *et al.*, 1978; Moodie *et al.*, 1974; Neurath *et al.*, 1976). Cloning and sequencing of HBV DNA has provided evidence that the gene coding for HBV env proteins has the capacity to code for proteins larger than the 226 amino acid chain of P25 (Tiollais *et al.*, 1981) (Fig. 1), designated S protein, the translational product of the S region of the HBV env gene. Stibbe and Gerlich (1983) suggested that the minor HB_sAg components GP33 and GP36 have the same sequence as P25 plus 55 additional amino acids at the *N*-terminus encoded by the pre-S (pre-S2) region of the env gene. They also reported that these components are probably not important for hepatitis B vaccines.

Further investigations using monoclonal antibodies, synthetic peptides with sequences predicted from the nucleotide sequences of the pre-S region of the env gene, and pre-S polypeptide fusion products expressed in *Escherichia coli*, established that the protein moieties of HBV (HB_sAg) components with >29,000 M_r correspond to the following sequences: (1) the middle protein, representing S protein with 55 additional amino acids at the *N*-terminus encoded for by the pre-S region [amino acid residues pre-S (120–174)]; and (2) the large protein, consisting of the middle protein with an additional 108–119 *N*-terminal amino acids [depending on the antigenic subtype; amino acid residues pre-S (1–119) or pre-S (12–119) (designated pre-S1) (Neurath and Kent, 1985)] and

A. R. Neurath and N. Strick • The Lindsley F. Kimball Research Institute of The New York Blood Center, New York, New York 10021. *S. B. H. Kent and K. Parker* • Division of Biology, California Institute of Technology, Pasadena, California 91125.

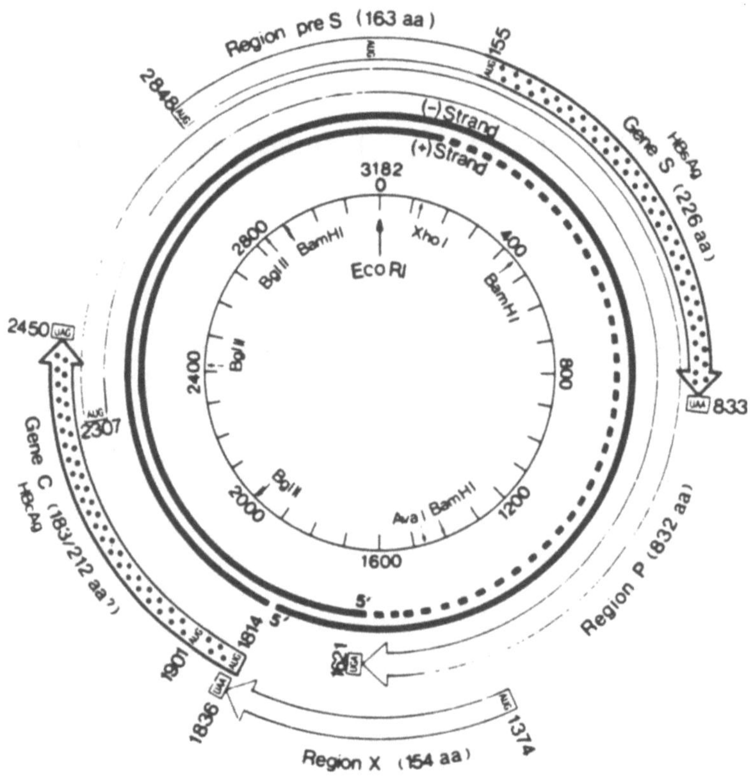

Figure 1. Genetic organization of HBV. The partial restriction map and numbering of the nucleotides indicated on the inner circle correspond to the *ayw* genome. The broad arrows surrounding the genome represent the four large open reading frames of the (−) strand transcript. These four potential coding regions are termed region S (= env gene = gene S + region pre-S), gene C, region P, and region X. The number of amino acids in the brackets corresponds to the length of the translational products. The two regions corresponding to genes S and C are represented by stippling. (Tiollais *et al.*, 1981.)

containing all amino acids encoded by the HBV env gene (pre-S and S regions) (Heermann *et al.*, 1984; Machida *et al.*, 1984; Michel *et al.*, 1984; Neurath *et al.*, 1984*a*, 1985*a*; Offensperger *et al.*, 1985; Okamoto *et al.*, 1985; Persing *et al.*, 1985; Wong *et al.*, 1985) (Fig. 2).

It has already been firmly established that HB_sAg lacking pre-S sequences (Neurath *et al.*, 1985*a*) elicits protective antibodies in immunocompetent recipients (Stevens *et al.*, 1984) and that monoclonal antibodies directed against an antigenic determinant on the S protein are virus neutralizing (Iwarson *et al.*, 1985). However, considerable differences between distinct hepatitis B vaccines were observed in efficacy trials in immunocompromised individuals. The vaccine lacking pre-S sequences was the least efficient. Before discussing the potential of synthetic peptide analogues of the HBV env protein, it is therefore necessary to consider the biologic role of the pre-S sequence and of the corresponding antibodies.

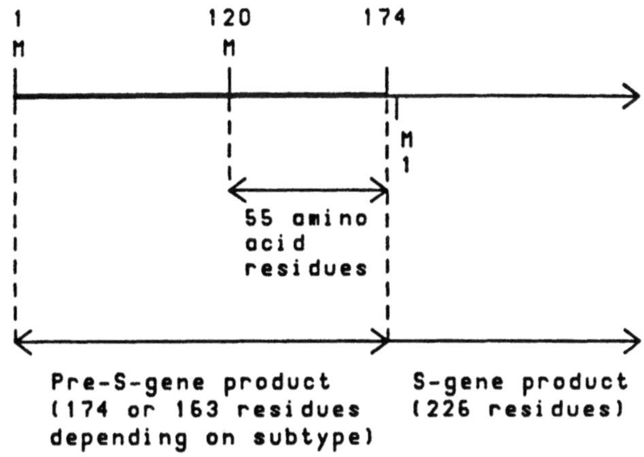

Figure 2. Schematic representation of the translational products of the open reading frame of the HBV DNA envelope gene. Position of N-terminal methionines (M) of the distinct translational products is indicated on top for the pre-S region and on the bottom of the line for the S-region. The three translational products are (1) the S-gene product containing 226 amino acids; (2) the S-gene product + 55 amino acids corresponding to pre-S (120–174), containing 281 amino acids (= middle protein); and (3) the S-gene + pre-S gene product, containing 389 to 400 amino acid residues (= large protein).

II. ANTIGENICITY, IMMUNOGENICITY, AND BIOLOGIC ROLE OF PRE-S SEQUENCES OF THE HBV ENV MIDDLE AND LARGE PROTEINS

This section summarizes the evidence that sequences coded for by the pre-S region of the HBV env gene (1) are highly immunogenic; (2) are recognized by antibodies produced as the result of HBV infection or immunization with HB_sAg (HBV); (3) play an important role in the life cycle of HBV; and (4) are expected to increase the efficacy of hepatitis B vaccines.

A. Results of Experiments in Rabbits

1. Antisera prepared by immunization of rabbits with HBV recognize at high dilution synthetic peptides spanning the pre-S sequence. The highest dilution endpoints were observed with peptides near the *N*-terminal portions of the HBV env middle and large proteins (Fig. 3). These peptides were also the most efficient in eliciting antibodies reacting with HBV (Fig. 4).
2. Rabbit anti-HBV also reacted with a fusion protein (expressed in *E. coli*) containing the sequences of 55 amino acids of the HBV middle protein encoded by the pre-S2 region of the HBV env gene (Offensperger *et al.*, 1985) (Fig. 5). However, this fusion protein had a poor immunogenicity.
3. Rabbit antisera to some of the synthetic peptides inhibited the attachment of HBV (HB_sAg) to hepatocytes (hepatoma cells) (Neurath *et al.*, 1985*a*). The anti-HBV serum (from which antibodies to the S protein had been removed) had similar effects, while anti-protein S serum displayed no activity in this respect (Table 1). These results establish the essential role of pre-S sequences in the first step of HBV-cell interactions, the attachment of the virus to hepatocytes.
4. Rabbit antiserum to the synthetic peptide pre-S (120–145) neutralized the infectivity of HBV (Fig. 6).

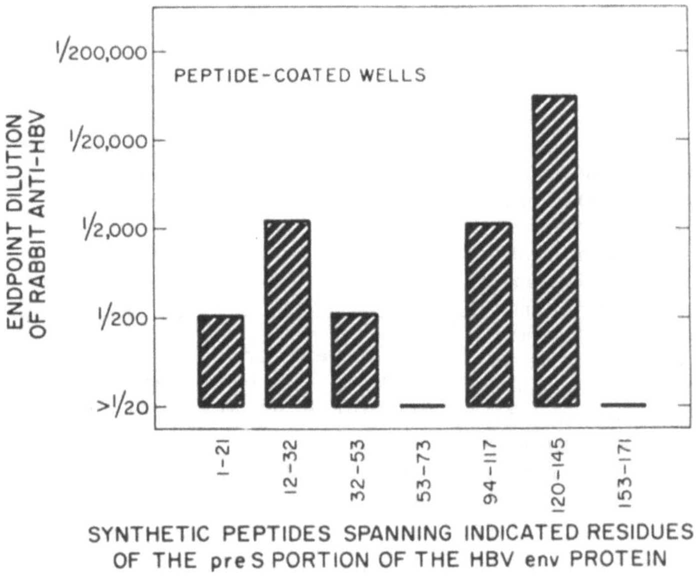

Figure 3. Scanning of the pre-S portion of HBV env proteins for regions optimally recognized by rabbit anti-HBV. Wells of polystyrene plates were coated with the synthetic peptides defined on the abscissa and reacted with serial dilutions of anti-HBV. The rabbit IgG attached to the wells was detected by ^{125}I-labeled anti-rabbit IgG. RIA ratio units were determined by dividing the number of counts per minute (cpm) corresponding to diluted anti-HBV by cpm corresponding to equally diluted control sera. The endpoint dilutions correspond to the highest dilution at which the RIA ratios were greater than 2.1. (Reprinted from Neurath *et al.*, 1986*a*.)

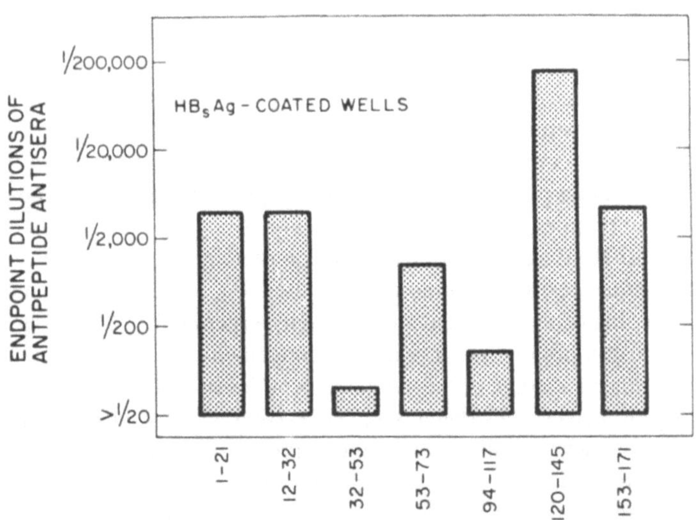

Figure 4. Scanning of the pre-S portion of HBV env proteins for regions eliciting the highest levels of antibodies reacting with HBV (HB$_s$Ag). Rabbits were immunized with synthetic peptides defined on the abscissa. The antisera were screened by RIA tests using HB$_s$Ag-coated wells of 96-well polystyrene plates. Procedures indicated for Fig. 3 were used. (Reprinted from Neurath *et al.*, 1986*a*.)

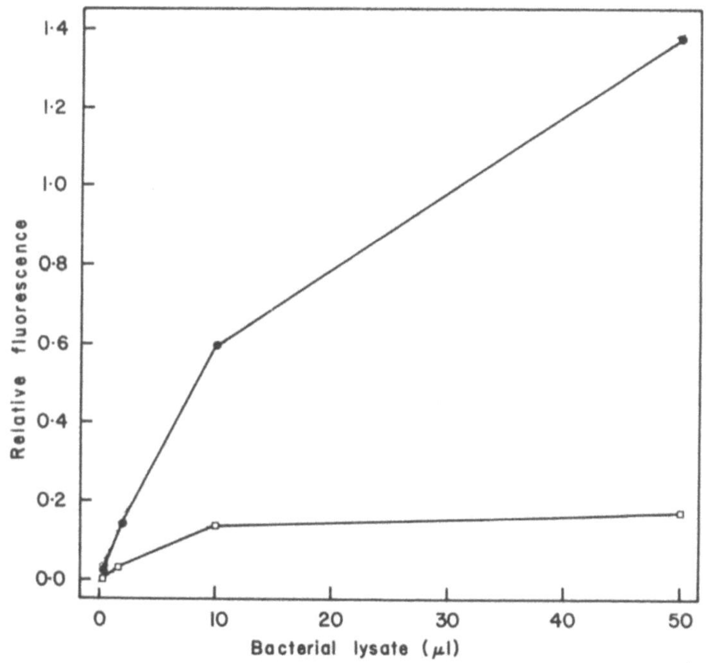

Figure 5. Immunoassay of pre-S(120–174)-β-galactosidase fusion protein expressed in *E. coli*. Results obtained with anti-HBV (●——●) and with control normal rabbit serum (□——□). (Offensperger *et al.*, 1985).

Table 1. Interaction of HBV (HB_sAg) with Human Hepatoma Cells

Cells used	Antibodies added	Percentage of cells attached to HB_sAg–cellulose
HeLa	None	10–20
Rat hepatocytes	None	10–20
HepG2 (human hepatoma)	None	O (BSA–cellulose)
HepG2	None	80–95 (16 experiments)
HepG2	Anti-HBV serum devoid of anti-S protein	8
HepG2	Anti-HB_s anti-S protein-specific	90
HepG2	Anti-pre-S (1–21)	70
HepG2	Anti-pre-S (12–32)	30
HepG2	Anti-pre-S (32–53)	6
HepG2	Anti-pre-S (53–73)	94
HepG2	Anti-pre-S (94–117)	65
HepG2	Anti-pre-S (120–145)	60
HepG2	Anti-pre-S (153–171)	84
HepG2	Anti-pre-S (12–32) + Anti-pre-S (120–145)	20

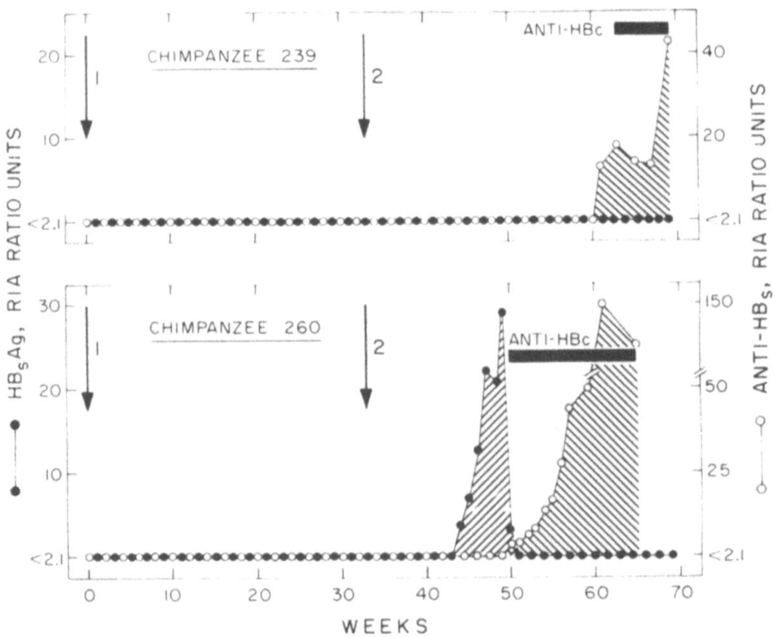

Figure 6. Neutralization of HBV by rabbit anti-pre-S (120–145). HBV subtype *adw* [6 × 10^3 chimpanzee infectious doses (CID50)] was mixed with rabbit anti-pre-S (120–145). After sterile filtration, the mixture was incubated for 1 hr at 37°C and stored at −70°C until used for inoculation. Two 3-year-old chimpanzees were injected intravenously each with one third of the HBV-anti-pre-S (120–145) mixture at time = 0 (indicated by arrows 1). Sera were tested for HB_sAg S-protein, anti-HB_s and antibodies to hepatitis B core antigen (anti-HB_c). Thirty-three weeks after injection with the HBV-anti-pre-S (120–145) mixture (arrows 2), the chimpanzees were challenged with live HBV using a mixture containing normal rabbit serum instead of anti-pre-S (120–145). (Reprinted from Neurath *et al.*, 1986*a*.)

B. Results of Experiments in Mice

1. Studies on the genetic restriction of the immune response in inbred strains of mice to the S protein and to synthetic peptides corresponding to the pre-S sequence indicate that antibody responses to S and pre-S determinants are regulated by distinct genes (Neurath *et al.*, 1983; Neurath *et al.*, 1985*b*) (Fig. 7a,b). Similar conclusions were reached using HB_sAg particles with a high content of pre-S sequences (Milich *et al.*, 1985).
2. Preimmunization with the synthetic peptide pre-S (120–145) of a strain of mice nonresponsive to S protein resulted in an enhanced response to S protein after subsequent immunization with HB_sAg particles containing both S protein and pre-S sequences (Neurath *et al.*, 1985*b*). Thus, nonresponsiveness to S protein is circumvented by the presence of pre-S sequences in HB_sAg in agreement with data presented by Milich *et al.* (1985, 1986). The presence of pre-S sequences in HB_sAg particles enhances the antibody response to S protein (Coursaget *et al.*, 1985) (Fig. 8).
3. Pre-S sequences are substantially more immunogenic than S protein and the

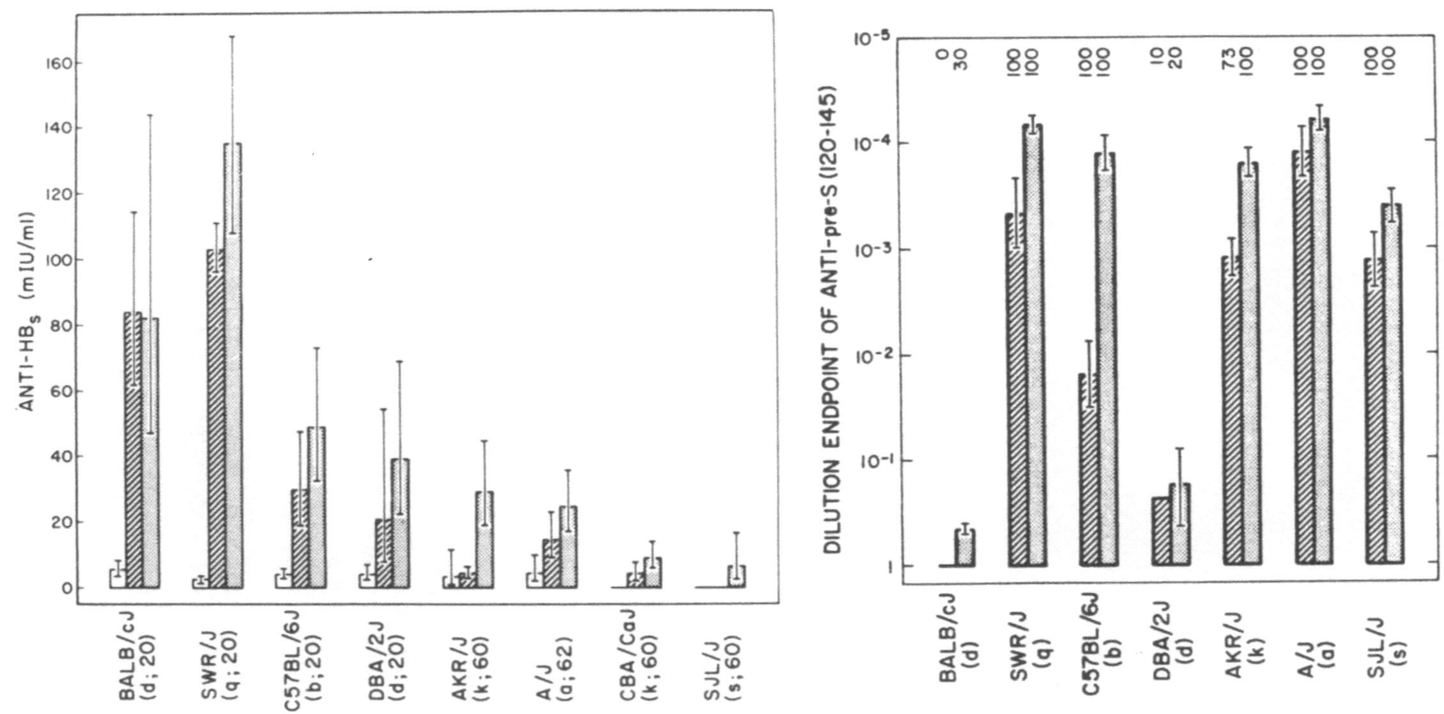

Figure 7. (A) Geometric mean levels of anti-HB_s and (B) geometric mean dilution endpoints for anti-pre-S (120–145) sera from distinct inbred mouse strains after immunization with one (□), two (▨), or three (□) doses of pepsin-treated HBsAg (A) and the synthetic peptide, respectively (B). Length of vertical lines for each bar indicates 95% confidence limits of the means, which were calculated only for mice responding to immunization. The percentage of responders is indicated on top of the bars for (B). Mouse strains and their H-2 haplotypes are indicated on the abscissa. (Neurath *et al.*, 1983, 1985*b*.)

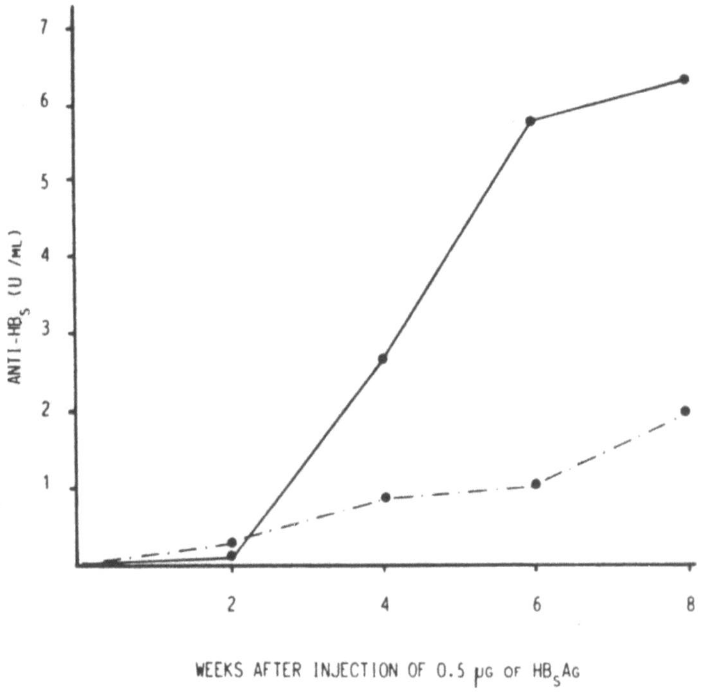

Figure 8. Immune response in mice to two Pasteur vaccines: one with (●——●) and one without (●- -●) detectable pre-S sequences. Pre-S sequences were detected on the basis of reactivity with polymerized serum albumin (Coursaget *et al.*, 1985). [More sensitive and more specific methods for detection of pre-S sequences have now become available (Neurath *et al.*, 1986*c*).]

antibody response to pre-S sequences is elicited more rapidly than the response to S protein (Milich *et al.*, 1985) (Fig. 9).

These findings suggest that the presence of pre-S sequences at high enough levels in vaccines containing S protein will: (1) decrease the proportion of nonresponders to the vaccines; (2) enhance the immune response to S protein; and (3) result in an earlier immunity to HBV, which may be essential for the high efficacy of vaccines expected to prevent perinatal transmission of hepatitis B.

C. Results of Investigations with Sera of Humans Who Had Been Infected with HBV or Who Were Vaccinated with Hepatitis B Vaccines

1. IgG from pooled human anti-HB$_s$-positive sera (= HBIgG) contains antibodies specifically recognizing the HBV env middle protein (Neurath *et al.*, 1984*a*) (Fig. 10).
2. Humans infected with HBV develop antibodies specific for pre-S sequences of the HBV env middle and large proteins (Neurath *et al.*, 1985a) (Figs. 11 and 12).

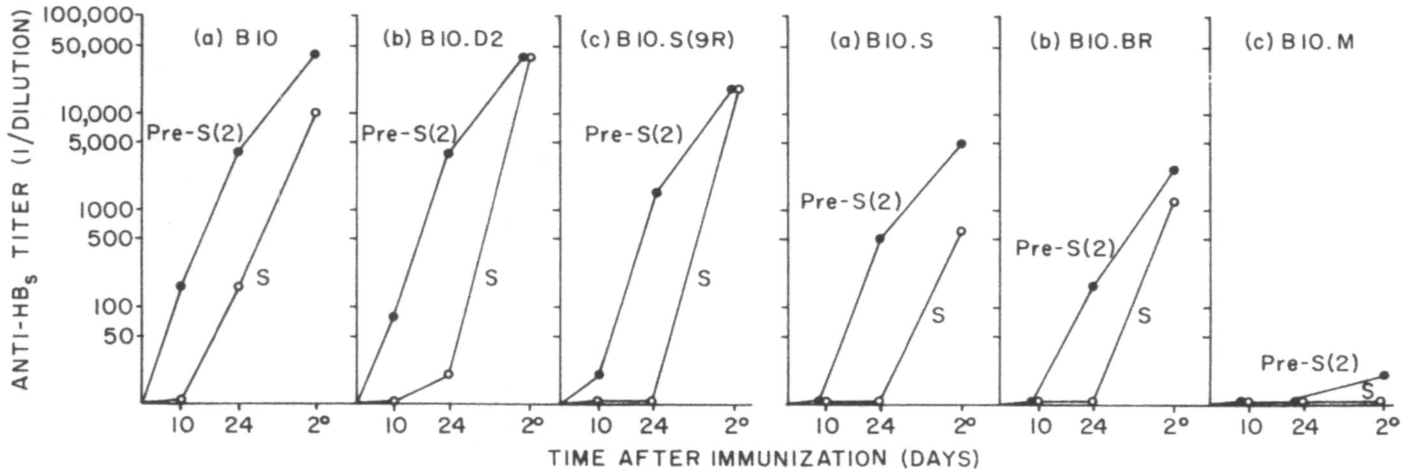

Figure 9. Comparison of production of antibodies to the S and pre-S(2) region of HBsAg *in vivo*. Groups of 5 mice of the indicated H-2 congenic strains were immunized intraperitoneally (i.p.) with 1 μg of HB_sAg/p34 particles (Michel *et al.*, 1984). The relative amount of p34 in the CHO-derived HB_sAg particles was 35 percent (Michel *et al.*, 1984) and the pre-S(2) region accounts for approximately 25% by weight of p34. Therefore, a 1-μg dose of HB_sAg/p34 particles is equivalent to 0.913 μg of S region protein and 0.087 μg of pre-S(2) region protein. The IgG antibody responses specific for the S region were determined by solid-phase RIA 10 and 24 days after primary immunization and 2 weeks after secondary immunization (2°). Antisera were also examined in a pre-S(2) region-specific assay. The antibody titers are expressed as the highest dilution to yield 2.5 times the counts of preimmunization sera. The pre-S(2) region corresponds to sequences pre-S(120–174). (Reprinted from Milich *et al.*, 1985.)

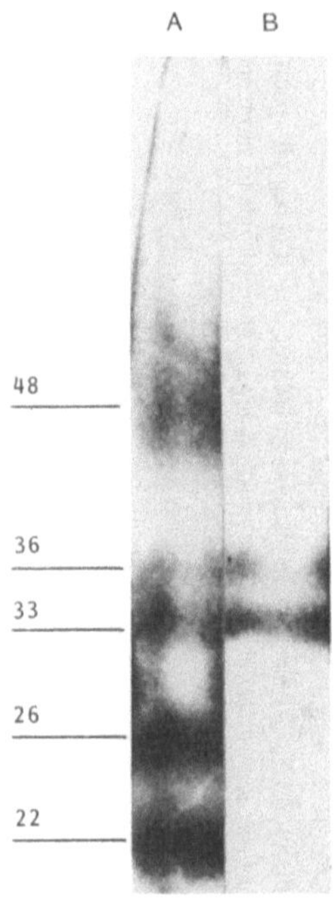

Figure 10. Polyacrylamide gel electrophoresis of HB_sAg. Separated HB_sAg polypeptides were stained with silver (A) or transferred to nitrocellulose and reacted with ^{125}I-labeled human anti-HBs (B). (Reprinted from Neurath *et al.*, 1984*a*.)

The sera preferentially recognize epitopes near the *N*-terminus of the HBV env middle and large proteins (Fig. 13). These antibodies appear early, sometimes preceding the appearance of HB_sAg in serum, and thus in some cases represent the earliest marker for HBV infection. Anti-pre-S antibodies thus represent a portion of the repertoire of antibodies elicited as the result of HBV infection.

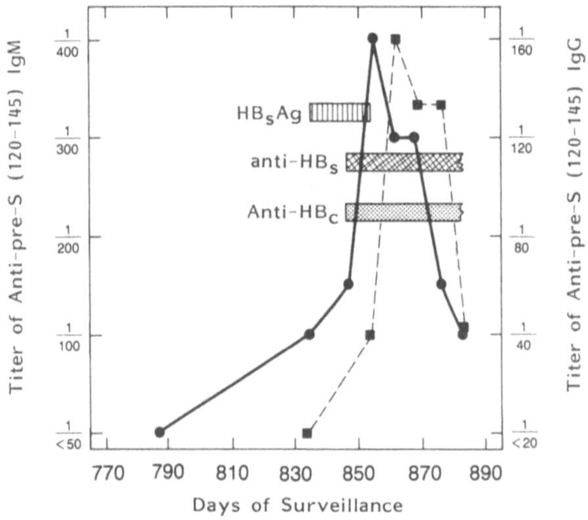

Figure 11. Development of antibodies to the pre-S gene-encoded proteins of HBV during acute hepatitis B infection. IgM (●——●) and IgG (■- - -■) antibodies to pre-S (120–145) were quantified; HB_sAg, anti-HB_s and anti-HB_c were assayed using commercial test kits (Abbott Laboratories). The broken line at the end of the bars corresponding to the different markers of HBV infection indicates their presence at the end of surveillance. Antibody titers represent the highest dilution of serum at which radioactivity counts corresponding to the specimens, divided by counts corresponding to equally diluted control serum, were greater than or equal to 2.1. (Reprinted from Neurath *et al.*, 1985*a*.)

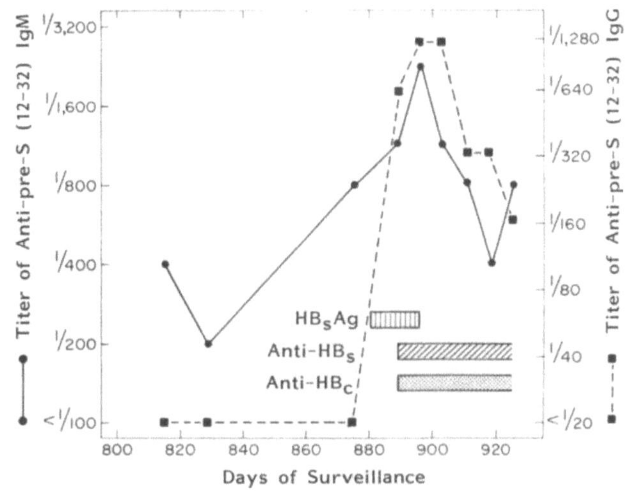

Figure 12. Development of antibodies to the pre-S gene-coded proteins of HBV during acute hepatitis B infection. Antibodies to pre-S (12–32) were quantified. This assay is specific for the HBV env large protein; HB_sAg, anti-HB_s and anti-HB_c were assayed using commercial test kits (Abbott Laboratories). The broken line at the end of the bars corresponding to the different markers of HBV infection indicates their presence at the end of surveillance. Antibody titers represent the highest dilution of serum at which radioactivity counts corresponding to the specimens, divided by counts corresponding to equally diluted control serum, were greater than or equal to 2.1.

3. Several experimental or licensed hepatitis B vaccines, i.e., the NIH vaccine (McAuliffe *et al.*, 1982), the Pasteur-Hevac B, and the HB vaccine developed by the Central Laboratory of the Netherlands Red Cross Blood Transfusion Service (CLB), have the potential to elicit anti-pre-S specific antibodies. These antibodies preferentially recognize the *N*-terminal portions of the HBV middle and large proteins (Figs. 14 and 15). Although vaccine lots used for immunization were not available, tests on other lots indicated that the Pasteur vaccine had a higher level of pre-S sequences (20 ng/μg) than the CLB vaccine (0.2 ng/μg). The Merck, Sharp and Dohme (MSD) vaccine lacked detectable pre-S sequences (Neurath *et al.*, 1985*a*) and failed to elicit anti-pre-S responses.
4. The Pasteur vaccine generally elicited anti-pre-S responses in a higher proportion of preselected vaccine recipients than the CLB vaccine, however, a variability

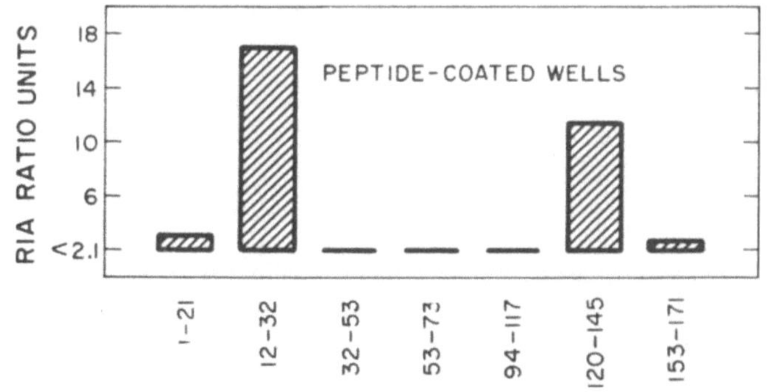

Figure 13. Scanning of the pre-S portion of HBV env proteins for regions optimally recognized by IgG antibodies elicited in humans as a result of HBV infection. For experimental details, see text to Fig. 3. Anti-human IgG was used as second antibody in the assays. (Reprinted from Neurath *et al.*, 1986*a*.)

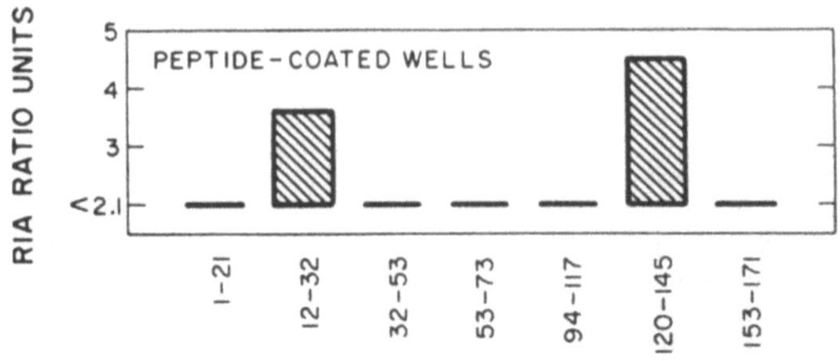

Figure 14. Scanning of the pre-S portion of HBV env proteins for regions optimally recognized by IgG antibodies elicited in humans after vaccination with an experimental hepatitis B vaccine (McAuliffe *et al.*, 1982). For experimental details, see text to Figs. 3 and 12. (Reprinted from Neurath *et al.*, 1986*a*.)

between different vaccine batches was noticed. Healthy individuals were better anti-pre-S responders than were hemodialysis patients (Table 2).

5. Pre-S sequences are highly immunogenic in humans (responses were observed with nanogram quantities of the immunogen).
6. It is possible that the reported lack of efficacy of the MSD vaccine in hemodialysis patients (Stevens *et al.*, 1984), contrasted with the efficacy of the Pasteur and CLB vaccines in a similar population (Desmyter *et al.*, 1983; Desmyter and Colaert, 1984), was due to the absence or presence of pre-S sequences in the respective vaccines. In this respect, it is of interest that hemodialysis patients who became infected with HBV, despite vaccination with the MSD vaccine, developed anti-pre-S antibodies as a consequence of infection. They were thus immunocompetent to respond to pre-S-specific epitopes.
7. The content of pre-S sequences in current serum-derived HB_sAg vaccines is low. The full potential of immunity to pre-S sequences can be realized only by vaccination with immunogens having a higher level of pre-S sequences. Synthetic peptides appear to be the tools of choice for preparation of such immunogens.

III. SYNTHETIC PEPTIDES AS POTENTIAL HEPATITIS B VACCINES

In considering the potential of synthetic peptide analogues of the HBV env protein, one has to strive at least for a "golden standard" that may already have been established by a serum-derived vaccine:

1. Prevention of perinatal transmission of hepatitis B (Fig. 16)
2. Induction of high levels of antibodies against the HBV env protein(s), resulting in long-term immunity (Fig. 17).

To have a potential for active immunization against hepatitis B, the synthetic analogues of the HBV env protein(s) should have the following properties:

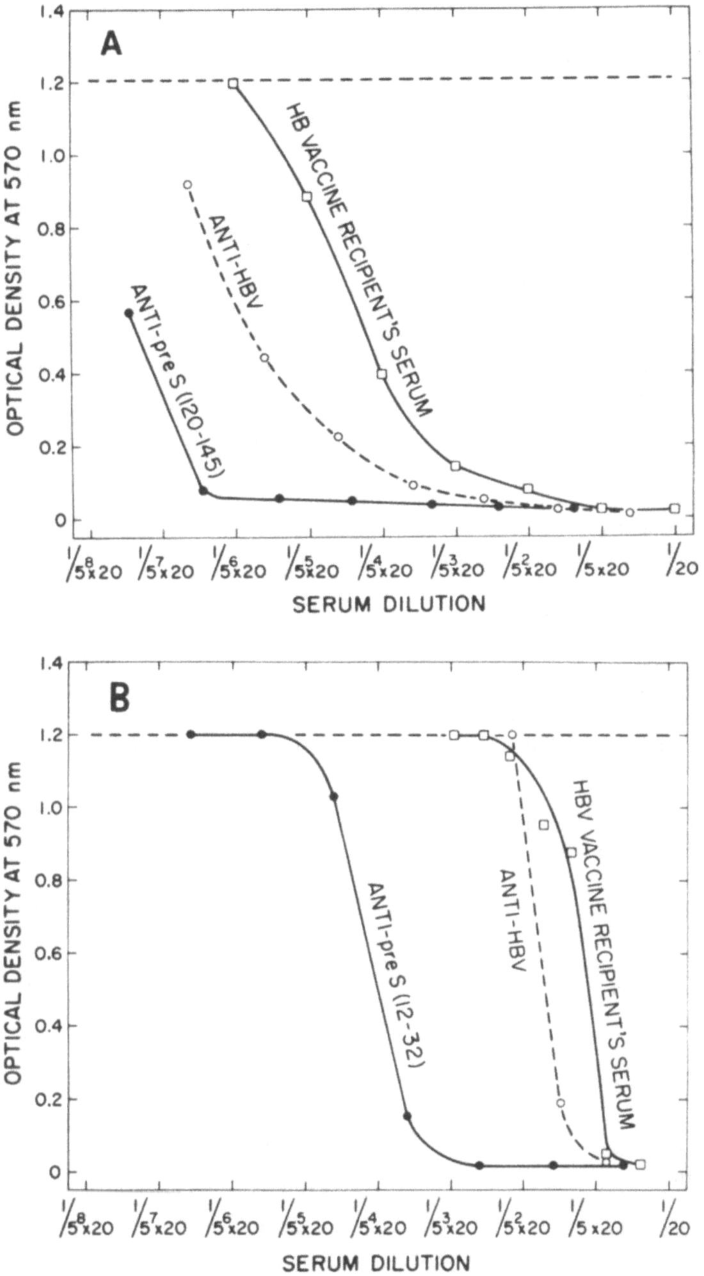

Figure 15. (A) Results of ELISA tests with serial dilutions of rabbit anti-HBV; of an antiserum to the synthetic peptide pre-S (120–145); and of a serum pool from recipients of a hepatitis B vaccine (McAuliffe *et al.*, 1982). Dashed horizontal line indicates optical density (OD_{570}) corresponding to substrate. Antibody-positive samples [recognizing the sequence pre-S (120–145)] cause a decolorization of substrate in this assay. The dilution endpoint of the serum pool from recipients of the NIH vaccine (McAuliffe *et al.*, 1982) was ~1/25,000. The dilution endpoints for pooled sera from anti-pre-S responders belonging to groups 3,4,5 (Table 2) were ~1/5,000, ~1/200 and ~1/50, respectively. (B) Results of similar ELISA tests for antibodies recognizing the sequence pre-S (12–32), i.e., specific for antibodies recognizing the large HBV env protein. The dilution endpoint of the serum pool from recipients of the NIH vaccine (McAuliffe *et al.*, 1982) was ~1/130. (Reprinted from Neurath *et al.*, 1986*d*.)

Table 2. Results of Screening Selected Sera from Persons Who Had Been Infected with HBV or Who Were Vaccinated with Distinct Hepatitis B Vaccines for Anti-Pre-S-Specific Antibodies[a]

Characterization of population	Source of vaccine and lot number (if available)	Number of anti-pre-S-positive sera per total number of sera tested	Level of anti-HB_s (anti-S-protein antibodies) (mIU/ml)	
			Range	Geometric mean
Healthy individuals after receiving three doses of vaccine	Pasteur[b] 02	10/10	2,250–10,500	4,460
Healthy individuals after receiving three doses of vaccine	Pasteur[b] 1005	0/12	600–12,000	7,170
Hemodialysis personnel 1 month after 4th dose of vaccine	Pasteur[b]	15/15	53,000–533,000	152,000
Hemodialysis patients 1 month after 4th dose of vaccine	Pasteur[b]	14/20	49–15,520	646
Subdivision of group 4	Pasteur[b]	12/12	364–15,250	2,150
Subdivision of group 4	Pasteur[b]	2/8	49–291	55
Hemodialysis personnel 1 month after 4th dose of vaccine	CLB[c]	11/15	50,000–690,000	133,800
Hemodialysis patients 1 month after 4th dose of vaccine	CLB[d]	2/20	3,200–88,700	9,030
Hemodialysis patients 1 month after 5th dose of vaccine	CLB[d]	0/20	10,400–119,600	40,710
Healthy individuals after receiving 3 doses of vaccine[e]	CLB[d]	10/25	610–7,800	1,990
Hemodialysis personnel 1–5 months after 3rd dose of vaccine	MSD[g]	0/10	—	13,000[f]
Homosexual men who acquired anti-HB_s after HBV infection	—[g]	13/20	280–1,020	595
Hemodialysis patients in the course of transient hepatitis B	—[g]	10/10[h]	—	—

[a] From Neurath *et al.* (1986*b*).
[b] Vaccine dose = 5 μg.
[c] Vaccine dose = 3 μg.
[d] Vaccine dose = 27 μg.
[e] Different batches of vaccine were used.
[f] The pool of 10 sera was assayed by the AUSAB test.
[g] Vaccine dose = 20 μg.
[h] Six of these patients had received the MSD vaccine but were not protected.

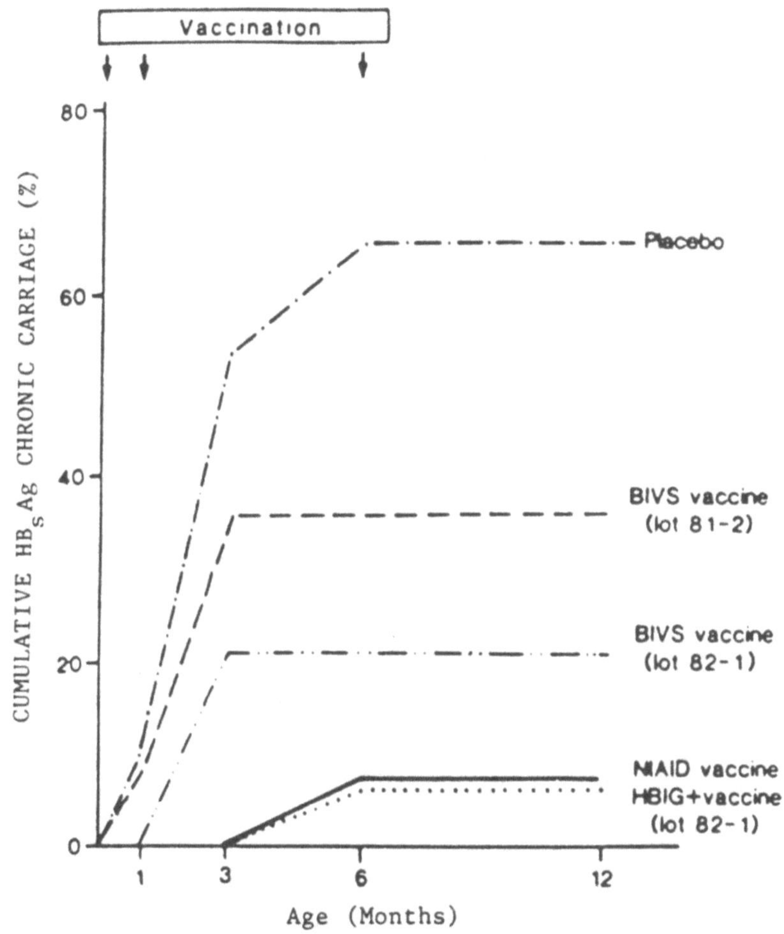

Figure 16. Efficacy of distinct vaccines in prevention of perinatal transmission of hepatitis B and in establishment of a chronic carrier state in babies born to HBV-infected mothers. The NIAID (National Institute of Allergy and Infectious Diseases, U.S.A.) vaccine [known to have a potential to elicit high levels of anti-pre-S2 specificity (Neurath *et al.*, 1986*b*)] was highly effective alone without administration of antibodies to HB_sAg (= HBIg). The other vaccine manufactured under conditions known to remove pre-S sequences (Neurath *et al.*, 1985*a*) had low efficacy which could be enhanced by administration of HBIg. Whether the high efficacy of the NIAID vaccine was due only to the presence of pre-S sequences or due also to other factors is not known. (Reprinted from Xu *et al.*, 1985.)

1. High immunogenicity (i.e., the capacity to elicit high levels of antibodies recognizing the HBV env protein in its native form) is desirable. The chances that a peptide meets this criterion are higher if the peptide is recognized by antiviral antibodies (anti-HBV) and the binding constants of the anti-HBV-peptide and anti-HBV-env protein reactions are similar. Only a few of the synthesized analogues of the S protein were reported to react with anti-HBV. Most of these peptides have to be in a cyclic form (cyclization by disulfide bonds) for the reaction to occur (Fig. 18). On the other hand, many of the linear peptide

analogues corresponding to the pre-S sequence are recognized by anti-HBV (Figs. 19 and 20). Such recognition depends on the location of the peptide along the S or pre-S sequence and on the length of the peptide. The recognition is probably influenced by conformational features of the peptide as they relate to the conformation of the homologous sequence within the HBV env protein.

2. Antibodies elicited by the synthetic peptides should neutralize the infectivity of HBV and should be protective in species susceptible to HBV infection. Because of limitations imposed by the fact that only chimpanzees can be used for the relevant experiments, only a few HBV env protein analogues were evaluated by this criterion (Figs. 6, 18, 19).
3. A protein carrier antigenic in humans should not be required to confer immunogenicity on the synthetic peptides for the following reasons:

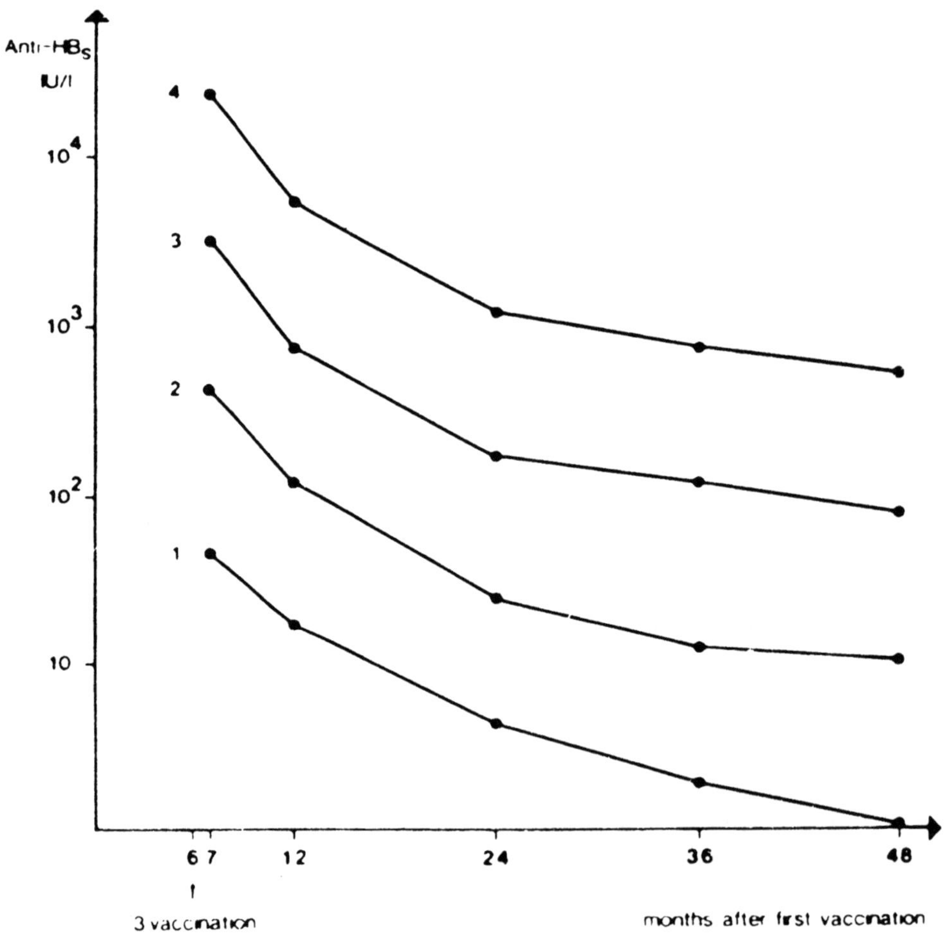

Figure 17. Decrease in anti-HB_s levels in serum of immunized subjects with different initial antibody titers after three doses of hepatitis B vaccine. These results indicate that the duration of immunity is related to the initial antibody titer. (Reprinted from Deinhardt and Zuckerman, 1985.)

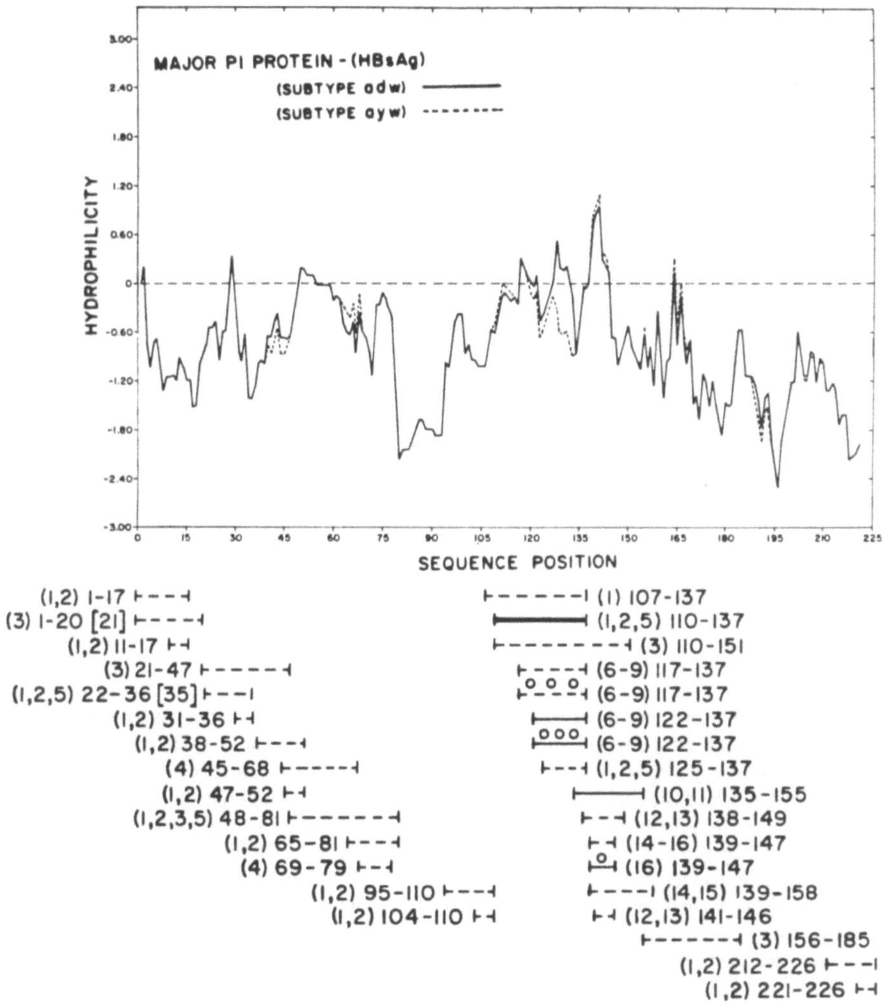

Figure 18. Schematic representation of results of studies on synthetic analogs of the HBV S-protein. All synthetic peptides eliciting anti-S protein antibodies are represented by dashed bars; peptides which, in addition, were reported to react with anti-HB_s in RIA or ELISA assays are represented by solid bars. The thicker solid bar indicates the peptide reported to elicit protective antibodies in chimpanzees. Numbers to the left or right of the bars indicate N- and C-terminal ends of the peptides in the scale (abcissa) of the superimposed hydrophilicity plot (Hopp and Woods, 1981). Circles on top of the bars indicate cyclic peptide. All peptides listed were immunogenic when bound to protein carriers and administered with complete and incomplete Freund's adjuvant. Immunogenicity without protein carriers and Freund's adjuvant was reported only for peptides S (122–137) (cyclic; Sanchez *et al.*, 1982) and S (135–155) (Neurath *et al.*, 1984*d*). Numbers in parentheses indicate references: (1) Alexander and Lerner, 1984; (2) Lerner *et al.*, 1981; (3) Shih *et al.*, 1984; (4) Neurath *et al.*, 1984*b*; (5) Gerin *et al.*, 1983; (6) Dreesman *et al.*, 1982; (7) Dreesman, 1984; (8) Sancheez *et al.*, 1982; (9) Ionescu-Matiu *et al.*, 1983; (10) Neurath *et al.*, 1982; (11) Neurath *et al.*, 1984*c*; (12) Hopp, 1981; (13) Prince *et al.*, 1982; (14) Bhatnagar *et al.*, 1982; (15) Vyas *et al.*, 1983; (16) Howard *et al.*, 1984.

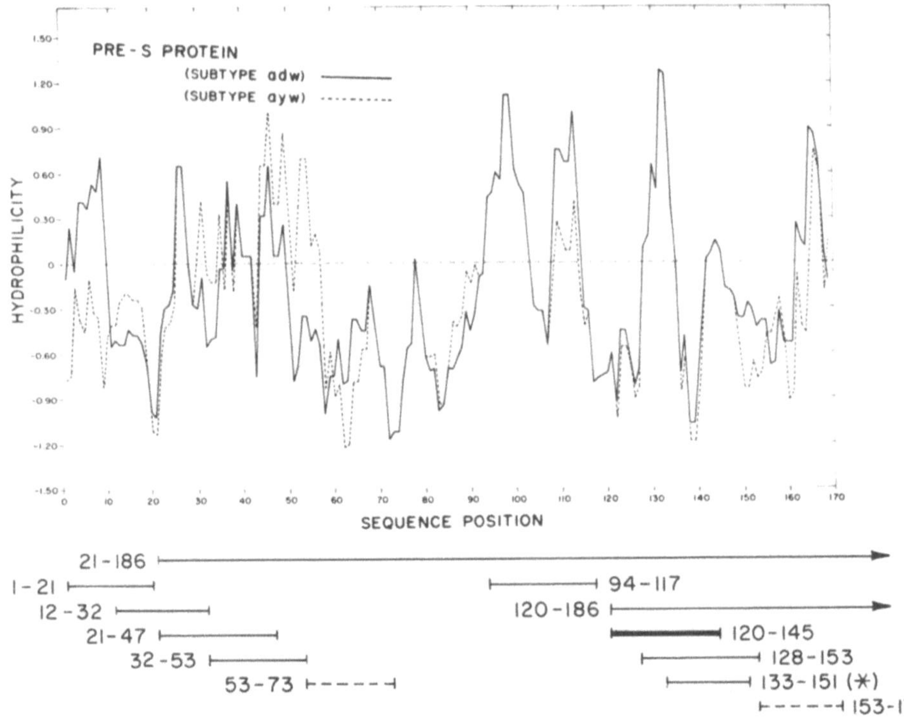

Figure 19. Schematic representation of results of studies on synthetic analogs of the pre-S portion of the HBV env large protein. All synthetic peptides eliciting anti-HBV env protein antibodies are represented by dashed bars; peptides also recognized by anti-HBV are represented by solid bars. The thicker solid bars indicate the peptide reported to induce virus neutralizing antibodies. Numbers to the left or right of the bars indicate N- and C-terminal ends of the peptides in the scale (abscissa) of the superimposed hydrophilicity plot (Hopp and Woods, 1981). All the peptides were immunogenic without protein carriers [except pre-S (133–151) which was tested only in the form of a conjugate with ovalbumin]. Pre-S (12–32) and pre-S (120–145) elicited antibodies when injected without complete and incomplete Freund's adjuvant (other peptides were not assayed for immunogenicity under these conditions). (Data compiled from Neurath *et al.*, 1984*a*, 1985*a*,*c*, 1986*a*,*d*; Okamoto *et al.*, 1985; and from unpublished data of the authors.)

a. The use of protein carriers may result in suboptimal immune responses to the synthetic peptides (Neurath *et al.*, 1985*c*).
b. It may be prohibitive for regulatory or economic reasons. Therefore, nonantigenic carriers are being developed or free peptides are tested as immunogens (Neurath *et al.*, 1985*c*).

4. Adjuvants approved for human use should confer high immunogenicity to the synthetic peptides.
5. The synthetic immunogens should prime the immune systems of the recipient, so that subsequent encounters with HBV will result in increased levels of protective antibodies.

These requirements can be fulfilled if the synthetic immunogens have in addition to antibody binding sites, regions reacting with appropriate T-cell receptors for the HBV env

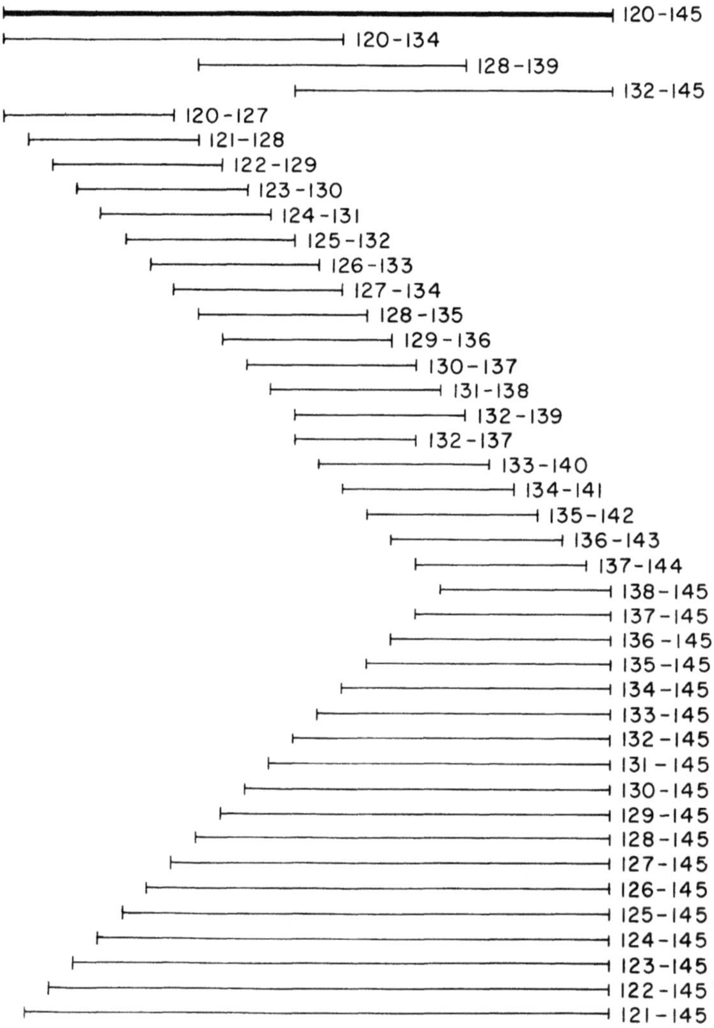

Figure 20. Schematic representation of results of studies on synthetic analogs of the pre-S (120–145) portion of the HBV env protein. Studies with the series of related pre-S peptides of differing chain length is expected to allow the precise localization of the immunodominant epitope on the pre-S2 sequence. For further explanations, see Fig. 19. There were considerable differences in antigenicity and immunogenicity between these peptides (data to be published).

protein as well as with Ia antigens (Berzofsky, 1985*a;* Berzofsky, 1985*b;* Watts *et al.*, 1985).

Only a few of the synthetic analogues of the HBV env protein(s) have been reported to elicit antibodies if administered without either a protein carrier and/or complete (incomplete) Freund's adjuvant (not permissible for use in humans) (Figs. 18 and 19). Two linear synthetic peptides eliciting protective [S(110–137)] or virus-neutralizing [pre-S (120–145)] antibodies (Gerin *et al.*, 1983; Neurath *et al.*, 1986*a*) were reported to have both T-cell and B-cell determinants (Milich and Chisari, 1984; Milich *et al.*, 1986). Data

Table 3. Pattern of Progress in Vaccine Production[a]

Vaccine type	Relies on
Conventional	Empirical knowledge
	Living attenuated and crude killed vaccines
Improved	Large-scale culture (including mammalian cell culture)
	Purification of antigens
	Genetic basis for virulence
	Immunologic understanding
Single-protein	Monoclonal antibodies
	Pure antigens
	Gene cloning and sequencing
	Improved adjuvants
Simpler synthetic peptide	Immunochemistry
	Computer chemistry
	Understanding of basis of adjuvanticity
Simple synthetic peptide, active orally	Medicinal chemistry

[a] From Vane and Cuatrecasas (1984).

concerning T-cell binding sites on promising cyclic peptide analogues of the S protein, S(122–137) and S(139–147) (Dreesman, 1984; Howard *et al.*, 1984), have not been reported.

IV. CONCLUSIONS

Peptide vaccines represent the ultimate step towards the development of fully synthetic "designer vaccines" (Table 3). Synthetic analogues of the HBV env proteins have become invaluable in (1) defining the biologic function of the pre-S sequence (Neurath *et al.*, 1984, 1985*a,b,* 1986*a,d*), (2) developing pre-S-specific diagnostic assays (Neurath *et al.*, 1986*b,c*), (3) defining the fine specificity of B- and T-cell responses to HBV (HB_sAg) (Milich *et al.*, 1985, 1986; Milich and Chisari, 1984), (4) characterizing idiotype-anti-idiotype interactions in immunity to HB_sAg (Dreesman, 1984), and (5) defining antigenic subtypes of HBV (Dreesman, 1984; Gerin *et al.*, 1983). The application of some of these analogues to active prophylaxis of hepatitis B in the future will require additional knowledge concerning the immunochemistry of the HBV env proteins and the selection of optimal delivery systems (carriers and adjuvants).

ACKNOWLEDGMENTS

Human serum samples for the reported studies were provided by L. Baker, J. Desmyter, M. Girard, F. B. Hollinger, P. N. Lelie, H. W. Reesink, C. E. Stevens, P. Taylor, and F. Tron. We thank T. Huima and J. Krestik for assistance in preparation of the manuscript.

REFERENCES

Alberti, A., Diana, S., Sculard, G. H., Eddleston, S. L. W. F., and Williams, R. (1978). *Br. Med. J.* **1986,** 1056–1058.

Alexander, H., and Lerner, R. A. (1984). In *Advances in Hepatitis Research* (F. V. Chisari, ed.), pp. 230–237, Masson, New York.

Berzofsky, J. (1985*a*). In *Modern Approaches to Vaccines* (R. Chanock, R. Lerner, and F. Brown, eds.), pp. 21–23, Cold Spring Harbor Laboratory, Cold Spring Harbor, New York.

Berzofsky, J. (1985*b*) *Science* **229,** 932–940.

Bhatnagar, P. K., Papas, E., Blum, H. E., Milich, D. R., Nitecky, D., Karels, M. J., and Vyas, G. N. (1982). *Proc. Natl. Acad. Sci. USA* **79,** 4400–4404.

Coursaget, P., Barres, J. L., Chiron, J. P., and Adamowicz, P. (1985). *Lancet* **1,** 1152–1153.

Deinhardt, F., and Zuckerman, A. J. (1985). *J. Med. Virol.* **17,** 209–217.

Desmyter, J., and Colaert, J. (1984). In *Viral Hepatitis and Liver Diseases* (G. N. Vyas, J. L. Dienstag, and J. F. Hoofnagle, eds.), pp. 709–710, Grune & Stratton, Orlando, Florida.

Desmyter, J., Colaert, J., DeGroote, G., Reynders, M., Reerink-Brongers, E. E., Dees, P. J., Lelie, P. N., Reesink, H. W., and the Leuven Renal Transplantation Collaborative Group. (1983). *Lancet* **2,** 1323–1328.

Dreesman, G. R. (1984). In *Advances in Hepatitis Research* (F. V. Chisari, ed.), pp. 216–222, Masson, New York.

Dreesman, G. R., Sanchez, Y., Ionescu-Matiu, I., Sparrow, J. T., Six, H. R., Peterson, D. L., Hollinger, F. B., and Melnick, J. L. (1982). *Nature (Lond.)* **295,** 158–160.

Gerin, J. L., Alexander, H., Shih, J. W.-K., Purcell, R. H., Dapolito, G., Engel, R., Green, N., Sutcliffe, J. G., Shinnick, T. M., and Lerner, R. A. (1983). *Proc. Natl. Acad. Sci. USA* **80,** 2365–2369.

Heermann, K. H., Goldmann, U., Schwartz, W., Seyffarth, T., Baumgarten, H., and Gerlich, W. (1984). *J. Virol.* **52,** 396–402.

Hopp, T. P., and Woods, K. R. (1981). *Proc. Natl. Acad. Sci. USA* **78,** 3824–3828.

Howard, C. R., Brown, S. E., Hogben, D. N., Zuckerman, A. J., Murray-Lyon, I. M., and Steward, M. W. (1984). In *Viral Hepatitis and Liver Disease* (G. N. Vyas, J. L. Dienstag, and J. H. Hoofnagle, eds.), pp. 561–572, Grune & Stratton, Orlando, Florida.

Ionescu-Matiu, I., Kennedy, R. C., Sparrow, J. T., Culwell, A. R., Sanchez, Y., Melnick, J. L., and Dreesman, G. R. (1983). *J. Immunol.* **130,** 1947–1951.

Iwarson, S., Tabor, E., Thomas, H. C., Goodall, A., Waters, J., Snoy, P., Shih, J. W.-K., and Gerety, R. J. (1985). *J. Med. Virol.* **16,** 89–96.

Koistinen, V. V. (1980). *J. Virol.* **35,** 20–23.

Lerner, R. A., Green, N., Alexander, H., Liu, F.-T., Sutcliffe, J. G., and Shinnick, T. M. (1981). *Proc. Natl. Acad. Sci. USA* **78,** 3403–3407.

Machida, A., Kishmoto, S., Ohnuma, H., Baba, K., Ito, Y., Miyamoto, H., Funatsu, G., Oda, K., Usuda, S., Togami, S., Nakamura, T., Miyakawa, Y., and Mayumi, M. (1984). *Gastroenterology* **86,** 910–918.

McAuliffe, V. J., Purcell, R. H., Gerin, J. L., and Tyeryar, F. J. (1982). In *Viral Hepatitis* (W. Szmuness, H. J. Alter, and J. E. Maynard, eds.), pp. 425–435, Franklin Institute Press, Philadelphia.

Michel, M.-L., Pontisso, P., Sobczak, E., Malpiece, Y., Streeck, R. E., and Tiollais, P. (1984). *Proc. Natl. Acad. Sci. USA* **81,** 7708–7712.

Milich, D. R., and Chisari, F. V. (1984). In *Advances in Hepatitis Research* (F. V. Chisari, ed.), pp. 91–109, Masson, New York.

Milich, D. R., Thornton, G. B., Neurath, A. R., Kent, S. B., Michel., M., Tiollais, P., and Chisari, F. V. (1985). *Science* **228,** 1195–1199.

Milich, D. R., Thornton, G. B., McNamara, M. K., McLachlan, A., and Chisari, F. V. (1986). In *New Approaches to Immunization* (F. Brown, R. M. Chanock, and R. A. Lerner, eds.), pp. 377–382, Cold Spring Harbor Laboratory, Cold Spring Harbor, New York.

Moodie, J., Stannard, L. M., and Kipps, A. (1974). *J. Gen. Virol.* **24,** 375–379.

Neurath, A. R., and Kent, S. B. H. (1985). In *Immunochemistry of Viruses. The Basis for Serodiagnosis and Vaccines* (M. H. V. van Regenmortel and A. R. Neurath, eds.), pp. 325–266, Elsevier, Amsterdam.

Neurath, A. R., Trepo, L., Chen, M., and Prince, A. M. (1976). *J. Gen. Virol.* **30,** 277–285.

Neurath, A. R., Kent, S. B. H., and Strick, N. (1982). *Proc. Natl. Acad. Sci. USA* **79,** 7871–7875.
Neurath, A. R., Stark, D., Strick, N., and Sproul, P. (1983). *J. Med. Virol.* **12,** 277–285.
Neurath, A. R., Kent, S. B. H., and Strick, N. (1984*a*). *Science* **224,** 392–395.
Neurath, A. R., Kent, S. B. H., and Strick, N. (1984*b*). *Virus Res.* **1,** 321–331.
Neurath, A. R., Strick, N., and Kent, S. B. H. (1984*c*). *J. Virol. Methods* **9,** 341–346.
Neurath, A. R., Kent, S. B. H., and Strick, N. (1984*d*). *J. Gen. Virol.* **65,** 1009–1014.
Neurath, A. R., Kent, S. B. H., Strick, N., Taylor, P., and Stevens, C. E. (1985*a*). *Nature (Lond.)* **315,** 154–156.
Neurath, A. R., Kent, S. B. H., Strick, N., Stark, D., and Sproul, P. (1985*b*). *J. Med. Virol.* **17,** 119–125.
Neurath, A. R., Strick, N., and Kent, S. B. H. (1985*c*). In *Vaccines 85* (R. A. Lerner, R. M. Chanock, and F. Brown, eds.), pp. 185–189, Cold Spring Harbor Laboratory, Cold Spring Harbor, New York.
Neurath, A. R., Kent, S. B. H., Parker, K., Prince, A. M., Strick, N., Brotman, B., and Sproul, P. (1986*a*). *Vaccine* **4,** 35–37.
Neurath, A. R., Kent, S. B. H., Strick, N., and Hepatitis B vaccine trial study groups (1986*b*). *J. Gen. Virol.* **67,** 453–461.
Neurath, A. R., Strick, N., Kent, S. B. H., Offensperger, W., Wahl, S., Christman, J. K., and Acs, G. (1986*c*). *J. Virol. Methods* **12,** 185–192.
Neurath, A. R., Kent, S. B. H., and Strick, N. (1986*d*). In *Synthetic Peptides in Biology and Medicine* (K. Alitalo, P. Partanen, and A. Vaheri, eds.), pp. 113–131, Elsevier, Amsterdam.
Offensperger, W., Wahl, S., Neurath, A. R., Price, P., Strick, N., Kent, S. B. H., Christman, J. K., and Acs, G. (1985). *Proc. Natl. Acad. Sci. USA* **82,** 7540–7544.
Okamoto, H., Imai, M., Usuda, S., Tanaka, E., Tachibana, K., Mishiro, S., Machida, A., Nakamura, T., Miyakawa, Y., and Mayumi, M. (1985). *J. Immunol.* **134,** 1212–1216.
Persing, D. H., Varmus, H. E., and Ganem, D. (1985). *Proc. Natl. Acad. Sci. USA* **82,** 3440–3444.
Peterson, D. L., Paul, D. A., Gavilanes, F., and Achord, D. T. (1984). In *Advances in Hepatitis Research* (F. Chisari, ed.), pp. 30–39, Masson, New York.
Prince, A. M., Ikram, H., and Hopp, T. P. (1982). *Proc. Natl. Acad. Sci. USA* **79,** 579–582.
Sanchez, Y., Ionescu-Matiu, I., Sparrow, J. T., Melnick, J. L., and Dreesman, G. R. (1982). *Intervirology* **18,** 209–213.
Shih, J. W.-K., Yajima, H., Gerety, R. J., and Liu, T. Y. (1984). In *Viral Hepatitis and Liver Disease* (G. N. Vyas, J. L. Dienstag, and J. H. Hoofnagle, eds.), pp. 659, Grune & Stratton, Orlando, Florida.
Stevens, C. E., Alter, H. J., and Taylor, P. E., Zang, E. A., Harley, E. J., and Szmuness, W. (1984). *N. Engl. J. Med.* **311,** 496–501.
Stibbe, W., and Gerlich, W. H. (1983). In *Developments in Biological Standardization* (G. Papaevangelou and W. Hennessen, eds.), Vol. 54, pp. 33–43, S. Karger, Basel.
Tiollais, P., Charnay, P., and Vyas, G. N. (1981). *Science* **213,** 406–411.
Vane, J., and Cuatrecasas, P. (1984). *Nature (Lond.)* **312,** 303–305.
Vyas, G. N., Bhatnagar, P. K., Blum, H. E., Expose, J., and Heldebrandt, C. M. (1983). In *Developments in Biological Standardization* (G. Papaevangelou and W. Hennessen, eds.), Vol. 54, pp. 93–102, S. Karger, Basel.
Watts, T., Gariepy, J., Schoolnik, G., and McConnell, H. (1985). *Proc. Natl. Acad. Sci. USA* **82,** 5480–5484.
Wong, D. T., Nath, N., and Sninsky, J. J. (1985). *J. Virol.* **55,** 223–231.
Xu, Z-Y., Liu, C-B., Francis, D. P., Purcell, R. H., Gun, Z-L., Duan, S-C., Chen, R-J., Margolis, H. S., Huang, C-H., Maynard, J. E., and the United States–China Cooperative Study Group on Hepatitis B. (1985). *Pediatrics* **76,** 713–718.

10

Chemical Synthesis of Poliovirus Peptides and Neutralizing Antibody Responses

Michael G. Murray and Eckard Wimmer

I. INTRODUCTION

Poliomyelitis, a disease caused by poliovirus, remains a serious health problem throughout the world. Only in developed countries where the existing vaccines are regularly administered is the incidence of the disease low. However, occasional outbreaks of poliomyelitis in developed countries, albeit rare, are evidence of the present danger of infection with neurovirulent strains that may be imported from other places in the world. Moreover, the existing live vaccines produce a very low level of disease in vaccine recipients and/or persons coming in contact with vaccine recipients. During 1975–1984, the incidence of vaccine related poliomyelitis in the United States of America was one case per 3.22×10^6 doses of live vaccine administered to immunologically normal recipients (CDC, 1986). Even such low incidence has aroused much public discussion recently (Chasan, 1986). Organizations concerned with public health, such as the World Health Organization (WHO), are interested therefore in developing programs that may eventually lead to the worldwide eradication of poliomyelitis. It should be stressed that the existing live (OPV) and inactivated (IPV) vaccines have proven to be excellent vaccines whenever applied properly. However, investigation to identify and eliminate those properties of the OPV strains that are responsible for reversion to the neurovirulent phenotype or research to explore the possibility of augmenting a protective immune response by the development of synthetic antigens has become an important subject for the WHO: Although much effort may have to be expended, a synthetic vaccine may be developed that possesses the required potency and that can be produced at such low cost, that it can be considered as an alternative to those vaccines now in existence. This in turn may serve as a model to combat other enteroviral disease such as hepatitis caused by HAV.

Michael G. Murray and Eckard Wimmer • Department of Microbiology, School of Medicine, Health Sciences Center, State University of New York at Stony Brook, Stony Brook, New York 11794.

Here we will review briefly the antigenic properties of poliovirus that have recently been elucidated. The insight gained from studies of the antigenicity of poliovirus has helped to understand previous experiments with synthetic peptides and kindled renewed interest in the possibilities of the development of synthetic vaccines.

Poliovirus, a member of the family Picornaviridae, is one of the best characterized animal viruses (Koch and Koch, 1985; Davis *et al.*, 1980, Nicklin *et al.*, 1986, Kuhn *et al.*, 1986, Wimmer *et al.*, 1986). Picornaviruses have been found to be the etiologic agents of many human and animal diseases, including hepatitis, meningitis, conjunctivitis, enteritis, carditis, pericarditis, paralytic poliomyelitis, and foot-and-mouth disease. Poliovirus is about 310 Å in diameter and exists as a nonenveloped icosohedrally shaped particle composed of 60 copies each of four capsid proteins (VP1, VP2, VP3, and VP4) (Hogle *et al.*, 1985). Biochemical studies (Wetz and Habermehl, 1979, and references cited therein) and the recently determined three-dimensional structures of human rhinovirus 14 and poliovirus type 1, Mahoney, have shown that VP1, VP2, and VP3 of these picornaviruses are exposed to the surface and considerably intertwined whereas VP4 is located entirely internal (Rossmann *et al.*, 1985; Hogle *et al.*, 1985). The naked capsid surrounds a single stranded plus-sense RNA molecule of about 7500 nucleotides, which is polyadenylated at its 3′ terminus (Yogo and Wimmer, 1975) and covalently attached, via a phosphodiester bond, to a virus-encoded polypeptide called VPg (Lee *et al.*, 1977; Nomoto *et al.*, 1977; Rothberg *et al.*, 1978).

Poliovirus exists in three stable serotypes. Each serotype is defined by its ability to elicit neutralizing antibodies that are incapable of neutralizing the other two serotypes. The process of antibody-mediated neutralization is poorly understood, but it is likely to occur by several different mechanisms (virus aggregation, inhibition of release of the viral RNA, or physical blockage of the receptor-binding site). The chemical structure and gene organization of poliovirus is known (Kitamura *et al.*, 1981), and the availability of the nucleotide sequence of all three neurovirulent serotype of poliovirus (type 1, Mahoney, type 2, Lansing, and type 3, Leon) as well as of the attenuated strains (Nomoto *et al.*, 1982; Toyoda *et al.*, 1984) has permitted an assessment of the genetic differences between these closely related genomes. This, in turn, has led to the elucidation of the antigenic structure of the poliovirion. Finally, the entire genome of the neurovirulent poliovirus type 1, Mahoney (PV-1) has been cloned into suitable plasmids that have been shown to produce infectious virus upon transfection into suitable mammalian cells (Racaniello and Baltimore, 1981*a*; Semler *et al.*, 1984*a*). The construction of infectious cDNA clones corresponding to the entire genome of the attenuated Sabin strain of PV-1 has also been recently reported (Omata *et al.*, 1984). These reagents will permit the construction of site-specific mutants and viral recombinants that, combined with the recently determined three-dimensional structure of the picornavirion, will afford the opportunity to correlate structure with biologic function.

Study of the antigenic structure of poliovirus is significant for several reasons. First, it is conceivable that the viral element which interacts with the cellular receptor may be identified. Second, the molecular identity of neutralization antigenic sites can aid a study of the mechanism(s) of neutralization. Finally, the expression of genetically engineered viral gene segments or the synthesis of peptides capable of mimicking virion surface structures may ultimately result in the production of new vaccines.

II. NEUTRALIZATION ANTIGENIC STRUCTURE OF POLIOVIRUS

Our laboratory and several others have been interested in the antigenic structure of poliovirus. A wealth of information regarding poliovirus neutralization antigenic sites, antibody binding, and induction of neutralizing antibodies by synthetic peptides has recently been reviewed (Wimmer *et al.*, 1986). To summarize very briefly, surface labeling and chemical cross-linking studies have shown VP1 to be the capsid protein that exhibits the highest degree of exposure on the surface of the virion, followed by VP3 and VP2 (Lonberg-Holm and Butterworth, 1976; Beneke *et al.*, 1977, Wetz and Habermehl, 1979), a result corroborated by the three-dimensional structure. All three surface-exposed capsid proteins carry neutralizing antigenic determinants. Each of the three surface-exposed capsid proteins isolated by SDS-PAGE or HPLC were found to be capable of eliciting a neutralizing response (van der Marel *et al.*, 1983; Dernick *et al.*, 1983). When two different purified monoclonal antibodies raised against PV-1 were covalently cross-linked to the virion surface, it was found that these antibodies bind specifically to capsid protein VP1 (Emini *et al.*, 1983*b*; Blondel *et al.*, 1983). Initial efforts to identify neutralization antigenic sites (N-Ag) concentrated therefore on this protein.

The deduced amino acid sequence of VP1 of PV-1 was subjected to hydrophilicity analysis in an attempt to predict the surface-exposed regions of the protein. Regions of high hydrophilicity were identified and compared with the corresponding regions of the other two serotypes, with special attention to the diversity in amino acid sequences. It was reasoned that hydrophilic regions that differ in amino acid sequence from serotype to serotype might represent neutralization antigenic sites. This approach did indeed lead to the identification of such sites with foot-and-mouth disease virus (Chapter 8, this volume) but was somewhat misleading in the case of poliovirus. Peptides corresponding in amino acid sequence to three hydrophilic regions of capsid protein VP1 of PV-1 (amino acids 93–103, 70–75, and 11–17) were synthesized, cross-linked to carrier proteins and injected into rabbits. The resulting anti-peptide antisera were tested for virus neutralizing activity. Only the 93–103 synthetic peptide elicited directly a type-specific neutralizing antibody response (Emini *et al.*, 1983*a;* Wychowski *et al.*, 1983), whereas the 70–75 peptide and the 11–17 peptide did not give such a response (Emini *et al.*, 1983*a*). All three peptides were, however, capable of priming the rabbit immune system for high titer neutralizing antisera (Emini *et al.*, 1983*a*) (see below, for a description of peptide priming). The 93–103 and the 70–75 peptides were shown in an enzyme-linked immunosorbent assay (ELISA) to bind to neutralizing monoclonal antibodies (N-mcAb) elicited by intact PV-1. The 93–103 peptide was also shown in ELISA to bind to an N-mcAb, the C3 antibody, elicited by heated PV-1 (Blondel *et al.*, 1983). These results led to the conclusion that all three peptides represented neutralizing antigenic sites (N-Ags) at the surface of the virion; a conclusion that must now be reconsidered. The recent elucidation of the X-ray crystal structure of PV-1 shows the 93–103 site to be particularly exposed on the surface of the virion, whereas the 11–17 and the 70–75 sequences, at least in the intact virion, are internal sequences. As we shall see, region 93–103 of VP1 is indeed an N-Ag in both poliovirus type 1 and poliovirus type 3 (PV-3), but the response of the immune systems of different animal species to this region is remarkably different. The use of a peptide corresponding to a hydrophilic region in VP2 of PV-1 has led to the identification

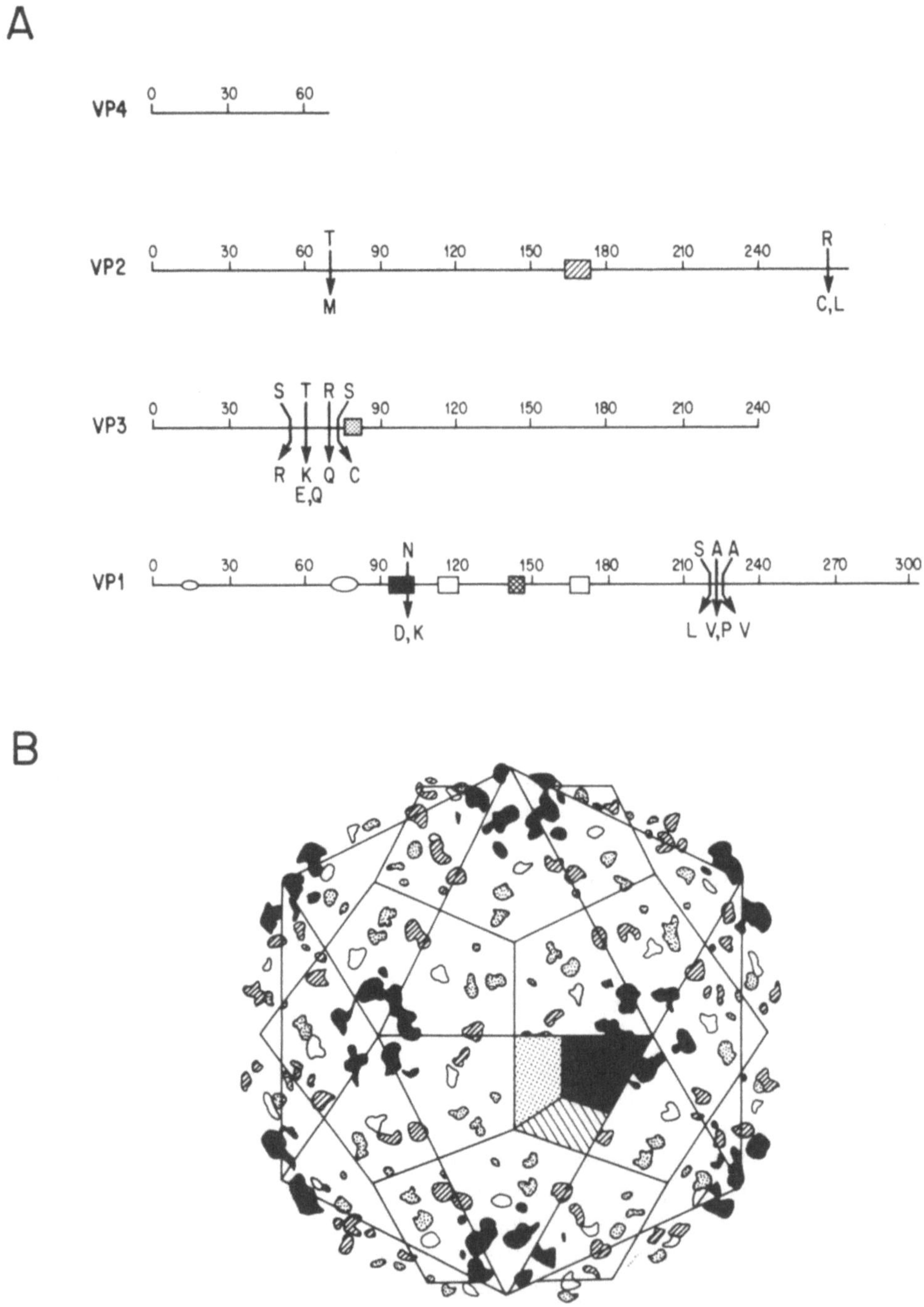

Figure 1. Known antigenic structure of poliovirus type 1, Mahoney. (A) The position within the four capsid proteins of known neutralization antigenic sites (N-Ag) and amino acid substitutions found in diverse N-mcAb escape mutants. N-Ags, identified through peptide immunizations and priming, are represented as boxes and ovals. Boxes represent peptides that can elicit directly neutralizing antibodies; these sites are exposed on the exterior surface of the virion. The ovals represent sites on the interior surface of the virion and peptides which

of an N-Ag in a capsid protein other than VP1 (Emini *et al.*, 1984). However, the most reliable strategy to localize N-Ags proved to be sequence analysis of neutralization escape mutants in combination with the three-dimensional structure of the virus.

Minor *et al.* (1983, 1985) and Evans *et al.* (1983) identified mutations in numerous PV-3 neutralization escape mutants that were selected with a large panel of PV-3-specific N-mcAbs. All but a few mutations mapped to region 89–100 of VP1; the few exceptions mapped in the C-terminal region (aa 286–288) of VP1. Diamond *et al.* (1985*a,b*) and Blondel *et al.* (1986) analyzed escape mutants of PV-1 and observed an entirely different pattern. Only a single mutation was found in region 93–103 of VP1 (Fig. 1A, solid box); all other mutations mapped to various regions within the amino acid sequence of all three capsid polypeptides (Fig. 1A). According to the three-dimensional structure, all mutations occur at the surface of the virion (Hogle *et al.*, 1985) and fall into three distinct sites (Fig. 1B). These results can be summarized as follows:

1. Poliovirus has three major N-Ags (termed N-AgI, N-AgII and N-AgIII; see Fig. 1B). Only N-AgI (Fig. 1A, solid box; region 93–103 in VP1) appears to be a linear sequence of amino acids; N-AgII and N-AgIII are composed of noncontiguous amino acid chains. However, region 162–174 of VP2 (Fig. 1A, hatched box, could serve as a linear N-Ag, since a corresponding peptide elicits a neutralizing antibody response (Emini *et al.*, 1984). Analogous to the situation in poliovirus, a corresponding region in human rhinovirus 2 (HRV2) is recognized by an N-mcAb; this N-mcAb will precipitate VP2 of HRV2 even if VP2 is denatured (E. Kuechler, personal communication).
2. The response of inbred mice to PV-1 and PV-3 is different. Whereas most N-mcAbs antibodies elicited by PV-3, Leon, recognize region 93–103, none of the N-mcAbs elicited by PV-1, Mahoney, recognizes this region. Instead, anti-PV-1 N-mcAbs bind to noncontiguous sites (N-AgII and N-AgIII) composed of segments of VP1, VP2, and VP3. Of interest is N-mcAb C3, elicited by heat-inactivated PV-1. C3 recognizes region 93–103 in native virus and in purified denatured VP1 (Blondel *et al.*, 1983).
3. The difference in immune response to PV-1 and PV-3 disappears when outbred animals are immunized (Icenogle *et al.*, 1986). It appears that in outbred animals N-AgII and N-AgIII are the dominant antigenic sites (Icenogle *et al.*, 1986).

---

elicit a neutralizing antibody response only through priming. (B) The solid, hatched, and stippled regions generally correspond to the similarly shaded regions described in (A):

Solid	VP1 residues 89–103, 168 and 154
Hatched	VP1 residues 222–224 and VP2 residues 166, 169, 170, and 270
Stippled	VP3 residues 58–60, 71–73, and VP2 residue 72 (this last residue has some influence on the previous site)
Open	VP1 residues 284–287

The relative positions of the capsid proteins are shown in one promoter; solid, VP1; hatched, VP2; stippled, VP3; VP4 is almost completely internal. (Compiled from computer graphics prepared by A. J. Olson, Research Institute of the Scripps Clinic.)

A. Interaction of Poliovirus-Specific Synthetic Peptides with the Immune System

Viral infections remain a major cause of human disease today. Commonly used antiviral chemotherapy is restricted to fewer than 10 drugs (Davis *et al.*, 1980; Stringfellow *et al.*, 1983). Thus, antiviral immunization remains the major strategy to lower the incidence of viral disease. Many potent viral vaccines have been developed, and poliovirus vaccines, both the killed virus and the live attenuated virus, have been models for subsequent vaccine research. Surprisingly little is known about how the immune system reacts to the viral antigens. The possibility of using synthetic peptides as vaccines has led to interesting new results in several picornavirus systems, yet the immune response to these peptides as representatives of "natural antigenic sites" is obscure. Synthetic peptides, corresponding in amino acid sequence to PV-1 VP1 amino acid numbers 61–80, 93–103, 91–109, 182–201, 222–241, and 244–264, have been shown to elicit a neutralizing antibody response in rats and rabbits (Emini *et al.*, 1983*a*, 1984; Chow *et al.*, 1985). However, the immune response elicited by these synthetic peptides is lower, by several orders of magnitude, than the immune response to intact virus and corresponds to that of isolated capsid polypeptides. The low virus-neutralizing immune response, elicited by peptides, is not surprising when one considers the probability that a linear synthetic peptide will assume a conformation close to the conformation of the same sequence present in intact virus. If the N-Ag consists of a composite of amino acids coming from different regions of a polypeptide or even from different polypeptide chains, the site can hardly be mimicked by a linear peptide.

Although synthetic peptides have the potential to be presented to the immune system as if they were part of an N-Ag, the absence of the structural constraints that exist on the corresponding sequences in the native virion capsid reduce this potential. One simple approach used to stabilize certain conformations, by restraining the linear peptide from assuming all possible conformations, is to circularize the sequence of amino acids representing a particular N-Ag site covalently. Another empirical approach toward constraining a synthetic peptide amino acid sequence which represents an N-Ag site is to oligomerize short sequences of amino acids. Both approaches have resulted in peptides that express epitopes differently from their linear counterparts (Dreesman *et al.*, 1982; Ferguson *et al.*, 1985).

Studies attempting to mimic native antigenic conformations with synthetic antigens are warranted in light of the advantages of a synthetic peptide vaccine. Chemically synthesized vaccines are well defined and are entirely free of the biologic hazards of inactivated or attenuated virus vaccines. The recent elucidation of the three-dimensional structure of PV-1 should greatly enhance this line of research. For example, residues 91–104 of VP1 form a loop at the peak of the fivefold axis that twists out from the surface and is particularly exposed. Residues 245–251 and residues 142–152 of VP1 form other exposed loops at this peak. Other exposed loops consist of residues 72–75, 160–170, and 240–244 of VP2 and residues 75–81 and 196–206 of VP3. A large projection is formed by residues 127–185 of VP2 and a somewhat smaller projection is formed by residues 53–69 of VP3 (Rossman *et al.*, 1985; Hogle *et al.*, 1985). With this information available, the engineering of synthetic peptide immunogens, that more closely mimic native conformations on the intact virus, becomes more feasible. It is hoped that this type of study will lead to the practical production of synthetic vaccines.

B. Priming the Immune System with Synthetic Peptide Conjugates: An Unexplained Phenomenon

The complete amino acid sequences of many viral surface proteins are known (Kitamura *et al.*, 1981; Racaniello and Baltimore, 1981*a*; Nomoto *et al.*, 1982; Stanway *et al.*, 1983; Toyoda *et al.*, 1984; Carroll *et al.*, 1984; Palmenberg *et al.*, 1984; Skern *et al.*, 1985; R. Colonno and E. A. Emini, unpublished observations, 1985). In spite of this knowledge, it has been difficult to elicit virus-neutralizing antibodies with carrier linked synthetic peptides. Synthetic peptides corresponding in amino acid sequence to a segment of the tobacco mosaic virus (TMV) capsid protein were used in the 1960s to produce neutralizing antibodies against TMV (Anderer and Schleumberger, 1965). However, only one system has been reported in which an immunization with synthetic peptide conjugates has offered definitive protection against subsequent challenge with virulent virus (foot-and-mouth disease virus, another picornavirus) in a whole animal system (DiMarchi *et al.*, 1986; Bittle *et al.*, 1982).

The potential of a synthetic peptide to act as a vaccine is assessed by the peptide's ability to elicit directly an immune response that will neutralize virus. Except for the protection against foot-and-mouth disease, synthetic peptides have been found to be impotent vaccines by this assessment. Results of many independent investigations have found that poliovirus-specific synthetic peptides elicit neutralizing antibody responses that are orders of magnitude below the levels of neutralizing antibodies found in hyperimmune sera and thus are highly unlikely to protect against disease. These low levels may be due to the unstable expression of epitopes by linear peptides.

A second possible explanation of the low level of virus-neutralizing response, elicited by a synthetic peptide, is that each peptide only represents a single antigenic site. This problem may be overcome by coupling several synthetic peptides, representing several immunogenic sites, to a single carrier molecule. We have coupled the PV-1, VP1 93–103, and VP2 162–173 synthetic peptides to a single carrier and have observed an increase in the neutralization titer resulting from this innoculum over either synthetic peptide conjugate alone (Jameson *et al.*, 1985).

Perhaps the potential of synthetic peptides as vaccines lies not in their ability to elicit directly an antiviral immune response but rather in their ability to prime an immune system for subsequent encounters with the viral pathogen. Emini *et al.* (1983*a*) reasoned that even in the absence of a virus-neutralizing response it was reasonable to assume that small peptides, when presented to an animal's immune system, would be "seen" in conformations similar to the conformations on the virion surface. These researchers therefore investigated whether various peptides were capable of priming by analyzing the sera of rabbits that had been inoculated first with the peptides and subsequently injected with a subimmunogenic dose (10^5 PFU) of intact virions (these experiments were carried out in the rabbit, an animal in which poliovirus is not infectious). At the end of the inoculation schedule with the peptide alone, animals exhibited an anti-peptide response;

but with the exception of the VP1 93–103 peptide, none of the sera were capable of neutralizing virus. However, animals inoculated with an ordinarily subimmunogenic dose of intact virus, after having been primed with the synthetic peptides, exhibited the production of a high titer neutralizing antibody response (Emini *et al.*, 1983*a*). The results of a typical experiment for the synthetic peptide representing PV-1 VP1 amino acid 70–75 are shown below, and similar results were obtained with other poliovirus specific synthetic peptides (Fig. 2). The neutralizing response observed after peptide priming is of high titer and comparable to that seen in hyperimmune serum.

Priming has been demonstrated in three animal systems using two viral systems to date. Priming with synthetic peptides of poliovirus and foot-and-mouth disease virus (FMDV) has been demonstrated in rabbits and guinea pigs, respectively (Emini *et al.*, 1983*a*; Francis *et al.*, 1985*a,b*). Priming in the rat system has been demonstrated with purified capsid proteins of poliovirus (van der Marel *et al.*, 1983; von Wetzel *et al.*, 1985). Although synthetic peptide priming in certain animals is a very reproducible phenomenon, it has never been demonstrated in mice. Moreover, it appears to be restricted to certain peptides in combination with certain animal species.

One can speculate that it would be possible to use peptide priming to alert the immune system, so that when the vaccine recipient is exposed to virulent virus, priming may lend the necessary advantage to prevent clinical disease by allowing the rapid production of neutralizing antibodies. Experiments to determine whether peptide mediated priming could protect primates against naturally acquired infection by neurovirulent

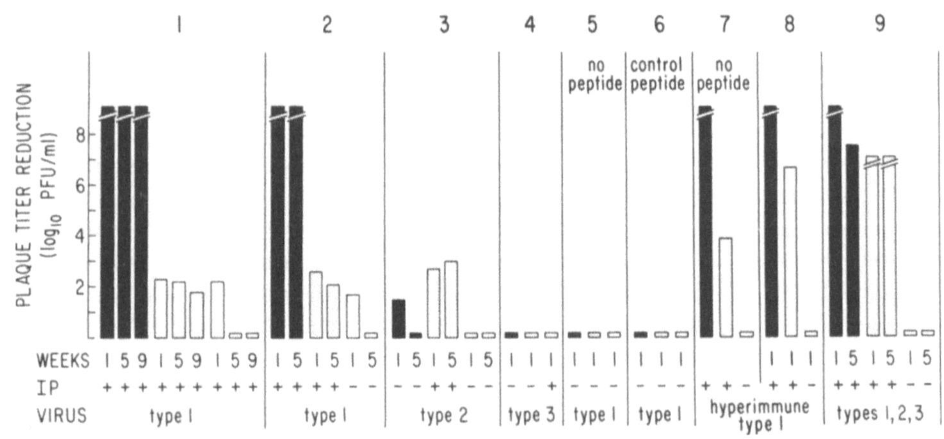

Figure 2. Priming of an anti-poliovirus immune response with a synthetic peptide representing PV-1 VP1 aa 70–75. Rabbits were inoculated with a bovine serum albumin (BSA)-conjugated peptide (1 mg per inoculation) followed by a single challenge with 10^5 PFU of poliovirus of either type 1, type 2, of type 3. Then, 1, 5, or 9 weeks later, immune sera were assayed for their ability to neutralize PV-1 (closed bars), PV-2 (shaded bars), or PV-3 (open bars), respectively, or tested whether they can immunoprecipitate the serotypes (IP). (1) Three inoculations with peptide, challenged with PV-1; (2) one inoculation with the peptide, challenge with PV-1; (3,4) three inoculations with the peptide, challenge with PV-2 and PV-3 respectively; (5) no inoculation of peptide, single inoculation with PV-1; (6) single inoculation with a peptide unrelated to poliovirus, challenge with PV-1; (7) no inoculation with peptide, but three inoculations with 50 μg each of PV-1; (8) three inoculations of 50 μg each of PV-1, followed by a single inoculation with peptide; (9) same as (1) but challenged with all three types of poliovirus. (For details, see Emini *et al.*, 1983*a*.)

A

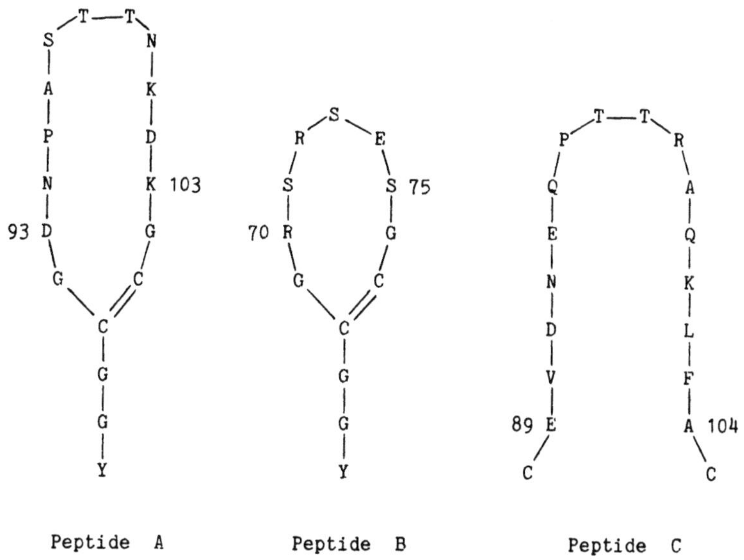

B

Plaque reduction (log_{10}pfu/ml)

peptide	rabbit	anti-peptide serum	anti-virus serum
A	1	1.7	0.7
	2	1.1	1.0
B	3	0.8	0.8
	4	<0.5	<0.5

Figure 3. (A) Synthetic peptides corresponding to PV-1 VP1 aa 93–103, PV-1 VP1 aa 70–75, and PV-3 VP1 aa 89–104. (B) Results of a neutralization assay (immune response elicited directly by the peptide and after injection of a subimmunogenic dose of virus).

poliovirus have not been performed to date. Since the molecular mechanisms of peptide mediated immunity and peptide priming are obscure, our observations warrant further study.

Before the three-dimensional structure of HRV 14 and PV-1 was reported, we hypothesized that if a hydrophilic sequence exists as a loop structure on the native virus, one should be able to engineer a synthetic peptide immunogen that would closely mimic the loop structure of the native virus. We synthesized peptides containing VP1 aa 93–103 and aa 70–75 in which the viral sequences were flanked with two glycine residues followed by two cystein residues and a glycine–glycine–tyrosine sequence at one end for the purposes of coupling to carrier proteins. Such a sequence would allow the circularization of the monomeric peptide by oxidation of the sulfhydryl groups of the cystein residues under dilute conditions. These peptides are shown in Fig. 3A. The peptides were conjugated to bovine serum albumin and used as an immunogen in rabbits following procedures published by Emini *et al.* (1983*a*). Antiserum elicited to each peptide was tested for its ability to neutralize virus. Sometime later, each rabbit was injected with a subimmunogenic dose of intact virus in order to test the priming effect of these peptides. The results of these experiments are shown in Fig. 3B.

Since the three-dimensional crystal structure of PV-1 provides evidence which supported our earlier assumption that the 93–103 site may exist as a loop in the native virus, it was hoped that peptide A would mimic the native loop conformation and thus elicit a virus-neutralizing response that would closely approximate the response elicited by native virus. This is clearly not the case. Peptide A does, however, elicit a virus-neutralizing response directly. The response is about the same as that elicited by the linear-version peptide (Emini *et al.*, 1983*a*). The most important result is that the circularization of these peptides has abolished the priming effect exhibited by the linear versions of these synthetic peptides.

Ferguson *et al.* (1985), following a similar approach, synthesized a peptide (Fig. 3A, peptide C) that corresponded to residues 89–104 of VP1 of PV-3. In addition, the peptide carried two extra cystein residues at its amino and carboxyl terminals to facilitate circularization. This peptide elicited a strong neutralization response in rabbits, but surprisingly, it did so in the polymerized form and not in the circular monomeric form (Ferguson *et al.*, 1985). A linear nonpolymerized version of the peptide (lacking the Cys residue at the *N*-terminal) or a shorter version [Cys-(92-100)-Cys] were very poor immunogens in regard to neutralizing titers. Moreover, the Cys-(89–104)-Cys did not induce a neutralizing response in guinea pigs or mice. These interesting, but disappointing, results underscore the empirical nature of studies that are aimed at the design of a synthetic peptide vaccine.

III. CONCLUSION

Synthetic peptides undoubtedly hold promise as immunogens that can mimic naturally occurring neutralization antigenic sites of pathogenic microorganisms and viruses and may eventually augment or even replace conventional vaccines. In spite of this rather optimistic statement, the current results with peptide immunogens are less than satisfactory, at least in the case of poliovirus-specific peptides.

Several major obstacles block the path to a peptide vaccine. First, our ignorance in regard to the interaction between carrier-conjugated or even unbound peptides with the immune systems of different animal species prevents us from "designing" a specific antigenic structure.

Second, the "design" of a specific antigenic structure requires that we can predict the preferred spatial structure of an amino acid chain. This cannot be achieved as yet, and thus even the knowledge of the three-dimensional structure of a virus is of little help at present. Third, the problem of antigen presentation (e.g., the nature of carrier molecules and adjuvants) has not been solved. It follows that the outcome of experiments with peptide immunogens during the past few years was unpredictable and usually harbored unexpected surprises. For example, FMDV-specific peptides deliver protective immunity in natural hosts, whereas peptides specific to poliovirus VP1 yielded detectable neutralizing immune response in rabbits and rats but not in monkeys (B. A. Jameson, A. Nomoto, and E. Wimmer, unpublished results).

The enormous potential of peptides as immunogens in medicine, however, must not be overlooked despite the initial disappointing results. An unexpected phenomenon may give us the necessary breakthrough to design effective peptide vaccines.

ACKNOWLEDGMENTS

This work is supported by grants AI-15122 and CA-28146 from the U.S. Public Health Services. The authors also acknowledge the expert photographic assistance by Christopher Helmke and excellent secretarial assistance by Lynn Zawacki.

REFERENCES

Anderer, F. A., and Schlumberger, H. D. (1965). *Biochim. Biophys. Acta* **97,** 503–509.

Beneke, T. W., Habermehl, K. O., Diefenthal, W., and Budiholtz, M. (1977). *J. Gen. Virol.* **34,** 387–390.

Bittle, J. L., Houghten, R. A., Alexander, H., Shinnick, T. M., Sutcliffe, Y. G., Lerner, R. A., Rowlands, D. Y., and Brown, F. (1982). *Nature (Lond.)* **298,** 30–33.

Blondel, B., Akacem, O., Crainic, R., Corillin, P., and Horodniceanu, F. (1983). *Virology* **126,** 707–710.

Blondel, B., Crainic, R., Fichot, O., Onfraisse, G., Candrea, A., Diamond, D. C., Girard, M., and Horand, F. (1986). *J. Virol.* **57,** 81–90.

Carroll, A. R., Rowlands, D. J., and Clarke, B. E. (1984). *Nucleic Acids Res.* **12,** 2461–2472.

Centers for Disease Control. (1986). *MMWR* **35,**(11), 180–182.

Chasan, D. J. (1986). *Science* **86,** 37–39.

Chow, M., Yabrov, R., Bittle, J., Hogle, J., and Baltimore, D. (1985). *Proc. Natl. Acad. Sci. USA* **82,** 910–914.

Davis, B. D., Dulbecco, R., Eisen, H. N., and Ginsberg, H. S. (1980). *Microbiology* 3rd ed., Harper and Row, New York.

Dernick, R., Heukeshoven, Y., and Hilbrig, M. (1983). *Virology* **130,** 243–246.

Diamond, D. C., Wimmer, E., and Emini, E. A. (1987). In *Synthetic Vaccines,* Vol. II (R. Arnon, ed.), CRC Press, Boca Raton, Florida, pp. 79–91.

Diamond, D. C., Jameson, B. A., Bonin, J., Kohara, M., Abe, S., Itoh, H., Komatsu, T., Arita, M., Kuge, S., Osterhaus, A. D. M. E., Crainic, R., Nomoto, A., and Wimmer, E. (1985). *Science* **229,** 1090–1093.

DiMarchi, R., Brooke, G., Gale, C., Cracknell, V., Doel, T., and Mowat, N. (1986) *Science* **232,** 639–641.

Dreesman, G. R., Sanchez, Y., Ionescu-Matiu, I., Sparrow, J. T., Six, H. R., Peterson, D. L., Hollinger, F. B., and Melnick, J. L. (1982). *Nature (Lond.)* **295,** 158–160.

Emini, E. A., Jameson, B. A., and Wimmer, E. (1983*a*). *Nature (Lond.)* **304,** 699–703.

Emini, E. A., Kao, S.-Y., Lewis, A. J., Crainic, R., and Wimmer, E. (1983*b*). *J. Virol.* **46,** 466–474.

Emini, E. A., Jameson, B. A., and Wimmer, E. (1984). *J. Virol.* **52,** 719–721.

Evans, D. M. A., Minor, P. D., Schild, G. C., and Almond, J. W. (1983). *Nature (Lond.)* **304,** 459–462.

Ferguson, M., Evans, D. M. A., Magrath, D. I., Minor, P. D., Almond, J. W., and Schild, G. C. (1985). *Virology* **143,** 505–515.

Francis, M. J., Fry, C. M., Rowlands, D. J., Brown, F., Bittle, J. L., Houghten, R. A., and Lerner, R. A. (1985*a*). *J. Gen. Virol.* **66,** 2347–2354.

Francis, M. J., Fry, C. M., Rowlands, D. J., Brown, F., Bittle, J. L., Houghten, R. A., and Lerner, R. A. (1985*b*). In *Vaccines 85* (R. A. Lerner, R. M. Chanock, and F. Brown, eds.), pp. 203–210, Cold Spring Harbor Laboratory, Cold Spring Harbor, New York.

Hogle, J. M., Chow, M., and Filman, D. J. (1985). *Science* **229,** 1358–1363.

Icenogle, J. P., Minor, P. D., Ferguson, M., and Hogle, J. M. (1986). *J. Virol.* **60,** 297–301.

Jameson, B. A., Bonin, J., Murray, M. G., Wimmer, E., and Kew, O. (1985). In *Vaccines 85* (R. A. Lerner, R. M. Chanock, and F. Brown, eds.), pp. 191–198, Cold Spring Harbor Laboratory, Cold Spring Harbor, New York.

Kitamura, N., Semler, B. L., Rothberg, P. G., Larsen, G. R., Adler, C. J., Dorner, A. J., Emini, E. A., Hanecak, R., Lee, J. J., van der Werf, S., Anderson, C. W., and Wimmer, E. (1981). *Nature (Lond.)* **291,** 547–553.

Koch, F., and Koch, G. (1985). *The Molecular Biology of Poliovirus*, Springer-Verlag, New York.

Kuhn, R. J., and Wimmer, E. (1986). In *The Molecular Biology of Positive Strand RNA Viruses* (D. J. Rowlands, B. W. J. Mahy, and M. Mayo, eds.) Academic Press, Orlando, Florida.

LaMonica, N., Merian, C., and Racaniello, V. R. (1986). *J. Virol.* **57,** 515–525.

Lee, Y. F., Nomoto, A., Detjen, B. M., and Wimmer, E. (1977). *Proc. Natl. Acad. Sci. USA* **74,** 59–63.

Lonberg-Holm, K., and Butterworth, B. E. (1976). *Virology* **71,** 207–216.

Minor, P. D., Schild, G. C., Bootman, J., Evans, D. M. A., Ferguson, M., Reeves, P., Spitz, M., Stanway, G., Cann, A. J., Haupmann, R., Clarke, L. D., Mountford, R. C., and Almond, J. W. (1983). *Nature (Lond.)* **301,** 674–679.

Minor, P. D., Evans, D. M. A., Ferguson, M., Schild, G. C., Westrop, G., and Almond, J. (1985). *J. Gen. Virol.* **65,** 1159–1165.

Nicklin, M. J. H., Toyoda, H., Murray, M. G., and Wimmer, E. (1986). *Biotechnology* **4,** 33–42.

Nomoto, A., Detjen, B., Pozzatti, R., and Wimmer, E. (1977). *Nature (Lond.)* **268,** 208–213.

Nomoto, A., Omata, T., Toyoda, H., Kuge, S., Horie, H., Kataoka, Y., Genba, Y., Nakano, Y., and Imura, N. (1982). *Proc. Natl. Acad. Sci. USA* **79,** 5793–5797.

Omata, T., Kohara, M., Sakai, Y., Kameda, A., Imura, N., and Nomoto, A. (1984). *Gene* **32,** 1–10.

Palmenberg, A. C., Kirby, E. M., Janda, M. R., Drake, N. L., Duke, G. M., Potratz, K. F., and Collett, M. S. (1984). *Nucleic Acids. Res.* **12,** 2969–2985.

Racaniello, V. R., and Baltimore, D. (1981*a*). *Science* **214,** 916–919.

Racaniello, V. R., and Baltimore, D. (1981*b*). *Proc. Natl. Acad. Sci. USA* **78,** 4887–4891.

Rossmann, M. G., Arnold, E., Erickson, J. W., Frankenberger, E. A., Griffith, J. P., Hecht, H. J., Johnson, J. E., Kamer, G., Luo, M., Mosser, A. G., Rueckert, R. R., Sherry, B., and Vriend, G. (1985). *Nature (Lond.)* **317,** 145–153.

Rothberg, P. G., Harris, T. J. R., Nomoto, A., and Wimmer, E. (1978). *Proc. Natl. Acad. Sci. USA* **75,** 4868–4872.

Semler, B. L., Dorner, A. J., and Wimmer, E. (1984). *Nucleic Acids. Res.* **12,** 5123–5141.

Skerm, T., Sommergruber, W., Blaas, D., Gruender, P., Fraundorfer, F., Pieler, C., Fogy, I., and Kuechler, E. (1985). *Nucleic Acids Res.* **13,** 2111–2126.

Stanway, G., Cann, A. J., Hauptmann, R., Hughes, P., Clark, L. D., Mountford, R. C., Minor, P. D., Schild, G. C., and Almond, J. W. (1983). *Nucleic Acids Res.* **11,** 5629–5643.

Stringfellow, D. A., Bale, J. F., Corey, L., Couch, R. B., Glasgow, L. A., Korn, E. R., Lennette, D. A., Lennette, E. T., Rapp, F., and Stanberry, L. R. (1983). *Virology* The Upjohn Company, Kalamazoo, Michigan.

Toyoda, H., Kohara, M., Kataoka, Y., Suganuma, T., Omata, T., Imura, N., and Nomoto, A. (1984). *J. Mol. Biol.* **174,** 561–585.

van der Marel, P., Hazendank, T. G., Henneke, M. A. C., and van Wezel, A. L. (1983). *Vaccine* **1,** 17–22.

Wetz, K., and Habermehl, K.-O. (1979). *J. Gen. Virol.* **44,** 525–534.

Wimmer, E., Emini, E. A., and Diamond, D. C. (1986). In *Concepts in Viral Pathogenesis* Vol. II, (A. L. Notkins and M. B. A. Oldstone, eds.), pp. 159–173, Springer-Verlag, New York.

Wychowski, C., van der Werf, S., Siffert, O., Crainic, R., Bruneau, P., and Girard, M. (1983). *EMBO J.* **2,** 2019–2024.

Yogo, Y., and Wimmer, E. (1975). *J. Mol. Biol.* **92,** 467–477.

11

Immune Responses to Synthetic Peptide Vaccines of Veterinary Importance

P. Frenchick, M. I. Sabara, M. K. Ijaz, and L. A. Babiuk

I. INTRODUCTION

Vaccination against virus diseases has been one of the major triumphs of modern medicine. However, even with great success in some areas, many viral diseases continue to plague human and animal populations. Some of the reasons for the inability to control viral infections are related to the epidemiology of the infection and the pathogenesis of the specific disease in question. Other impediments to producing effective vaccines are (1) it is difficult to grow sufficient quantities of virus *in vitro;* and (2) if they do grow, some viruses are extremely labile and may be inactivated prior to administration. A final problem with production of conventional vaccines is that the causitive agent may be dangerous to manipulate in the laboratory. The use of synthetic peptides as vaccines has been suggested as an alternate approach to induce immunity to a wide variety of different virus diseases while avoiding many of the problems mentioned above (Brown, 1984).

Acute infectious diseases caused by rotaviruses occur worldwide in many species and are responsible for enormous economic losses. These losses occur primarily in young animals and infants; however, recent reports also indicate severe rotavirus diarrhea in human adults (Chen *et al.*, 1985). Since diarrhea, especially in animal species, occurs during the first few weeks of life, it is difficult to induce adequate active immunity prior to exposure to the pathogen. Controlling of gastrointestinal (GI) infection requires local immunity whether acquired or active to prevent infection, which is difficult to produce in the young. Consequently, one of the most productive strategies for control of enteric viral infections involves vaccination of the mother, who then transfers protective antibody in the colostrum and milk. The young calf suckling a cow with high levels of milk antibody is protected as long as there is antibody in the milk and in the intestinal lumen (Saif *et al.*, 1983). Although there are licensed vaccines for immunizing cattle against rotavirus infections, the efficacy of these vaccines for parental administration to pregnant cows has been questioned (DeLeeuw *et al.*, 1980). The reported reason for the apparent poor immunologic response to this vaccine is that rotavirus does not grow to very large quantities *in*

P. Frenchick, M. I. Sabara, M. K. Ijaz, and L. A. Babiuk • Veterinary Infectious Disease Organization, Saskatoon, Saskatchewan, S7N 0W0 Canada.

vitro; therefore, the vaccines may not contain the required concentration of antigen for proper stimulation of immunity. Increasing the antigenic mass and incorporating the virus into appropriate adjuvants does induce high levels of antibody in the serum and in the milk of vaccinated dams (Saif and Smith, 1983), indicating that this approach to immunization and disease control is feasible.

Since the quantity of antigen required to produce adequate immunity is high, it may be more economical to identify the specific antigens involved in inducing protective immunity and to either clone the genes encoding them or to sequence a specific epitope and produce synthetic oligopeptides. These approaches appear to be especially relevant to the production of vaccines against viruses, such as rotavirus, which do not replicate to high levels in tissue culture and where high antigenic mass is required for induction and protection. In the case of rotavirus, we have identified a number of synthetic peptides that are important in inducing protection, making this approach feasible for vaccine production. However, it is also important to have an in-depth understanding regarding the factors involved in the formulation of such vaccines including carriers and adjuvants, so that the immune system can be appropriately stimulated with these vaccines. The present report identifies a cross-reactive neutralizing epitope on rotavirus in addition to describing various ways of formulating vaccines in an attempt to develop a simple economically feasible method of producing protective immune response against rotavirus infections.

II. IDENTIFICATION OF EPITOPE

The major outer shell protein of rotavrius is a glycoprotein, VP7 (gp41), with an approximate molecular weight of 38,200 in its unreduced form and 41,900 in its reduced form. This glycoprotein has been shown to be responsible for virus attachment to cells (Sabara *et al.*, 1985). It is also the major antigen responsible for inducing neutralizing antibodies to the virus (Bastardo *et al.*, 1981) and for serotype differences (Greenberg *et al.*, 1983). Although the VP7 glycoprotein from different serotypes does contain different antigenic determinants, accounting for serotype differences, amino acid sequence analysis indicates homology between serotypes ranges from 75 to 85% with several conserved regions (Richardson *et al.*, 1984; Dyall-Smith and Holmes, 1984). These conserved regions could be responsible for the cross-protection provided by the different serotypes (Woode *et al.*, 1976; Woode *et al.*, 1983).

Epitope mapping of VP7 (gp41), using neutralizing monoclonal antibodies, localized a neutralizing domain to a component peptide with an approximate molecular weight of 14,000. This 14,000-M_r fragment stimulated neutralizing antibodies when administered to mice and blocked absorption of virus to cells. Furthermore, this peptide contained some regions that were conserved between different rotaviruses, making it a useful region to identify potential universal rotavirus peptide vaccines. From the amino acid sequence of this region, several smaller peptides were selected for synthesis. Since a probable disulfide bond responsible for maintaining the conformational structure of the 14,000-M_r peptide could not be identified, no cyclic peptides were made. However, one peptide with a discontinuous structure was synthesized and examined for activity.

Figure 1 shows the activity of three different synthetic peptides derived from the 14,000-M_r fragment in blocking the binding of the radiolabeled bovine rotavirus to

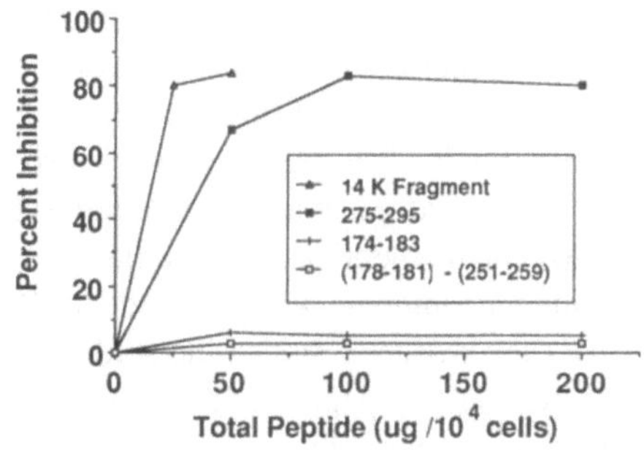

Figure 1. Inhibition of rotavirus binding to MA-104 cells by gp41 sequences. The isolated 14,000-M_r fragment or synthetic peptides in increasing amounts were mixed with MA-104 cells and radiolabeled virus. The cells were then pelleted and the binding evaluated.

MA-104 cells. The parent 14,000-M_r fragment was very efficient in preventing rotavirus binding, as was the peptide containing amino acids (aa) 275–295 (Fig. 2). This peptide was almost as efficient in its ability to block absorption as the parent 14,000-M_r peptide, even though it does not contain any disulfide bonds. By contrast, two other peptides from this fragment displayed little or no activity. Monoclonal antibodies that reacted with the peptide capable of inhibiting binding gp41 (275–295) also react with other serotypes, suggesting that this region may be conserved among rotaviruses.

In addition to the outer glycoprotein, the nucleocapsid protein (45,000 M_r) also induces neutralizing antibodies (Bastardo *et al.*, 1981). This may occur as a result of the protrusion of the nucleocapsid through the outer capsid in such a way that regions are exposed; alternately, some double-shelled particles have portions of the outer shell removed. Since the nucleocapsid protein is shared by most rotaviruses, this protein also appeared to have potential as a broad-spectrum vaccine. Thus, attempts were made at identifying a region on the nucleocapsid that would be recognized by neutralizing monoclonal antibodies. It was possible to identify such a region in the area of aa 40–60 (Fig. 2).

III. SYNTHESIS OF PEPTIDE AND CONJUGATE

The peptides were synthesized using 9-fluorenylmethyloxycarbonyl (Fmoc)-protected amino acids and solid-phase peptide synthesis (Cambridge Research Biologics, Cambridge, England; IAF Biochemical, Laval, Quebec, Canada). The peptide–keyhole

Ala-Pro-Gln-Thr-Glu-Arg-Met-Met-Arg-Ile-Asn-Trp-Lys-Lys-Trp-Trp-Gln-Val-OH

Amino Acids 275-294 from BRV gp41

Gly-Asn-Glu-Phe-Gln-Thr-Gly-Gly-Ile-Gly-Asn-Leu-Pro-Ile-Arg-Asn-Trp-Asn-OH

Amino Acids 40-60 from BRV p45

Figure 2. Sequence of synthetic peptides derived from bovine rotavirus proteins.

limpet hemacyanin (KLH) conjugates were produced using *N*-maleimido-benzoyl-*N*′-hydroxysuccinimide (MBS) ester-derivitized KLH and a cysteine added to the *N*-terminal of the peptide (Cambridge Research Biologics, Cambridge, England) or water-soluble carbodiimide. The peptides were also conjugated to the *Escherichia coli* pilin protein K99. Urea-extracted and ammonium sulfate-precipitated K99 was purified (deGraff *et al.*, 1980) and the peptides were conjugated to it using the carbodiimide method. This protocol yielded a conjugate with a peptide to K99 at a ratio of 3.5 : 1, as determined by ultraviolet (UV) spectroscopy and amino acid analysis and confirmed by gel electrophoresis. Following synthesis of peptides and conjugates, the products were assessed for immunoreactivity using an immunoblot enzyme-linked inmunosorbent assay (ELISA) developed with polyclonal anti-rotavirus serum.

IV. IMMUNE RESPONSES TO CONJUGATES

Synthetic peptides have been shown to be poorly immunogenic when used as "free" monomers. Approaches to improve the immune response has been to polymerize or aggregate peptides, either linking the peptides into large complexes or inducing micelles from the peptides (Jolivet *et al.*, 1983; Hopp, 1984; Sanchez *et al.*, 1982). However, these approaches have rarely produced strong long-lasting immunity. Alternatively, peptides have been conjugated to proteins, where they have become very large haptens, relying on the immune response against the carrier to assist in producing the antipeptide response (Muller *et al.*, 1982). These conjugates generally produce long-lasting immunity to both carrier protein and peptide. Most of the proteins examined for use as carriers have been large highly immunogenic proteins. The use of these proteins, such as KLH, heterologous albumins, tetanus toxoid, or cytochromes, is handicapped because they are expensive, may cause adverse responses, or may be poor immunogens in species other than mice. Recent reports suggest that preexisting immunity may also affect the response to the peptide haptens (Schultze *et al.*, 1985).

Our approach was designed to investigate the importance of carrier size in relationship to the antirotavirus response. Animals were immunized with 100 μg of either the KLH or the K99 conjugates in Freund's complete adjuvant (FCA) and the antirotavirus immune responses followed. The amount of the conjugates used in the immunizations contained approximately equal quantities of each of the rotavirus peptides. No difference in the response of animals to the conjugates was seen in three immunizations (Fig. 3). Although the two carriers differ greatly in molecular weight, KLH being approximately 3.5×10^6 and K99 approximately 19,000 M_r, they appear to be equally effective in generating antibodies to these peptide haptens. These results suggest that the size of a carrier is not the key factor in inducing immune responses to synthetic peptides.

In addition to the use of carriers in producing immune responses to peptides, many adjuvants have been investigated to increase these responses further. Unfortunately, most adjuvants powerful enough to increase the immune responses are not acceptable for use in humans or animals. The adjuvants chosen were the quatenary amine surfactant dimethyldioctadecylammonium (DDA) bromide and FCA. DDA was selected because of its reported activity as a powerful adjuvant toward haptens and the fact that it is approved for

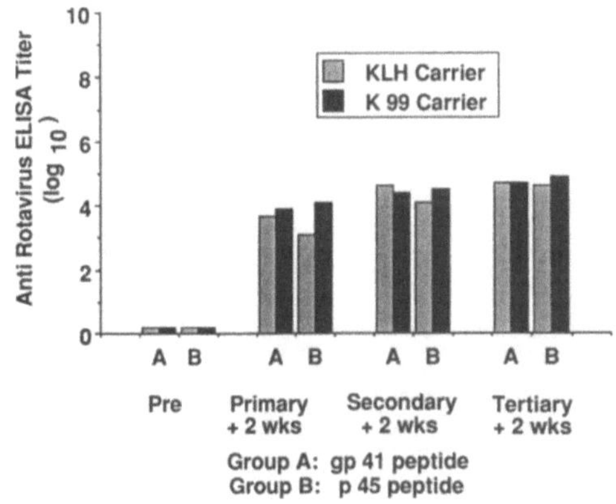

Figure 3. Effect of the carrier on the antirotavirus antibody response generated by synthetic peptides conjugated to them. Group A: gp41 275–295 peptide; group B: p45 40–60 peptide.

use in food animals. To compare the efficacy of these two adjuvants in synthetic peptide vaccines, they were mixed with rotavirus peptide gp41 (275–295) of VP7 with K99 as a carrier. Each mouse received 100 μg of conjugate with either 0.1 ml FCA or 100 μg of DDA. Following two immunizations, there was little difference in the antibody titer to rotavirus and only a slight difference in the anti-K99 titer. The levels of antirotavirus activity seen in this experiment approximate that seen following a protective immunization schedule using live rotavirus (Fig. 4).

The next series of experiments were designed to examine the dose of antigen needed to produce protective level of immunity. Three doses of conjugate—1, 10, and 100 μg/animal—with DDA as an adjuvant were used. In each case, antirotavirus and anti-K99 antibody levels were measured. As shown in Fig. 5, the conjugate produced dose-related responses to the K99 carrier protein, but only the 100-μg dose approached the antirotavirus antibody level produced by virus-immunized controls. Neither the 1- nor 10-μg doses of conjugate showed a secondary antirotavirus response following reimmunization, even though the carrier antibody levels increased above previous titers.

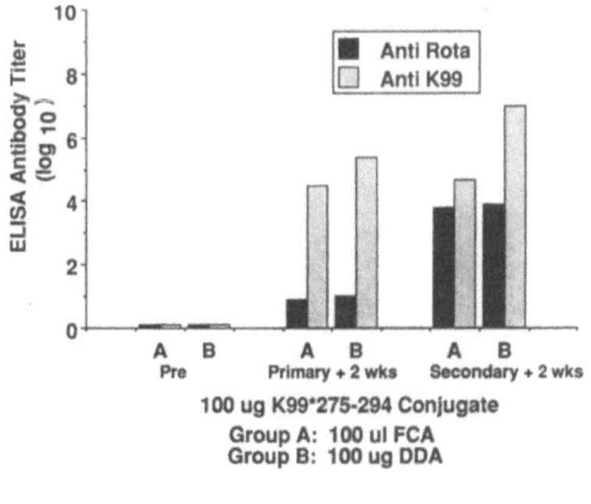

Figure 4. Effect of adjuvants on the antibody response to 100 μg K99–gp41 275–295 conjugate. Group A: Freund's complete adjuvant; Group B: dimethyldioctadecylammonium bromide.

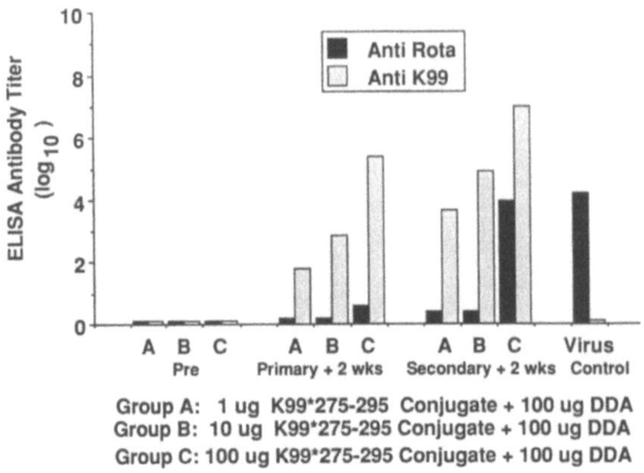

Figure 5. Effect of conjugate dose on the antibody response. Group A: 1 μg K99–gp41 (275–295); group B: 10 μg K99–gp41 (275–295); group C: 100 μg-g41 (275–295). The adjuvant for all groups was 100 μg dimethyldioctadecyl ammonium bromide.

V. IMMUNIZATION WITH FREE PEPTIDES

In general, free synthetic peptides generate only low levels of antibody of short duration. The generation of immunity to free peptides requires the use of strong adjuvants of limited practical value. However, peptides that form aggregates, either spontaneously or through the use of lipid side chains, induce significant immune responses (Hopp, 1984; Sanchez *et al.*, 1982). Peptides have also been used to prime the immune system such that subsequent exposure to a suboptimal dose of parent protein results in a secondary type of response (Emini *et al.*, 1983). Alternative approaches of immunizing with peptides might be possible by (1) associating the peptides with liposomes containing an adjuvant, (2) aggregating the peptides with the adjuvant DDA, and (3) combining peptides with FCA to prime the immune response. To investigate these possibilities, an equivalent amount of gp41-(275–295) peptide (100 μg) was combined with three adjuvants—FCA, DDA, and phosphatidylcholine–cholesterol–avridine (7 : 2 : 2) liposomes. The liposomes containing avridine and associated peptide were produced by reswelling a film containing avridine, dipalmitylphosphatidylcholine, and cholesterol with a buffer containing the synthetic peptide. The peptide in FCA and DDA was injected intramuscularly. The liposomal preparation was injected into the peritoneum of the mice with no attempt to separate the liposome-associated peptide from free peptide. Peptide in FCA and DDA produced low

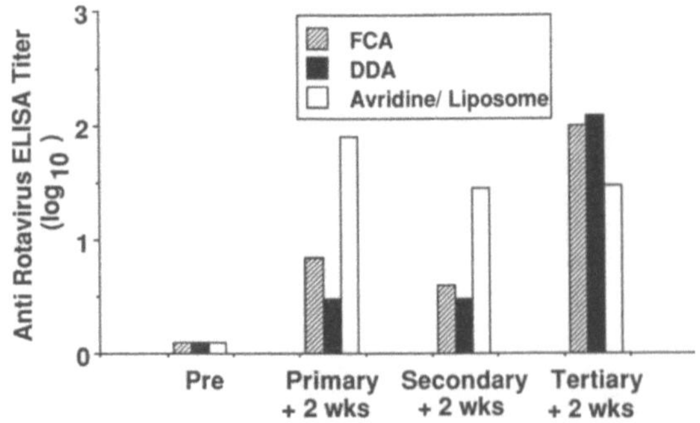

Figure 6. Effect of adjuvant on the anti-rotavirus antibody response following immunization with 100 μg gp-41 (275–295). The adjuvants used were Freund's complete adjuvant (FCA), dimethyldioctadecylammonium (DDA), and avridine-containing liposomes.

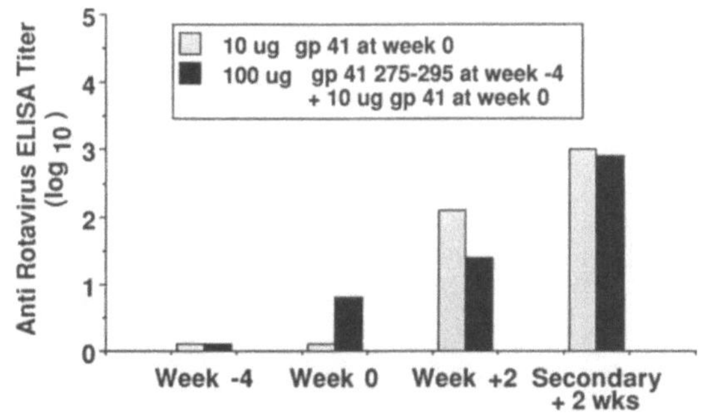

Figure 7. Effect of priming with free peptide gp41 275–295 on the antibody response to rotavirus following immunization with gp41.

levels of antibody after two injections but resulted in significant levels after the third injection (Fig. 6). By contrast, liposomes produced a response after only one injection, but no boosting of the response occurred after either the second or third immunization.

To determine whether the peptide/FCA-immunized mice were primed following one immunization of synthetic peptide, mice given 100 μg of free peptide in FCA at week −4 were then immunized with the authentic gp41 glycoprotein at week 0. The control animals received only gp41 at week 0. The single injection of synthetic peptide had no effect on the response to the gp41 (Fig. 7), suggesting that it did not prime the animals' immune system. Although there is a slight decrease in response after one immunization in the primed animals, their responses were equal following a second immunization with gp41.

VI. CARRIER-INDUCED SUPPRESSION

The use of carriers has been shown in numerous studies to be important in producing immune responses to synthetic peptides (Shinnick *et al.*, 1983). However, it has also been shown that preexisting immunity, either passive or acquired, to the carrier can limit the responses to classic haptens and the large peptide haptens as used in these studies (Schultze *et al.*, 1985; Sumida and Taniguchi, 1985; van Oirschot and DeLeeuw, 1985). On the basis of these findings, the type of potentially useful carrier would be limited and would prevent

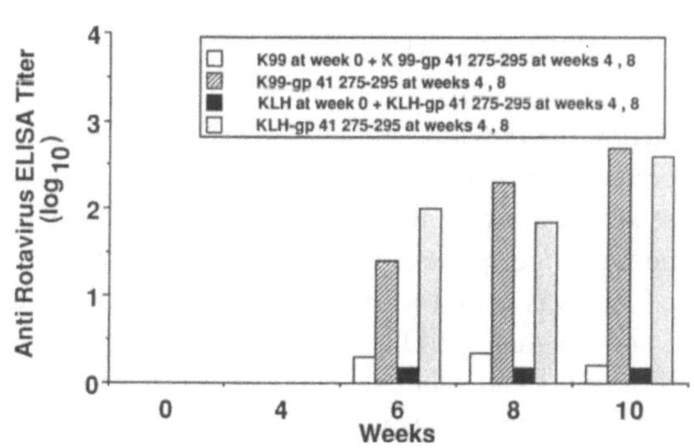

Figure 8. Carrier-induced suppression of the antirotavirus antibody response.

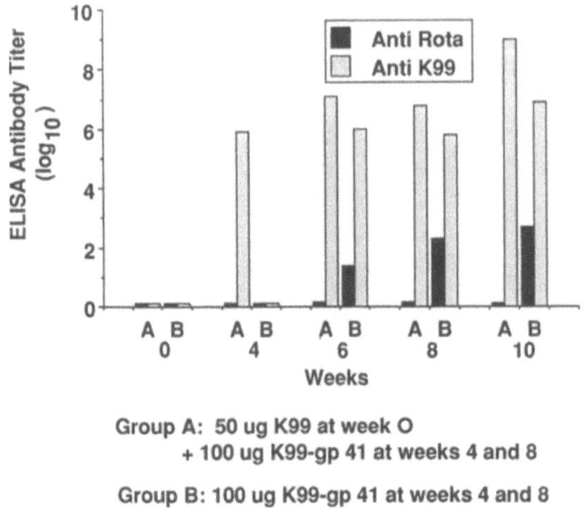

Figure 9. Antibody responses to rotavirus and carrier following carrier priming. Group A received 50 μg K99 (carrier) at week 0, then 100 μg K99 gp41 (275–295) at weeks 4 and 8. Group B received only 100 μg K99 gp41 (275–295) at weeks 4 and 8.

the use of a universal carrier for all synthetic peptides. To investigate this possibility, groups of mice were immunized with 50 μg of either K99 or KLH in FCA at week 0. Two groups were left as unimmunized controls. At weeks 4 and 8, all groups received 100 μg of the respective carrier–peptide conjugates. The antirotavirus response of the carrier-primed animals was suppressed compared with that of the controls (Fig. 8). Both carriers, K99 and KLH, produced this suppression with these peptides. This experiment was then repeated with K99 as the carrier, and the anti-K99 response was followed to investigate whether any suppression of the carrier response developed. Following the primary immunization with K99, the mice exhibited the expected anti-K99 response, which was boosted following each immunization with the peptide K99 conjugates (Fig. 9). The control group produced the expected antibody response to both rotavirus and K99 following immunizations with only the peptide K99 conjugate. Preimmunization with the carrier again effectively limited the antirotavirus response. The experiments were repeated with DDA as the adjuvant rather than Freund's complete adjuvant. Similar suppression was observed indicating the suppression is carrier related rather than adjuvant related (data not shown).

VII. PROTECTION FROM ROTAVIRUS CHALLENGE

The data generated in the previous studies showed that synthetic peptides conjugated to carriers could produce significant levels of anti-rotavirus antibodies in the serum of mice, similar to those seen following live virus innoculation. This finding encouraged us to use a rotavirus mouse model system in an attempt to protect neonates from disease following challenge. This was achieved by immunizing female mice three times before and during breeding and pregnancy. The last immunization was given 2 weeks prior to whelping. Each of the carriers and synthetic peptides was used in combination with FCA. In addition, the K99–peptide conjugates were also used with DDA. The mouse pups were allowed to suckle and challenged at 7 days of age with a bovine strain of rotavirus. The effectiveness of the immunizations was determined following challenge of the neonates in a single-blind experiment. In this model, diarrhea and morbidity is apparent within 3–5 hr following challenge. Morbidity, mortality, and severity of diarrhea were scored over a 48-hr period following the challenge (Table I).

Two of the peptide preparations, one each from the gp41 and p45 protein, provided protection equal to the virus preparation, whereas the other synthetic peptide preparations provided a lower level of protection but still substantially reduced morbidity. In this experiment, the K99–peptide conjugate did not provide as much protection as did the KLH–peptide conjugate. Further studies are currently under way to determine the reproducibility of these results as well as the effects of other carriers and adjuvants in providing protection from challenge. In addition, studies are in progress using different synthetic peptide preparations as well as investigating the effectiveness of other routes of immunizations.

The present results clearly demonstrate that synthetic peptides corresponding to regions of either gp41 or p45 of rotavirus were capable of inducing protective anti-rotavirus immunity in mice. These results support previous studies wherein peptides to other viruses have induced neutralizing antibodies *in vivo* (Brown, 1984; Shinnick *et al.*, 1983). Synthetic peptides have often been criticized as being too specific; also, slight mutations of the virus in the region in which the synthetic peptide is synthesized would reduce the effectiveness of immunity induced by the peptide. Furthermore, the specificity of the synthetic peptides often requires that different synthetic peptides be produced for different serotypes of virus. The synthetic peptide prepared and tested in the present discussion corresponds to a region that is conserved between different serotypes and appears to be an invariable region involved in viral pathogenicity. It appears plausible that this is a crucial site for maintaining virus viability, and therefore the probability of mutations at this site may be low. Thus, this type of peptide vaccine would be for the induction of immunity. In an effort to answer this question fully, we are currently

Table 1. Protection of Neonates from Rotavirus Challenge following Three Immunizations of Dams with 100 μg of Immunogen[a,b]

Immunogen	Adjuvant	Diarrhea score
KLH	FCA	4+
KLH–p45(40–60)	FCA	1+
KLH–gp41(275–295)	FCA	—
K99–p45(40–60)	FCA	1+
K99–gp41(275–295)	FCA	2+
K99–p45(40–60)	DDA	±
K99–gp41(275–295)	DDA	2+
Virus control	FCA	—

[a] DDA, dimethyldioctaldecylammonium bromide; FCA, Freund's complete adjuvant.

[b] Suckling 7-day old mice from immunized dams were challenged with bovine rotavirus; the morbidity and mortality scored no symptoms (−) or mild (1+) to severe (4+). All dams received three immunizations of 100 μg of the indicated immunogen.

attempting to determine whether mutations at this site reduce the viability or virulence of the virus.

Another criticism often levied against the use of synthetic peptides is that they are not very immunogenic on their own and must be linked to various carrier proteins. Although this is true, it appears that the size and nature of the carrier protein that is required for linkage to the peptide is not crucial. Therefore, a variety of carrier proteins may be designed that are both economical to use and easy to link the synthetic peptide. One factor that must be taken into consideration when chosing a carrier protein is that the animal should not be previously immune to that specific carrier because preexisting immunity to the carrier may suppress the ability to produce antibody responses to the peptides conjugated to that carrier (Schultze *et al.*, 1985).

We are currently investigating the immunologic mechanisms responsible for this carrier-induced suppression to synthetic peptides. If it is related to the specific carrier, it may still be possible to identify a universal carrier that does not cause immunosuppression and one that can be used for multiple synthetic peptides, sequentially. It is possible that the immunosuppression only occurs with specific carriers or peptides. If suppression is only associated with certain peptides, in the case of those specific peptides, one would have to ensure that a unique carrier to which the animal is not immune is used during the immunization protocols.

Most immunization studies in laboratory animals are conducted using FCA. Unfortunately, this adjuvant is not suitable for use in humans or animals. Therefore, it is important to identify adjuvants with equivalent activity to FCA. In most studies, it has been very difficult to find an adjuvant equivalent to FCA. The present studies have indicated that the adjuvant DDA, which is approved for veterinary use, will produce equivalent immunity to that induced by FCA. Thus, not only is it possible to induce protective immunity to synthetic peptides against rotavirus antigens, but this can be done using a commercially acceptable adjuvant.

In conclusion, we have shown that synthetic peptides conjugated to low-molecular-weight carriers, such as bacterial proteins, when used with acceptable adjuvants, will protect against challenge by the etiologic agent. However, our data suggest that one general protein carrier may not be applicable for all synthetic peptide immunizations. We suggest the use of small proteins from bacteria involved in the disease syndrome or from related syndromes, as carriers may prove to be a method that can avoid this carrier suppression while forming an effective bivalent vaccine.

REFERENCES

Bastardo, J. W., McKimm-Breschkin, J. L., Sonza, S., Mercer, L. D., and Holmes, I. H. (1981). *Infect. Immun.* **34,** 641–647.

Brown, F. (1984). *Annu. Rev. Microbiol.* **38,** 221–235.

Chen, G. M., Hung, T., Bridger, J. C., and McCrae, M. A. (1985). *Lancet* **2,** 1123.

deGraff, F. K., Klemm, P., and Gaastra, W. (1980). *Infect. Immun.* **33,** 877–883.

DeLeeuw, P. W., Ellens, D. J., Straver, P. J., van Balken, J. A. M., Moerman, A., and Baanvinger, T. (1980). *Res. Vet. Sci.* **29,** 135–141.

Dyall-Smith, M. L., and Holmes, I. H. (1984). *Nucl. Acid Res.* **12,** 3973–3982.

Emini, E. A., Jameson, B. A., and Wimmer, E. (1983). *Nature (Lond.)* **304,** 699–703.

Greenberg, H. B., Valdesuso, J., van Wyke, K., Midthun, K., Walsh, M., McAuliffe, V., Wyatt, R. G., Kalica, A. R., Flores, J., and Hoshino, Y. (1983). *J. Virol.* **47,** 267–275.
Hopp, T. P. (1984). *Mol. Immunol.* **21,** 13–16.
Jolivet, M., Audibert, F., Beach, E. H., Tartar, A., Gras-Masse, H., and Chedid, L. (1983). *Biochem. Biophys. Res. Commun.* **117,** 359–366.
Muller, G. M., Shapira, M., and Arnon, R. (1982). *Proc. Natl. Acad. Sci. USA* **79,** 569–573.
Richardson, M. A., Iwanoto, A., Ikegami, N., Nomato, A., and Furuichi, Y. (1984). *J. Virol.* **51,** 860–862.
Sabara, M., Gilchrist, J. E., Hudson, G. R., and Babiuk, L. A. (1985). *J. Virol.* **53,** 58–66.
Saif, L. J., and Smith, K. L. (1983). *Int. Symp. Neonatal Diarrhea (Saskatoon)* **4,** 394–423.
Saif, L. J., Redman, D. R., Smith, K. L., and Theil, K. W. (1983). *Infect. Immun.* **41,** 1118–1131.
Sanchez, Y., Ionescu-Matiu, I., Sparrow, J. T., Melnick, J. L., and Dreesmoth, G. R. (1982). *Intervirology* **18,** 209–213.
Schultze, M. P., LeClerc, C., Jolivet, M., Audibert, F., and Chedid, L. (1985). *J. Immunol.* **135,** 2319–2322.
Shinnick, T. M., Sutcliffe, J. G., Green, N., and Lernek, R. A. (1983). *Annu. Rev. Microbiol.* **37,** 425–446.
Sumida, T., and Taniguchi, M. (1985). *J. Immunol.* **134,** 3675–3681.
van Oirschot, J. T., and DeLeeuw, P. W. (1984). *Vet. Microbiol.* **10,** 401–408.
Woode, G. N., Bridger, J. C., Jones, J. M., Flewett, T. H., Bryden, A. S., Davies, H. A., and White, G. B. B. (1976). *Infect. Immun.* **14,** 804–810.
Woode, G. N., Kelso, N. E., Simpson, T. F., Gaul, S. K., Evans, L. E., and Babiuk, L. A. (1983). *J. Clin. Microbiol.* **18,** 358–364.

III

Quality Control, Potency, and Standardization of Viral Vaccines

12

Production and Quality-Control Testing of Virus Vaccines for Use in Humans

J. Furesz, D. W. Boucher, and G. Contreras

I. HISTORICAL BACKGROUND

Vaccination is the best means available for the prevention of viral diseases in humans. This procedure evolved from the use of variolation by the Chinese, centuries before the published work of Jenner, whose first experiment in humans consisted of transferring material from a cowpox lesion of a milkmaid to the arm of a boy (Jenner, 1798). When inoculated with smallpox virus 6 weeks later, the boy was immune to the disease. The global eradication of smallpox from 1967 to 1980 was accomplished with the use of this simply prepared vaccine: vaccinia virus propagated on calf skin. Table 1 describes the highlights in the development of virus vaccines.

The next giant step took place some 90 years later, after Jenner's ingenious discovery when Pasteur demonstrated that rabies virus could be attenuated by repeated intracerebral passage in rabbits. This virus, when partially inactivated, was subsequently shown to protect a boy severely bitten by a rabid dog (Pasteur, 1895). Pasteur's strain of fixed rabies virus became the seed virus for most subsequent vaccine preparations.

New concepts and methods were used for the development of a vaccine against yellow fever. This vaccine was established by repeated passage of the 17D virus strain, first in minced mouse embryonic tissue, then in minced whole chick embryo, and finally in minced chick embryo from which the brain and spinal cord had been removed. Following more than 160 passages in the latter type of culture, the attenuated 17D strain was shown to have lost its neurovirulence for monkeys (Theiler and Smith, 1937) and was found safe and effective in humans (Theiler, 1951). The discovery by Burnet (1940, 1941) that influenza virus could propagate in the amniotic sac and allantoic cavity of the chick embryo led to the development of inactivated influenza vaccines still in use today.

The next important breakthrough came in 1949, when Enders *et al.* (1949) demonstrated that poliovirus could multiply in non-nervous tissue of human origin. As a result of this discovery, Salk *et al.* (1953) introduced an inactivated poliovirus vaccine that had been prepared in monkey kidney cell cultures. This vaccine proved effective in reducing

---

J. Furesz, D. W. Boucher, and G. Contreras • Bureau of Biologics, Health Protection Branch, Department of National Health and Welfare, Ottawa, Ontario K1A 0L2, Canada.

Table 1. Highlights in the Development of Virus Vaccines

Highlights	Date	Vaccine
Discovery of vaccination; use of naturally existing attenuated virus strain	1798	Smallpox
Experimental attenuation *in vivo* of a pathogenic virus	1885	Rabies
Experimental attenuation *in vitro* of a pathogenic virus (tissues of mice and chick embryo)	1936	Yellow fever
Multiplication of viruses in living chick embryos	1937	Yellow fever, influenza
Multiplication of viruses in cell cultures	1949	Poliomyelitis, measles, mumps, rubella
Vaccine prepared from human blood	1971	Hepatitis B

the incidence of paralytic poliomyelitis in a large field trial directed by Francis *et al.* (1955). The use of inactivated vaccine was followed shortly by the development in primary monkey kidney cell cultures of an attenuated live oral poliovirus vaccine by Sabin (1955). This vaccine has since been used very successfully all over the world (Sabin, 1985).

The use of the new cell culture techniques led to the isolation of a host of viruses, consequently effective vaccines for measles, mumps and rubella were developed within a few years. Attempts to propagate hepatitis B virus in cell cultures failed and a new approach for the production of hepatitis B vaccine was needed. Krugman and associates (1971) demonstrated that the administration of heated human serum containing the surface antigen of hepatitis B virus conferred immunity to hepatitis B. This observation led to the development of vaccines from human blood. The surface antigen present in high concentration in the plasma of asymptomatic human carriers of the virus was harvested, purified, and inactivated (Buynak *et al.*, 1976, Barn *et al.*, 1978). Clinical studies in the United States have demonstrated that this vaccine was highly effective in preventing hepatitis B in high-risk groups (Szmuness *et al.*, 1980).

II. PROBLEMS IN PRODUCTION OF VIRUS VACCINES

Problems concerning the production and control procedures of virus vaccines present the greatest challenge to those involved in the control of biologic drugs. As knowledge in the field of virology continues to expand, the number, complexity, and cost of testing procedures have placed increasing demands on both the control agencies and the vaccine manufacturers. In order to understand the changing concepts of vaccine safety, one has to examine the history of vaccine development during the past 30 years. A major concern relates to the safety of the virus itself. If the vaccine contains killed or inactivated virus, can one be certain that the virus was completely killed? The importance of the Salk inactivated vaccine safety test for the detection of residual live poliovirus in monkeys and tissue cultures was already recognized in the 1950s. Nevertheless, the failure to detect live virus in some of the early batches of Salk vaccine, which caused paralytic poliomyelitis in some vaccine recipients (Nathanson and Langmuir, 1963), proved that larger volumes for testing and more stringent safety measures were required immediately.

There is also a critical need for ensuring the safety of the attenuated viruses used to prepare live virus vaccines. How genetically stable is the attenuated or weakened virus? The development of the concept of the seed virus system has helped establish consistent safety of live virus vaccines (Table 2). Here, a single primary seed virus lot is tested extensively to ensure that it is stable in laboratory tests and clinically safe and effective in humans. In addition, the evaluation of each production lot of live virus vaccine in the appropriate animal model provides additional evidence that the lot is similar to the accredited seed virus. In general, the number of viral passages in tissue cultures from the seed virus should be as low as possible and should not exceed five passages.

Considered no less a hazard than possible instability of the vaccine virus itself is the problem of extraneous microbial contamination of the cell substrates used in vaccine production. Such contamination can result from endogenous contamination of the primary cell cultures prepared from various mammalian and avian species or from exogneous contamination introduced during manufacture. In the past, there have been major problems in virus vaccine contamination from both endogenous and exogenous sources. Endogenous contamination resulted in the accidental inoculation of thousands of persons with simian virus 40 (SV40)-contaminated inactivated poliovirus vaccines (Mortimer *et al.*, 1981). This led to the development and introduction of many tests for detection of overt or latent endogenous viruses.

The most significant occurrence of exogenous contamination occurred during the early years of World War II in the manufacture of the live attenuated yellow fever virus vaccine. During production, human serum was added to the vaccine as a stabilizer. Hepatitis B virus was unknown at this time. Only after hepatitis had developed in several thousand vaccine recipients was it recognized that the human serum added to the vaccine during manufacture did in fact contain a human virus capable of causing hepatitis (Findlay and Martin, 1943; Sawyer *et al.*, 1944*a,b*). As a lesson from this unfortunate experience, human serum is no longer used in biologics manufacture.

Table 2. Scheme for Production of Virus Vaccines

Preparation of seed virus
Preparation of cell substrate primary cell cultures, cell bank—diploid, heteroploid
Embryonated eggs
Virus inoculation
Incubation
Virus harvest
Pooling, clarification
Live virus vaccine
Blending, stabilizing (bulk)
Filling
Inactivated virus vaccine
Concentration, purification, ultracentrifugation, ultrafiltration, chromatography, immunoadsorption, enzyme treatment
Inactivation
Blending, stabilizing (bulk)
Filling

Other exogenous contamination (e.g., with bacteria, *Mycoplasma,* or bacteriophage)—present often in the bovine serum used as a nutrient for the growth of cell cultures in vaccine manufacture—is also of concern but does not constitute a major problem.

Once the cell cultures—primary cells or continuous lines—are seeded and propagated in numerous small-size containers or large fermentation tanks, each step in the vaccine production is carried out under strictly controlled conditions (Table 2). Following virus inoculation, the cells are incubated at optimal temperatures to obtain high yield of uniform virus suspensions. Virus fluids are harvested one or more times, pooled, and clarified from cell debris by filtration or light centrifugation. While live virus vaccines are blended and stabilized with various chemicals and are ready for filling into final containers, inactivated vaccines must be concentrated and purified either before or after inactivation. Ultracentrifugation, ultrafiltration, chromatography, immunoadsorption, and enzyme treatment are the methods of choice for this purpose.

III. QUALITY-CONTROL TESTING OF VIRUS VACCINES

Before a new virus vaccine is approved for general use by the National Control Authority, the manufacturer must comply with a number of requirements (Table 3). The principles of these requirements are (1) to establish the consistency of manufacture by producing five consecutive lots with approved standard operational procedures for each manufacturing step; (2) to ensure the safety of the product (freedom from endogenous and exogenous contaminants) by testing in animals and cell cultures; (3) to provide evidence of purity (chemical and physical) and potency (*in vitro* with serologic assays and *in vivo* in cell cultures and animals) of the vaccine; (4) to document the lack of clinical adverse reactions (short and long term) in humans; and (5) to demonstrate the efficacy of the product and the duration of protective immunity in the vaccine recipients.

Table 3. Virus Vaccine Requirements[a]

Requirement	Test or observation
Establishment of consistency in production	Standard manufacturing and quality-control procedures, production of five consecutive lots
Safety testing in animals	Adult and suckling mice, guinea pigs, rabbits, monkeys, embryonated eggs
Safety testing in cell cultures	Primary, human diploid, heteroploid
Purity testing	Chemical, physical
Potency testing *in vitro*	HA, CF, ELISA, RIA, SRID
Potency testing *in vivo*	Infectivity in cell cultures, Antigenicity in animals (mice, monkeys, chicks, rats)
Clinical studies	Adverse reactions (local, systemic, short and long term) Antibody response Efficacy

[a]HA, hemagglutination; CF, complement fixation; ELISA, enzyme-linked immunosorbent assay; RIA, radioimmunoassay; SRID, single-radial immunodiffusion.

A. Quality-Control Tests on Primary Cell Cultures and Continuous Cell Lines

Originally the only freely available tissue for the production of inactivated poliovirus vaccine was monkey kidney. It was soon apparent that the tissues from these animals harbored simian RNA and DNA viruses, and about one-third of the tissues prepared from the kidneys of monkeys had to be discarded for this reason. One attempt at reducing the number of contaminating agents was to establish a mandatory quarantine of wild caught monkeys for a period of at least 6 weeks before their use in the country in which the vaccine was manufactured (Table 4). The introduction of various tests for the detection of extraneous agents rendered the quality-control testing of these vaccines cumbersome and expensive. Nevertheless, these tests improved vaccine safety considerably. For this purpose, 25% of the cell cultures obtained from each animal were used as control cultures, i.e., they were uninoculated and observed microscopically for viral cytopathic effects. These cultures were monitored for 1–2 weeks after virus harvest. Culture fluids were collected at various intervals from these control cells and were inoculated into additional new cultures of different cell types and observed for 2 weeks for viral cytopathic effect and the presence of hemadsorbing viruses (Table 4).

During the past 20 years, there have been two major developments in the use of cell substrates. The first development involved the use of closed colonies of animals (e.g., chickens and rabbits) freed from specific pathogens bred under isolated conditions. Cells from such colonies have been tested extensively for the presence of contaminating viruses; to date, none has been found. Similarly, kidneys from monkeys born in captivity or taken within a short time before expected birth have also been shown to be free from contaminating agents. The second development was the establishment of a cell bank prepared by the passage of fibroblast cells growing from human foetal lung tissue (Hayflick, 1961; Jacobs, 1971). Such cell banks stored in liquid nitrogen have been shown by extensive tests to be diploid, nontumorigenic, and free from contaminants. They have a finite life and have been used successfully over the past 20 years for the production of millions of doses of vaccine.

The obvious advantages of the use of well-characterized and thoroughly tested continuous diploid and heteroploid cell lines have been well established during the past 25 years. The manufacturer is required to prepare a working cell bank from such cell substrates and to test these cells for freedom of extraneous agents, tumorigenicity *in vitro*, and in animals and chromosomal characterization (Table 5). The tests for extraneous agents are performed in a variety of animals and cell cultures. The tests for tumorigenicity are a combination of sensitive *in vitro* and *in vivo* methods for detecting tumorigenic

Table 4. Quality-Control Tests on Primary Cell Cultures

Quarantine of animals
Monitoring of cell cultures for extraneous agents
Observation of uninoculated control cells
Subculturing of fluids onto various cell types
Test for hemadsorbing viruses

Table 5. Quality-Control Tests on Diploid and Heteroploid Cell Cultures

Working cell bank
Tests for extraneous agents
Adult and suckling mice, guinea pigs, rabbits
Embryonated eggs
Various types of cell cultures
Tumorigenicity
In vitro
Chick embryo skin cultures
Human muscle organ cultures
Anchorage independence in soft agar
In vivo
Nude (nu/nu) mice
Newborn antithymocyte-globulin-treated rats
Chromosomal characterization and monitoring
Cell cultures used in vaccine production
Tests for extraneous agents
Observation of uninoculated control cells
Subculturing of fluids onto various cell types
Test for hemadsorbing viruses
Identity test of cells

potential of the cellular substrates (Furesz *et al.*, 1985*b*). The newborn rat treated with antithymocyte globulin has been shown to be the most sensitive host for tumorigenic and metastatic potential of heteroploid cell cultures (van Steenis and van Wezel, 1982; Johnson *et al.*, 1982; Bather *et al.*, 1985; Furesz *et al.*, 1985*a;* Contreras *et al.*, 1985). The extensive testing of the working cell bank considerably reduces the number of tests to be conducted on cell cultures used in vaccine production and on the virus harvests as well.

B. Quality-Control Tests on Virus Harvest, Bulk Vaccine, and Final Product

Single virus harvests are tested for extraneous agents in tissue cultures and animals, for microbial sterility, for virus identity, and for potency (Table 6). The tests for extraneous agents are similar to those described for cell substrates. Following the pooling of virus harvests and the addition of a stabilizer, the bulk vaccine prior to filling is also tested for virus identity, microbial sterility, and potency. Live virus vaccines are tested for markers either in tissue culture assays or in animals to ensure that the vaccine virus has not undergone genetic changes during its several cycles of multiplication in vaccine manufacture. Inactivated bulk vaccines are tested for the absence of infectious virus in tissue cultures and/or animals to ensure that the virus inactivation procedure was complete and that no live virus remained. For this purpose, the testing of large volumes of the inactivated bulk vaccine is required. Inactivated vaccines are also assayed for purity (protein nitrogen content).

Tests on final products (Table 7) include identity, microbial sterility, potency, gener-

Table 6. Quality-Control Tests on Virus Harvests and Bulk Vaccines

Virus harvests
Tests for extraneous agents
Animals
Cell cultures
Embryonated eggs
Tests for bacteria, fungi, and *Mycoplasma*
Tests for *M. tuberculosis*
Virus identity
Potency
Bulk (pool of virus harvests)
Virus identity
Tests for bacteria and fungi
Potency
Genetic marker tests (for live virus vaccines)
Absence of live virus (for inactivated virus vaccines)
Purity (for inactivated virus vaccines)

al safety in guinea pigs and mice, pyrogenicity in rabbits, content of added substances (preservatives, adjuvants, inactivating agents, stabilizers), and moisture for freeze-dried products. The general safety test serves for the detection of chemical or microbial contaminants that would include acute toxic reaction for guinea pigs and mice. Long-term stability assays are required for the validation of the expiration date of the product.

IV. VACCINES IN CURRENT USE

An ideal vaccine should confer long-lasting, preferably lifelong, protection against the disease. It should be inexpensive enough for large-scale use and stable enough to remain potent during storage and shipping and should have no adverse effect on the recipient. Most virus vaccines do meet these criteria; the routine administration of live and inactivated virus vaccines play an important role in the control of many infectious diseases. A brief overview of the status of virus vaccines in current use is presented in Table 8.

Table 7. Quality-Control Tests on Final Product

Identity	Added substances
Tests for bacteria and fungi	Preservatives
Potency	Adjuvants
General safety in animals	Inactivating agents
Guinea pigs	Stabilizers
Mice	Moisture (freeze-dried products)
Pyrogenicity	Stability

Table 8. Vaccines in Current Use

Virus	Substrate (cells)
Live (attenuated) virus	
Polio (Sabin)	Monkey kidney, human diploid
Measles	Chick embryo
Mumps	Chick embryo
Rubella	Human diploid
Yellow fever	Chick embryo[a]
Adeno	Human diploid
Killed (inactivated) virus	
Polio (Salk)	Monkey kidney, human diploid, Vero
Rabies	Human diploid, hamster kidney, Vero
Hepatitis B	Human plasma
Influenza	Chick embryo[a]
Japanese B encephalitis	Chick embryo, hamster kidney
Tickborne encephalitis	Chick embryo

[a] Embryonated eggs.

A. Live Virus Vaccines

Of the live virus vaccines (in addition to smallpox vaccine used in the past), Sabin's trivalent oral poliovirus vaccine has been the most extensively administered in many countries. The vaccine has proved safe and very effective in the elimination of poliomyelitis from several countries (Sabin, 1985). Current vaccines are produced either in monkey kidney or in human diploid cell cultures.

Immunization with the attenuated live measles, mumps, and rubella viruses propagated in chick embryo or human diploid cell cultures resulted in a dramatic decline of these diseases and in an apparently long-lasting immunity of the vaccine recipients (Weibel *et al.*, 1980). In some countries, these vaccines are combined in a single preparation; in others, they are used singly. One of the greatest concerns of measles infection is involvement of the central nervous system (CNS), i.e., acute encephalitis, and persistence of the virus in the brain, which can lead to subacute sclerosing panencephalitis (SSPE). In countries in which mass vaccination against measles has been successful, the incidence of SSPE has declined markedly (CDC, 1982, 1985).

Because yellow fever vaccine has a record of being one of the safest and most effective, most manufacturers have failed to change their production method to take advantage of modern manufacturing processes. Nonetheless, some production problems should be addressed. The vaccine available in many countries is made from embryonated chick eggs contaminated with avian leukosis viruses. The stability of some vaccines was found not satisfactory. The menace of widespread yellow fever epidemics persists, generating a need for larger emergency stocks of a more stable vaccine. Successful attempts to improve the production and stability of yellow fever vaccine were recently reported (Georges *et al.*, 1985). It is anticipated that all vaccines will be prepared from chick embryos of specific pathogen-free flocks, and the 17D vaccine virus will be propagated in chick embryo cell cultures.

The use of adenovirus vaccine appears to be primarily indicated for the protection of closed populations, such as military recruits, since the incidence of adenovirus disease is generally low in the general population. The combined use of live oral adenovirus vaccines (types 4 and 7), prepared in human diploid cells, was shown to be highly effective in U.S. military personnel in interrupting epidemic acute respiratory disease as well as in preventing epidemics caused by these types of adenoviruses (Top, 1975). In view of the good results obtained in recent clinical studies conducted in Canadian military recruits (Chaloner-Larsson *et al.*, 1986), the Canadian Armed Forces introduced a routine immunization program with live oral adenovirus vaccines (types 4 and 7) among military personnel.

B. Inactivated Virus Vaccines

Technology developed during the past decade led to the production of inactivated poliovirus vaccines of high purity and potency. Replacement of primary monkey kidney cells by subcultured monkey kidney and human diploid or Vero cells for virus propagation and the introduction of the microcarrier culture technique have made cell and virus cultivation in large fermentors feasible at economically acceptable costs (van Wezel *et al.*, 1978). The use of gel filtration and ion-exchange chromatography resulted in a highly purified and potent vaccine. Clinical studies demonstrated that two doses of this vaccine stimulated antibody production in all immunized infants. The frequency and titer of the antibody response of recipients was similar in widely different geographic regions, including both developed and developing countries (Salk *et al.*, 1982; Stoeckel *et al.*, 1984).

Classic rabies vaccines are prepared from the brains of infected animals. Although effective, these vaccines are generally of low or variable potency and are usually administered to exposed subjects in a large number of doses. The high frequency of neurologic complications associated with these vaccines requires that vaccines containing neuronal tissue be replaced as soon as possible by cell-culture vaccines. Inactivated vaccines prepared from virus grown in human diploid fibroblast cells (Wiktor *et al.*, 1964; Bahmanyar *et al.*, 1976) and in certain other cell cultures of nonhuman origin, such as hamster kidney (Fenje, 1960) and dog kidney (van Wezel and van Steenis, 1978), stimulate antibody production efficiently and are very effective in preexposure and postexposure treatments. However, the cost of such cell-culture vaccines is high, and the recent development of less expensive vaccines prepared in continuous cell lines, such as Vero cells, is promising (Fournier *et al.*, 1985).

The safety, immunogenicity, and high protective efficacy of hepatitis B vaccines prepared from the plasma of asymptomatic human carriers have been described. Current influenza virus vaccines contain influenza A and B virus strains that are propagated in embryonated chick eggs and are available as whole virus or as split-virus vaccines. Influenza vaccines are recommended to persons at high risk. Type A vaccines confer protection in approximately 70% of recipients for 1 year after vaccine administration, if the prevalent strain remains the same or has only minor antigenic drift; however, a history of vaccination within a year does not preclude the need to revaccinate for optimal protection (CDC, 1985).

Inactivated vaccines against Japanese B encephalitis prepared from mouse brain or in chick embryo cell cultures have been used extensively in endemic regions with an average protective efficacy of 80% (Takaku *et al.*, 1968; Kanamitsu *et al.*, 1970). Controlled trials of the hamster kidney vaccine showed an efficacy of 76–96% (Darwish *et al.*, 1967). A live attenuated vaccine also prepared in hamster kidney is being tested in China with some success (Huang, 1982).

Inactivated tickborne encephalitis vaccine produced in cell culture is used in some countries (Heinz *et al.*, 1980; Kunz *et al.*, 1980). In the U.S.S.R., immunization of populations in high-risk endemic areas has been practiced for many years, using a vaccine of mouse brain (Smorodintsev, 1958). Second-generation inactivated tickborne encephalitis vaccines have been developed in Austria and in the U.S.S.R., using purified chick embryo cell culture-grown virus (Heinz *et al.*, 1980). Immunogenicity trials have shown seroconversion of 85–100% of antibody-negative patients after a single subcutaneous dose (Kunz *et al.*, 1980).

V. PROSPECTS FOR NEW VIRUS VACCINES

During the past few years, several new virus vaccines were prepared with conventional methods; i.e., viruses were propagated in cell cultures and used either as live attenuated or inactivated vaccines (Table 9). Some of these vaccines are in the developmental stage; others are being evaluated in clinical studies.

The successful propagation of hepatitis A virus in diploid cell cultures of rhesus monkey origin resulted in a live attenuated vaccine that demonstrated protection of marmosets and chimpanzees against challenge with wild hepatitis A virus (Provost *et al.*, 1982, 1983).

Table 9. Possible Future Virus Vaccines

Vaccines produced in cell cultures
Hepatitis A
Varicella
Herpes simplex
Cytomegalovirus
Epstein–Barr
Respiratory syncytial
Parainfluenza
Rota
Dengue
Yellow fever
Venezuelan equine encephalitis
Recombinant and/or synthetic peptide vaccines
Hepatitis A
Hepatitis B
Hepatitis non-A, non-B
HTLV-III/LAV
Polio
Rabies
Influenza
Herpes simplex

Ten years of experience with a live attenuated varicella vaccine developed by Takahashi and associates in 1974 in Japan has shown that the vaccine is safe. Data concerning the protective efficacy in normal and leukemic children have been encouraging (Gershon, 1985).

As a consequence of the growing appreciation of the pathogenic role of cytomegalovirus (CMV), there have been two major efforts at developing a live virus vaccine (Elek and Stern, 1974; Plotkin *et al.*, 1976). These efforts have yielded mixed results following studies in animal models and clinical trials, mainly in renal transplant patients (Osborn, 1981; Plotkin *et al.*, 1984; Sachs *et al.*, 1984). Concern still exists regarding the possible latency and genetic instability of the vaccine virus in recipients. A subunit CMV vaccine, free of nucleic acid, might be more acceptable for human vaccination, and efforts are being made to purify the virus envelope antigen (Hudecz *et al.*, 1985).

Attempts have been made in the past to prevent reactivation of herpes simplex virus infection in humans with the administration of multiple doses of an inactivated subunit vaccine prepared in rabbit kidney or chick embryo cells (Hilleman *et al.*, 1981). Although the subunit vaccine is well tolerated and induces antibody and cell-mediated immune responses in human subjects (Mertz *et al.*, 1984), it has not proved beneficial in the treatment of recurrent genital herpes (Hilleman, 1985).

A membrane antigen (gp340) of Epstein-Barr virus produced in marmoset lymphoblastoid cells (North *et al.*, 1982) was tested in cottontop tamarins in attempts to develop protection against Epstein-Barr virus (EBV)-associated malignant disease. The gp340-purified antigen in liposomes included neutralizing antibodies in animals. In preliminary trials, these tamarins resisted a challenge with the virus (Epstein *et al.*, 1985).

Attempts to produce an effective inactivated vaccine with respiratory syncytial virus (RSV) and parainfluenza virus have been largely unsuccessful because (1) immunity to these viruses is probably correlated with secretory antibody, and (2) early inactivated vaccines induced high levels of circulating but no secretory antibody (Fulginiti *et al.*, 1969; Chin *et al.*, 1969). In addition, some of the recipients of the inactivated vaccine, when subsequently infected with wild RSV, developed a severe illness complicated with bronchiolitis (Kim *et al.*, 1969). A recent double-blind placebo-controlled field trial in children with a live attenuated experimental RSV vaccine showed no efficacy in providing protection against subsequent infection (Belshe *et al.*, 1982).

Human and animal rotaviruses have been found to share a common group antigen (Wyatt *et al.*, 1975; Vesikari *et al.*, 1983); in addition, mucosal immunity of the intestinal tract plays a major role in providing protection (Kapikian *et al.*, 1980). These studies led to the development of live attenuated virus vaccines. Two such experimental vaccines have been evaluated in humans. One contained an attenuated bovine rotavirus, strain RIT-4237 (Vesikari *et al.*, 1983) that has been administered orally to more than 500 infants; a protection rate of 80% was demonstrated against rotavirus diarrhea (Vesikari, 1985). In the other vaccine, an attenuated rhesus monkey rotavirus (MMU 18006) was used. The results of a limited clinical study are promising (Losonsky *et al.*, 1986).

The development of dengue vaccines is hampered by the poor growth of the virus in cell cultures, by the lack of an animal model, and by the potential immunopathologic consequences of immunization. Attempts to develop attenuated live dengue virus vaccines are under way in several laboratories. Results of initial clinical studies have shown that immunity to yellow fever is associated with a greater seroconversion rate in recipients of the attenuated dengue-2 vaccine (Scott *et al.*, 1983; Bancroft *et al.*, 1984). An experimental attenuated vaccine for dengue virus type 4, tested in five volunteers immune to yellow fever virus, was considered unacceptable because of lack of satisfactory attenuation (Eckels *et al.*, 1984).

An attenuated vaccine to Venezuelan equine encephalitis was developed to immunize horses and was successfully used in limiting an epizootic in 1971; it has also been used to immunize laboratory workers and military personnel (Edelman *et al.*, 1979). Attempts have been made to improve this vaccine for humans (Trent *et al.*, 1979; Mecham and Trent, 1982). An inactivated vaccine made from the same vaccine virus strain (TC83) proved effective in humans and produced minimal side effects (Cole *et al.*, 1974; Edelman *et al.*, 1979).

The rapid growth of our knowledge in the fields of monoclonal antibody research, nucleic acid sequencing, and capsid protein analysis will result in new approaches in virus vaccine technology. The information obtained for many viruses using monoclonal antibodies has contributed to the identification of the critical protein sites required for effective artificially designed immunogens. Monoclonal antibodies have also been helpful in investigating the potential use of antiidiotypic antibodies as viral vaccines. Anti-idiotypic antibodies are prepared against monoclonal antibodies in an attempt to generate a population of internal image antibodies, which would mimic the epitope on the native viral antigen to which the monoclonal antibody is directed (Jerne, 1974). Induction of antiviral immune response by anti-idiotypic antibodies has been investigated for various viruses and has shown positive results in animal models (Gurish *et al.*, 1985). For instance, anti-idiotypic antibodies prepared in rabbits inoculated with mouse monoclonal antibodies specific for rabies virus glycoprotein were able to induce rabies neutralizing antibodies when injected in mice (Reagan *et al.*, 1983).

The increasing availability of complete or partial nucleic acid sequences for a variety of viruses, such as poliomyelitis (Racaniello and Baltimore, 1981; Nomoto *et al.*, 1982), herpes simplex (Watson *et al.*, 1982), hepatitis B (Valenzuela *et al.*, 1979), HTLV-III (Wain-Hobson *et al.*, 1985; Ratner *et al.*, 1985; Muesing *et al.*, 1985), and yellow fever (Rice *et al.*, 1985), will result in a better understanding of the immunogens probably responsible for the protective efficacy of vaccines. The knowledge of virus nucleic acids is complemented by recently obtained information at the atomic level of the spatial configurations and relationships of the capsid proteins of picornaviruses (Hogle *et al.*, 1985; Rossman *et al.*, 1985). These recent advances will contribute further in identifying not only sequences but spatial configurations and interactions of the virus polypeptides essential for generating the immune response.

Recombinant DNA technology can be used to modify host cell genomes (*Escherichia coli,* yeast, mammalian cells) by the insertion of genes in suitable vectors that may permit large-scale production by the host cell of selected proteins or oligopeptides for use as vaccines. This approach has led to the large-scale production of hepatitis B surface antigen in yeast (Valenzuela *et al.*, 1979, 1982; Scolnick, 1984). Expression of hepatitis B proteins in mammalian (human, simian, and rodent) cells has also been achieved (Macnab *et al.*, 1976; McAleer *et al.*, 1983; Liu *et al.*, 1982; Crowley *et al.*, 1983; Dubois *et al.*, 1980; Patzer *et al.*, 1984). In addition to hepatitis B, encouraging results have been obtained by cloning the genomes of influenza, rabies, and polio viruses (Smith *et al.*, 1983; Lecocq *et al.*, 1985; Racaniello and Baltimore, 1981; van der Werf *et al.*, 1981).

Synthetic peptide immunogens overcome many of the problems encountered in recombinant DNA technology because they do not require the propagation of the pathogenic organism and are chemically pure and stable under average storage conditions. For instance, it has been shown that specific synthetic peptides copying the amino acid sequence of selected antigenic sites of the capsid proteins induce a neutralizing immune response against polio, hepatitis A and herpes simplex viruses (Jameson *et al.*, 1985; Emini *et al.*, 1985; Dietzschold *et al.*, 1985). The key question remains whether a synthetic peptide vaccine can confer immunity. For this purpose, the peptide needs to be incorporated into a safe and effective adjuvant suitable for use in humans.

Vaccinia virus is being proposed as a vector of multiple immunogens (Paoletti *et al.*, 1985). Rabbits inoculated with recombinant vaccinia virus containing sequences for HB_sAg, HSV glycoprotein D, and influenza virus hemagglutinin produced antibodies to all three foreign antigens (Perkus *et al.*, 1985). The low cost and ease of administration of such vaccine could be important factors in determining the success of immunization programs in the developing world (Beale, 1983). However, some difficult problems remain to be solved before this type of vector can be used in humans. During the smallpox global erradication campaign, a high proportion of the world population was immunized with vaccinia virus; the immunity acquired to vaccinia virus could restrict the growth of the vector, consequently limiting the immunogenicity of its cloned foreign antigens. A second major problem is the occurrence of adverse reactions due to vaccinia virus in the recipients. Primary immunization of children under the age of 12 months with vaccinia virus used to carry the risk of five deaths per million; other severe and more frequent complications (e.g., high fever, encephalitis, generalized vaccinia infection) were observed in recipients of all ages, as well as an incidence of accidental vaccinia infection as high as 53 in 100,000 following primovaccination was reported (CDC, 1971). It is obvious that further research in molecular genetics is required to manipulate the vaccinia virus to modify its virulence in the human host or to select other animal viruses as vectors that are nonpathogenic for humans.

Despite the many excellent conventional vaccines currently available, it is difficult to escape the conclusion that vaccines consisting of individual proteins or peptides or attenuated viruses of defined nucleic acid composition will have replaced them within the next decade. Indeed, the increasing availability of nucleic acid sequences and spatial protein configurations for a variety of viruses means that a better understanding of the factors responsible for effective vaccines is now attainable.

VI. CONCLUSION

The past 30 years can be described as the renaissance of virology and the birth of modern virus vaccine technology. The development of complex procedures of vaccine production led to the need for new approaches to the quality-control testing of these products. A number of stringent tests were introduced (1) to ensure the safety of the cell substrate and the vaccine during manufacture, and (2) to verify in the laboratory that each batch of the product has the appropriate characteristics. The excellent safety and efficacy record of the new vaccines in the field has amply proved that the technology of production and quality control was sound and effective.

In view of the ever-expanding array of new biologic drugs other than virus vaccines, one has to reassess the criteria that would be reasonable in establishing the acceptability of various cell substrates and in developing appropriate testing procedures. Based on scientific data generated from modern purification technology and excellent biologic tracer studies during manufacture, the following products have been approved as biologic drugs for clinical use: interferon prepared in human lymphoblastoid cells, hybridoma-derived monoclonal antibodies, and plasminogen activator produced in Chinese hamster ovary (CHO) cells. Highly purified inactivated virus vaccines (polio, rabies, hepatitis B) prepared in heteroploid continuous cell cultures are now under clinical investigation.

In the coming years, the rapid technologic development in virus vaccine manufacture will require a great deal of flexibility in terms of regulatory requirements for appropriate production and testing procedures. From past experience, we have learned to keep the door open to a variety of options and not to hamper the regulatory control of biologic drugs with regulations that might become obsolete before they could be published. Notwithstanding the continuous emphasis on *in vitro* and *in vivo* laboratory testing of the new products, the ultimate assessment of their safety will be judged by their use in humans. For this purpose, short-term clinical trials will have to be supplemented by long-term clinical studies organized and financed by industry, university, and government. A combination of good science, common sense, and a responsible regulatory approach will assist industry in the development of new products and in protecting the public from unsafe drugs.

REFERENCES

Bahmanyar, M., Fayaz, A., Nour-Saheli, S., Mohammadi, M., and Koprowski, H. (1976). *JAMA* **236,** 2751–2754.

Bancroft, W. H., Scott, R. McN., Eckels, K. H., Hoke, C. H., Jr., Simms, T. E., Jesrani, K. T. D., Summers, P. L., Dubois, D. R., Tsoulos, D., and Russel, P. K. (1984). *J. Infect. Dis.* **149,** 1005–1010.

Barin, F., Goudreau, A., Coursaget, P., and Maupas, P. (1978). *Ann. Microbiol. (Paris)* **129B,** 87–100.

Bather, R., Becker, B. C., Contreras, G., and Furesz, J. (1985). *J. Biol. Standard.* **13,** 13–22.

Beale, J. (1983). *Nature (Lond.)* **302,** 476–477.

Belshe, R. B., Van Voris, L. P., and Mufson, M. A. (1982). *J. Infect. Dis.* **145,** 311–319.

Burnet, F. M. (1940). *Aust. J. Exp. Biol. Med. Sci.* **18,** 353–360.

Burnet, F. M. (1941). *Aust. J. Exp. Biol. Med. Sci.* **19,** 291–292.

Centers for Disease Control (1971). *MMWR* **20,** 339–345.

Centers for Disease Control (1982). *Measles Surv. Rep.* **11,** 1977–1981.

Centers for Disease Control (1985*a*). *MMWR* **34,** 366–370.

Centers for Disease Control (1985*b*). *MMWR* **34,** 261–275.

Chaloner-Larsson, G., Contreras, G., Furesz, J., Boucher, D. W., Krepps, D., Humphreys, G. R., and Mohanna, S. M. (1986). *Can. J. Publ. Health* **77,** 367–370.

Chin, J., Magoffin, R. L., Shearer, L. A., Schieble, J. H., and Lennette, E. H. (1969). *Am. J. Epidemiol.* **89,** 449–463.

Cole, F. E., Jr., May, S. W., and Eddy, G. A. (1974). *Appl. Microbiol.* **2,** 150–153.

Contreras, G., Bather, R., Furesz, J., and Becker, B. C. (1985). *In Vitro Cell. Dev. Biol.* **21,** 649–652.

Crowley, C. W., Liu, C.-C., and Levinson, A. D. (1983). *Mol. Cell. Biol.* **3,** 44–55.

Darwish, M. A., Hammon, W. McD., and Sather, G. E. (1967). *Am. J. Trop. Med.* **16,** 364–370.

Dietzschold, B., Heber-Katz, H., Hudecz, F., Hollosi, M., Fasman, G., Eisenberg, R. J., and Cohen, G. H. (1985). In *Vaccines 85* (R. A. Lerner, R. M. Chanock, F. Brown, eds.), pp. 227–234. Cold Spring Harbor Laboratory, Cold Spring Harbor, New York.

Dubois, M. F., Pourcel, C., Rousset, S., Chany, C., and Tiollais, P. (1980). *Proc. Natl. Acad. Sci. USA* **77,** 4549–4553.

Eckels, K. H., Scott, R. McN., Bancroft, W. H., Brown, J., Dubois, D. R., Summers, P. L., Russell, P. K., and Halstead, S. B. (1984). *Am. J. Med. Hyg.* **33,** 684–689.

Edelman, R., Ascher, M. S., Oster, C. N., Ramsburg, H. H., Cole, F. E., and Eddy, G. A. (1979). *J. Infect. Dis.* **140,** 708–715.

Emini, E., Berger, J., Hughes, J. V., Mitra, S. W., and Linemeyer, D. L. (1985). In *Vaccines 85* (R. A. Lerner, R. M. Chanock, F. Brown, eds.), pp. 217–220, Cold Spring Harbor Laboratory, Cold Spring Harbor, New York.

Enders, J. F., Weller, T. H., and Robbins, F. C. (1949). *Science* **109,** 85–87.

Epstein, M. A., Morgan, A. J., Finerty, S., Randle, B. J., and Kirkwood, J. K. (1985). *Nature (Lond.)* **318,** 287–289.

Fenje, P. (1960). *Can. J. Microbiol.* **6,** 605–610.

Findlay, G. M., and Martin, H. N. (1943). *Lancet* **1,** 678–680.

Fournier, P., Montagnon, B., Vincet-Falquet, J. C., Ajjan, N., Drucker, J., and Roumiantzeff, M. (1985). In *Improvements in Rabies Post-exposure Treatment* (I. Vodopija, K. G. Nicholson, S. Smerdel, U. Bijok, eds.), pp. 115–121, Liber University Press, Zagreb, Yugoslavia.

Francis, T., Jr., Napier, J. A., Voight, R. B., Hemphill, F. M., Wenner, H. A., Korns, R. F., Boisen, M., Tolchinsky, E., and Diamond, E. L. (1957). In *Poliomyelitis Vaccine Evaluation Center* pp. i–xiv; 1–63. University of Michigan, Ann Arbor, Michigan.

Fulginiti, V. A., Eller, J. J., Seiber, O. F., Jayner, J. W., Minamitani, M., and Meiklejohn, G. (1969). *Am. J. Epidemiol.* **89,** 435–448.

Furesz, J., Armstrong, R. E., Contreras, G., and Wachmann, B. (1984). *Rev. Infect. Dis.* **6**(Suppl. 6), S528–S534.

Furesz, J., Bather, R., Contreras, G., and Becker, B. C. (1985*a*). *In Vitro Cell. Dev. Biol.* **6,** 57–69.

Furesz, J., Levenbook, I., Contreras, G., Elisberg, B., Bather, R., and Petricciani, J. (1985*b*). *In Vitro Cell. Dev. Biol.* **6,** 70–73.

Georges, A. G., Tible, F., Meunier, D. M. Y., Gonzales, J.-P., Beraud, A. M., Sissoko-Dybdahl, N.-R., Abdul-Wahid, S., Fritzell, B., Girard, M., and Georges-Courbot, M.-C. (1985). *Vaccine* **3,** 313–315.

Gershon, A. A. (1985). *J. Infect. Dis.* **152,** 859–862.

Gurish, M. F., and Nisonoff, A. (1985). In *High-Technology Route to Virus Vaccines* (G. R. Dressman, J. G. Bronson, R. C. Kennedy, eds.), pp. 103–108, American Society for Microbiology, Washington, D.C.

Hayflick, L., and Moorhead, P. S. (1961). *Exp. Cell Res.* **25,** 585–621.

Heinz, F. X., Kung, C., and Fauma, H. (1980). *J. Med. Virol.* **6,** 213–221.

Hilleman, M. R., Larson, V. M., Lehman, E. D., Salerno, R. A., Conrad, P. G., and McLean, A. A. (1981). In *The Human Herpesviruses. An Interdisciplinary Perspective* (W. R. Dowdle, A. J. Nahmias, R. F. Schinazi, eds.), pp. 503–506, Elsevier, New York.

Hilleman, M. R. (1985). *J. Infect. Dis.* **151,** 407–419.

Hogle, J. M., Chow, M., and Filmon, D. J. (1985). *Science* **229,** 1358–1365.

Hollinger, F. B., Troisi, C. L., and Pepe, P. E. (1986). *J. Infect. Dis.* **153,** 156–159.

Huang, C. H. (1982). *Adv. Virus Res.* **27,** 71–101.

Hudecz, F., Gonczol, E.. and Plotkin, S. (1985). *Vaccine* **3,** 300–304.

Jacobs, J. P., Jones, C. M., and Baille, J. P. (1970). *Nature (Lond.)* **227,** 168–170.

Jameson, B. A., Bonin, J., Murray, M. G., Wimmer, E., and Kew, O. (1985). In *Vaccines 85* (R. A. Lerner, R. M. Chanock, and F. Brown, eds.), pp. 191–198. Cold Spring Harbor Laboratory, Cold Spring Harbor, New York.

Jenner, E. (1798). Pamphlet, Vol. 4232, Army Medical Library, Washington, D. C.

Jerne, N. K. (1974). *Ann. Immunol. (Paris)* **125C,** 373–389.

Kanamitsu, M., Hashimoto, N., Urasawa, S., Katsurada, M., and Kimura, H. (1970). *Biken J.* **13,** 313–328.

Kapikian, A. Z., Wyatt, R. G., Greenberg, H. B. Kalica, A. R., Kim, H. W., Brandt, C. D., Rodriguez, W. J., Parrott, R. M., and Chanock, R. M. (1980). *Rev. Infect. Dis.* **2,** 459–469.

Kaplan, G., Lubinski, J., Dasgupta, A., and Racaniello, V. R. (1985). *Proc. Natl. Acad. Sci. USA* **82,** 8424–8428.

Kim, H. W., Canchola, J. G., Brandt, C. D., Pyles, G., Chanock, R. M., Jensen, K., and Parrott, R. H. (1969). *Am. J. Epidemiol.* **89,** 422–434.

Krugman, S., Giles, J. P., and Hammond, J. (1971). *JAMA* **217,** 41–45.

Kunz, C., Heinz, F. X., and Hofman, H. (1980). *J. Med. Virol.* **6,** 103–109.

Lecocq, J. P., Kieny, M. P., and Lemoine, Y. (1985). In *World's Debt to Pasteur* (H. Koprowski and S. A. Plotkin, eds.), pp. 259–271, Liss, New York.

Liu, C.-C., Yansura, D., and Levinson, A. D. (1982). *DNA* **1,** 213–222.

Losonsky, G. A., Rennels, M. B., Kapikian, A. Z., Midthun, K., Ferra, P. J., Fortier, D. N., Hoffmann, K. M., Baig, A., and Levine, M. M. (1986). *Pediatr. Infect. Dis.* **5,** 25–29.

Macnab, G. M., Alexander, J. J., Lecatsas, G., Bey, E. M., and Ubranocvicz, J. M. (1976). *Br. J. Cancer* **24,** 509–515.

McAleer, W. J., Markus, H. Z., and Bailey, J. F. (1983). *J. VIrol. Methods* **7,** 263–271.

Mecham, J. O., and Trent, D. W. (1982). *J. Gen. Virol.* **63,** 121–129.

Mertz, G. J., Peterman, G., Ashley, R., Jourden, J. L., Salter, D., Morrison, L., McLean, A., and Corey, L. (1984). *J. Infect. Dis.* **150,** 242–249.

Montagnon, B. J., Fanget, B., and Vincet-Falquet, J. C. (1984). *Rev. Infect. Dis.* **6**(Suppl. 2), S341–S344.

Mortimer, E. A., Lepow, M. L., Gold, E., Robbins, F. C., Burton, G. J., and Fraumeni, J. F. (1981). *N. Engl. J. Med.* **305,** 1517–1518.

Muesing, M. A., Smith, D. H., Cabradilla, C. D., Benton, C. V., Lasky, L. A., and Capon, D. J. (1985). *Nature (Lond.)* **313,** 450–458.

Nathanson, N., and Langmuir, A. D. (1963). *Am. J. Hyg.* **78,** 16–81.

Nomoto, A., Omata, T., Toyoda, H., Kuge, H., Horie, Y., Kataoke, Y., Genba, Y., Nakano, Y., and Imura, N. (1982). *Proc. Natl. Acad. Sci. USA* **79,** 5793–5797.

North, J. A., Morgan, A. J., Thompson, J. L., and Epstein, M. A. (1982). *J. Virol. Methods* **5,** 55–65.

Osborn, J. E. (1981). *J. Infect. Dis.* **143,** 618–630.

Paoletti, E., Perkus, M. E., Piccini, A., Wos, S. M., Lipinskas, B. R., and Mercer, S. R. (1985). In *Vaccines 85* (R. A. Lerner, R. M. Chanock, and F. Brown, eds.), pp. 147–150, Cold Spring Harbor Laboratory, Cold Spring Harbor, New York.

Pasteur, L. (1895). *C. R. Hebd. Seances Acad. Sci.* **101,** 765–767.

Patzer, E. J., Nakamura, G. R., and Yaffe, A. (1984). *J. Virol.* **51,** 346–353.

Perkus, M. E., Piccini, A., Lipinskas, R. R., and Paoletti, E. (1985). *Science* **229,** 981–984.

Plotkin, S. A., Friedman, H. M., Fleisher, G. R., Dafoe, D. C., Grossman, R. A., Smiley, M. L., Starr, S. E., Wlodaver, C., Friedman, A. D., and Barker, C. F. (1984). *Lancet* **1,** 528–530.

Provost, P. J., Baker, F. S., Giesa, P. A., McAleer, W. J., Buynak, E. B., and Hilleman, M. R. (1982). *Proc. Soc. Exp. Biol. Med.* **170,** 8–14.

Provost, P. J., Conti, P. A., Giesa, P. A., Baker, F. S., Buynak, E. B., McAleer, W. J., and Hilleman, M. R. (1983). *Proc. Soc. Exp. Biol. Med.* **172,** 357–363.

Racaniello, V. R., and Baltimore, D. (1981). *Proc. Natl. Acad. Sci. USA* **78,** 4887–4891.

Ratner, L., Haseltine, W., Patarca, R., Livak, J. L., Starcich, B., Josephs, S. F., Doran, E. R., Rafalski, J. A., Whitehorn, E. A., Baumeister, K., Ivanoff, L., Petteway, S. R., Jr., Pearson, M. L., Lautenberger, J. A., Papas, T. S., Ghrayeb, J., Chang, N. T., Gallo, R. C., and Wong-Staal, F. (1985). *Nature (Lond.)* **313,** 277–284.

Reagan, K. J., Wunner, W. H., Wiktor, T. J., and Koprowski, H. (1983). *J. Virol.* **48,** 660–666.

Rice, C. M., Lenches, E. M., Eddy, S. R., Shin, S. J., Sheets, R. L., and Strauss, J. H. (1985). *Science* **229,** 726–733.

Rossman, M. G., Arnold, E., Erickson, J. W., Frankenberger, E. A., Griffith, J. P., Hecht, H-J., Johnson, J. E., Kamer, G., Luo, M., Mosser, A. G., Rueckert, R. R., Sherry, B., and Vriend, G. (1985). *Nature (Lond.)* **317,** 145–153.

Sabin, A. B. (1955). *Ann. NY Acad. Sci.* **61,** 924–938.

Sabin, A. B. (1985). *J. Infect. Dis.* **151,** 420–436.

Sachs, G. W., Simmons, R. L., and Balfour, H. H., Jr. (1984). *Vaccine* **2,** 215–218.

Salk, J. E. (1953). *JAMA* **151,** 1081–1089.

Salk, J. E., Stoechel, P., van Wezel, A. L., Lapinleimu, K., and van Steenis, G. (1982). *Ann. Clin. Res.* **14,** 204–212.

Sawyer, W. A., Meyer, K. F., Eaton, M. D., Bauer, J. H., Putnam, P., and Schwentker, F. F. (1944*a*). *Am. J. Hyg.* **39,** 337–340.

Sawyer, W. A., Meyer, K. F., Eaton, M. D., Bauer, J. H., Putnam, P., and Schwentker, F. F. (1944*b*). *Am. J. Hyg.* **40,** 35–107.

Scolnick, E. M., McLean, A. A., West, D. J., McAleer, W. J., Miller, W. J., and Buynak, E. B. (1984). *JAMA* **251,** 2812–2815.
Scott, R. McN., Eckels, K. H., Bancroft, W. H., Summers, P. L., McCown, J. M., Anderson, J. H., and Russell, P. K. (1983). *J. Infect. Dis.* **148,** 1055–1060.
Smith, G. L., Murphy, B. R., and Moss, B. (1983). *Proc. Natl. Acad. Sci USA* **80,** 7155–7159.
Smorodintsev, A. A. (1958). *Prog. Med. Virol.* **1,** 210–248.
Stoeckel, P., Schulmberger, M., Parent, G., Maire, B., van Wezel, A., van Steenis, G., Evans, A., and Salk, D. (1984). *Rev. Infect. Dis.* **6**(Suppl. 2), S463–S466.
Szmuness, W., Stevens, C. E., Harley, E. J., Zang, E. A., Oleszko, W. R., William, D. C., Sadovsky, R., Morrison, J. M., and Kellner, A. (1980). *N. Engl. J. Med.* **303,** 833–841.
Takahashi, M., Otsuka, T., Okuno, Y., Asano, Y., Yazaki, T., and Isomura, S. (1974). *Lancet* **2,** 1288–1290.
Takaku, K., Yamashita, T., Osnai, T., Yoshida, I., Kato, M., Goda, H., Takagi, M., Hirota, T., Amano, T., Fukai, K., Kunita, S., Inoue, K., Shoji, K., Igarashi, A., and Ito, T. (1968). *Biken J.* **11,** 25–39.
Theiler, M. (1951). In *Yellow Fever* (G. K. Strode, ed.), pp. 39–136, McGraw-Hill, New York.
Theiler, M., and Smith, H. H. (1937). *J. Exp. Med.* **65,** 787–800.
Top, F. H. (1975). *Yale J. Biol. Med.* **48,** 185–195.
Trent, D. W., Clewley, J. P., France, J. K., and Bishop, D. H. L. (1979). *J. Gen. Virol.* **43,** 365–381.
Valenzuela, P., Gray, P., Quiroga, M., Zaldivar, J., Goodman, H. M., and Rutter, W. J. (1979). *Nature (Lond.)* **280,** 815–819.
Valenzuela, P., Medina, A., Rutter, W. J., Amerer, G., and Hall, B. D. (1982). *Nature (Lond.)* **298,** 347–350.
van der Werf, S., Bregegere, F., Kopecka, H., Kitamura, N., Rothberg, P. G., Kourilsky, P., Wimmer, E., and Girard, M. (1981). *Proc. Natl. Acad. Sci. USA* **80,** 7155–7159.
van Wezel, A. L., van Steenis, G., Hannick, C. A., and Cohen, H. (1978). *Dev. Biol. Stand.* **41,** 159–168.
van Wezel, A. L., and van Steenis, G. (1978). *Dev. Biol. Stand.* **40,** 69–75.
Vesikari, T., Isolauri, E., Delem, A., D'Hondt, E., André, F. E., and Zissis, G. (1983). *Lancet* **2,** 807–811.
Vesikari, T. (1985). *Pediatr. Infect. Dis.* **4,** 612–614.
Wain-Hobson, S., Sonigo, P., Danos, O., Cole, S., and Alizon, M. (1985). *Cell* **40,** 9–17.
Watson, R. J., Weiss, J. H., Salstrom, J. S., and Enquist, L. W. (1982). *Science* **218,** 381–384.
Weibel, R. E., Buynak, E. B., McLean, A. A., Roehm, R. R., and Hilleman, M. R. (1980). *Proc. Soc. Exp. Biol. Med.* **165,** 260–263.
Wiktor, T. J., Fernandez, M. V., and Koprowski, H. (1964). *J. Immunol.* **93,** 353–366.
Wiktor, T. J., Macfarlan, R. I., Reagan, K. J., Dietzschold, B., Curtis, P. J., Wunner, W. H., Kien, M-P., Lathe, R., Lecocq, J-P., Mackett, M., Moss, B., and Koprowski, H. (1984). *Proc. Natl. Acad. Sci. USA* **81,** 7194–7198.
Wyatt, R. G., Mebus, C. A., Yolken, R. H., Kalica, A. R., James, Jr., H. D., Kapikian, A. Z., and Chanock, R. M. (1979). *Science* **203,** 548–550.

13

Potency Assay of Inactivated Influenza and Poliovaccine

Applications of Single-Radial Immunodiffusion

J. M. Wood, G. C. Schild, P. D. Minor, D. I. Magrath, Jennifer Mumford, and R. G. Webster

I. INTRODUCTION

Influenza virus naturally infects humans, pigs, horses, seals, and many different wild and domestic avian species (Hinshaw and Webster, 1982). Inactivated vaccines against influenza are used in humans and in horses and experimental vaccines have been studied in swine and domestic poultry. Vaccination in humans has been shown to induce serum antibody to influenza hemagglutinin (HA), which is capable of virus neutralization and prevention of infection (Dowdle *et al.*, 1974; Hobson *et al.*, 1979). It is therefore important to measure vaccine HA antigen concentration in order to standardize the dose required for effective immunization. Several years ago, single-radial diffusion (SRD) techniques were developed for *in vitro* potency testing of human influenza vaccines (Wood *et al.*, 1977). The SRD test was shown to be technically simple, reproducible, and sensitive and produced vaccine potency data that related to antigenicity of vaccines (Wood *et al.*, 1977).

Potency tests based on SRD have also been developed for inactivated polio vaccine (IPV) (Schild *et al.*, 1980). Vaccines contain each of the three poliovirus serotypes (1, 2, and 3), and each serotype is represented by two major populations of poliovirus particles (D and C) that can be separated antigenically and by sedimentation on sucrose gradients (Mayer *et al.*, 1957; Minor *et al.*, 1980). Clinical trials in children have shown that poliovirus D antigen concentration is directly related to IPV immunogenicity (Salk *et al.*, 1982); therefore, potency tests of IPV should be capable of quantifying D antigen. This

J. M. Wood, P. D. Minor, and D. I. Magrath • Division of Virology, National Institute for Biological Standards and Control, Potters Bar, Herts. EN6 3QG, England. *G. C. Schild* • National Institute for Biological Standards and Control, Potters Bar, Herts. EN6 3QG, England. *Jennifer Mumford* • Equine Virology Unit, Animal Health Trust, Lanwades Park, Kennett, Newmarket, Suffolk CB8 7DW, England. *R. G. Webster* • Department of Virology and Molecular Biology, St. Jude Children's Research Hospital, Memphis, Tennessee 38101.

chapter reviews the application of SRD to standardization of human and veterinary influenza vaccines and discusses *in vitro* potency testing of IPV.

II. HUMAN INFLUENZA VACCINES

The SRD test is based on the reaction between influenza HA and specific antiserum to HA in an agarose gel (Wood *et al.*, 1977). Figure 1 illustrates an SRD test of A/Victoria/3/75 (H3N2) HA. Antiserum to purified A/Vic/75 HA (Brand and Skehel, 1972) was incorporated in the agarose; after disruption with 1% (w/v) Zwittergent 3-14 detergent (Calbiochem. Behring Corp.), vaccine dilutions were added to wells in the agarose. The size of the resulting precipitation ring, seen after Coomassie blue staining, was proportional to the amount of HA antigen in the vaccine that could be calibrated in micrograms by the use of appropriate reference antigens.

The SRD test is influenza subtype specific so that each component in a trivalent vaccine can be assayed separately. Stable freeze-dried reference reagents (calibrated antigen and specific anti-HA serum) are available from the National Institute for Biological Standards and Control (NIBSC) in the United Kingdom and the Office of Biologics, Research and Review, Food and Drug Administration (FDA), in the United States for each influenza virus strain included in vaccines. In practice, the SRD test is simple to perform and is highly reproducible (coefficient of variation 8%), as was demonstrated by an international collaborative study (Wood *et al.*, 1981).

Clinical trials have been carried out to establish the relationship between vaccine HA concentration and serologic responses in humans. In one such study (Nicholson *et al.*, 1979), volunteers were immunized with graded doses (5–94 μg HA) of A/USSR/92/77 (H1N1) whole virus and subunit vaccine. The responses of young seronegative adults to whole virus vaccine are illustrated in Fig. 2. There was a good correlation between hemagglutination-inhibition (HI) antibody and vaccine HA concentration after both one- and two-vaccine doses (Fig. 2). In this and similar studies (Potter *et al.*, 1980), it has been established that two doses of vaccine containing 8–15 μg of HA stimulate protective levels of antibody (HI titers of $\geq 1:40$) in most seronegative vaccinees, whereas only one such dose is required for antigenically primed vaccinees. It is important to establish the protective dose of vaccine for each type of vaccine and for each new virus strain included in vaccines.

III. EQUINE INFLUENZA VACCINES

Epizootics caused by equine influenza viruses occur periodically in many countries and may have serious financial consequences in thoroughbred training stables (Simpson, 1973). Inactivated vaccines are widely available, and vaccination is now mandatory in some countries. There is evidence, however, that vaccination does not protect against recently circulating influenza viruses (Klingeborn *et al.*, 1980; Hinshaw *et al.*, 1983). Two of the possibilities for lack of vaccine efficacy are that (1) vaccines contain antigenically outdated virus strains, and (2) vaccine potency is inadequate. Conventionally, equine vaccine potency is measured by hemagglutination tests and by tests in laboratory

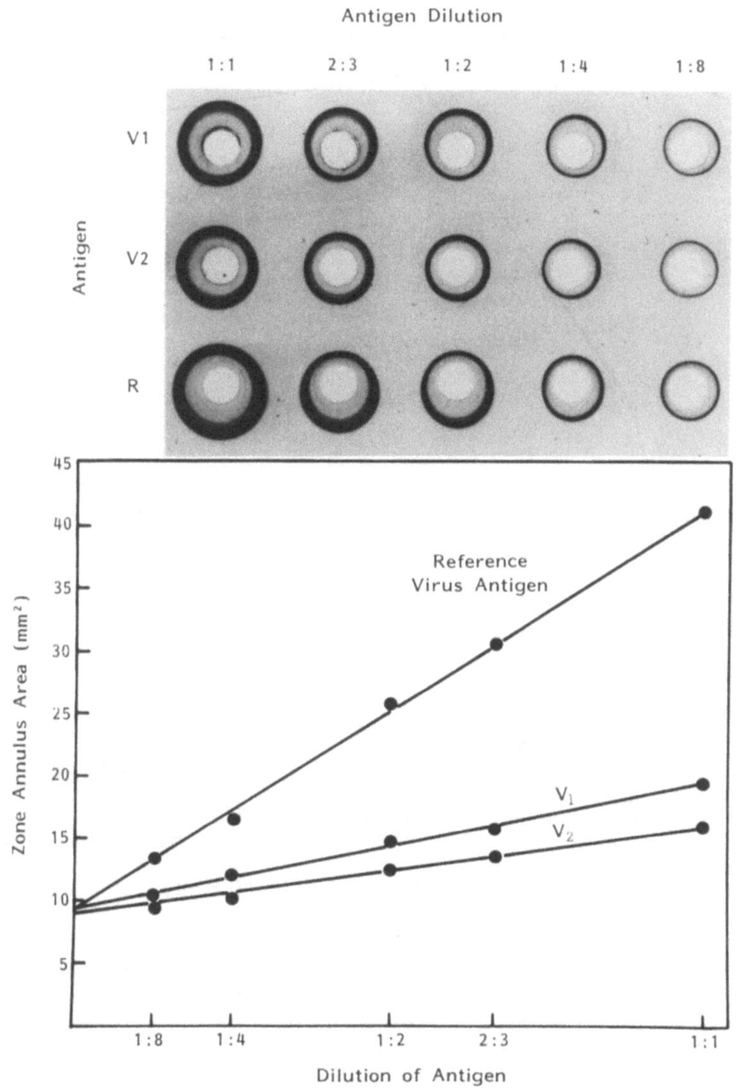

Figure 1. Single-radiodiffusion (SRD) potency assay of inactivated influenza vaccine. Vaccines (V_1 and V_2) and calibrated reference antigen (R_1, 80 μg HA/ml) prepared from A/Victoria/3/75 (H3N2) virus were treated with Zwittergent 3-14 detergent, dilutions were prepared, and 20-μl volumes applied to wells in an agarose gel containing sheep antiserum to A/Victoria/3/75 HA. SRD zones and resulting dose–response curves are illustrated. Each antigen showed a linear dose response with responses meeting at a common intercept. The HA concentration of V_1 and V_2 is calculated by comparing dose–response slopes with the slope of *R*.

animals. Experience with human influenza vaccine has demonstrated the inadequacies of the hemagglutination test (Schild *et al.*, 1975); recently, SRD techniques have been applied to standardization of equine influenza vaccines (Wood *et al.*, 1983*a*). A series of bivalent vaccines containing graded doses of β-propiolactone-inactivated A/equine/Prague/56 (H7N7) and A/equine/Miami/63 (H3N8) viruses was prepared for studies in

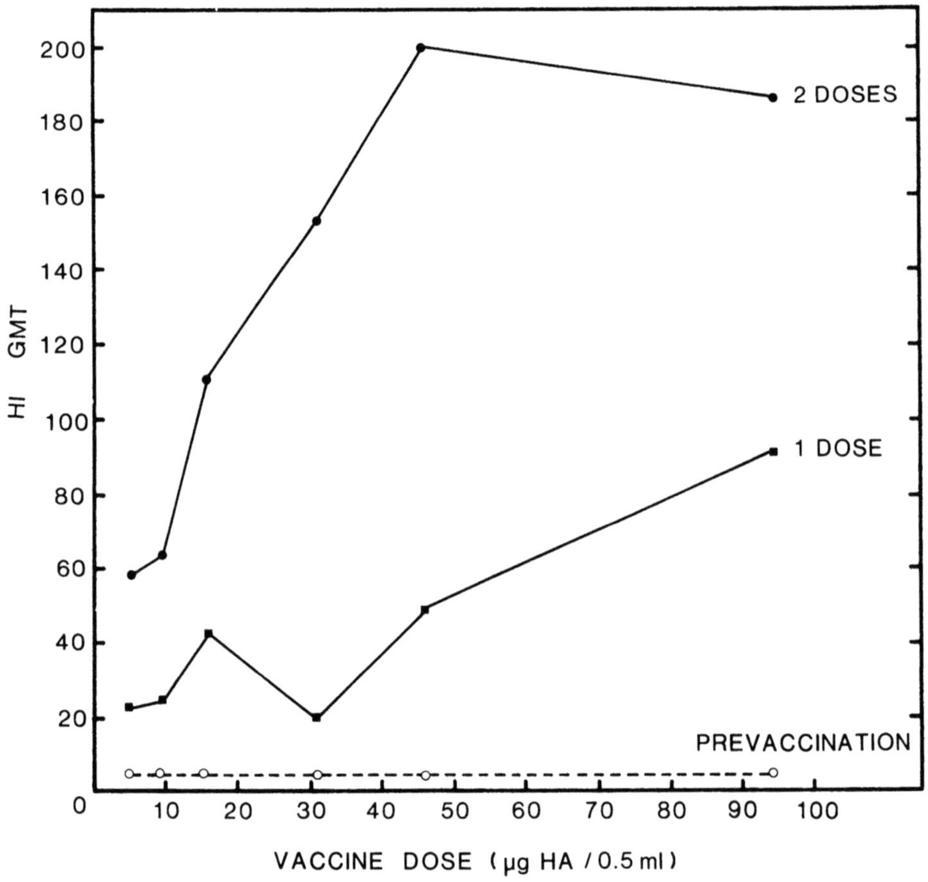

Figure 2. Serologic responses of young adults (aged 12–15 years) to graded doses of β-propiolactone-inactivated A/USSR/92/77 (H1N1) aqueous whole virus vaccine. Vaccine was given by deep subcutaneous injection (0.5 ml) at 4-week intervals and serum samples taken 4 weeks after the first and second vaccine doses were tested for hemagglutination-inhibition (HI) antibody to A/USSR/77 virus. The 175 vaccinees were initially seronegative. After one dose the dose-related antibody response was shallow, and after two doses the dose-related response was much steeper.

ponies (Wood *et al.*, 1983*b*). Three aqueous vaccines and one adjuvant-treated vaccine containing 4–56 μg of HA per strain per dose were used to immunize seronegative ponies and the single-radial hemolysis (SRH) serologic responses to A/equine/Prague/56 virus are shown in Fig. 3. There were vaccine dose-related antibody responses after both one and two doses of aqueous vaccine, whereas the adjuvant-treated vaccine containing two- and four-fold less antigen than the most potent aqueous vaccine produced highest antibody levels. Although the SRH antibody levels were high, there was a rapid decline in antibody, so that by 18 weeks after the second vaccine dose, serum antibody had fallen to low or undetectable levels. Similar data were obtained for antibody responses to A/equine/Miami/63 virus. The ponies were subsequently challenged with a representative of recent equine H3N8 viruses, A/equine/Newmarket/79, and protection was assessed by virus excretion, febrile responses, and antibody responses (Mumford *et al.*, 1983).

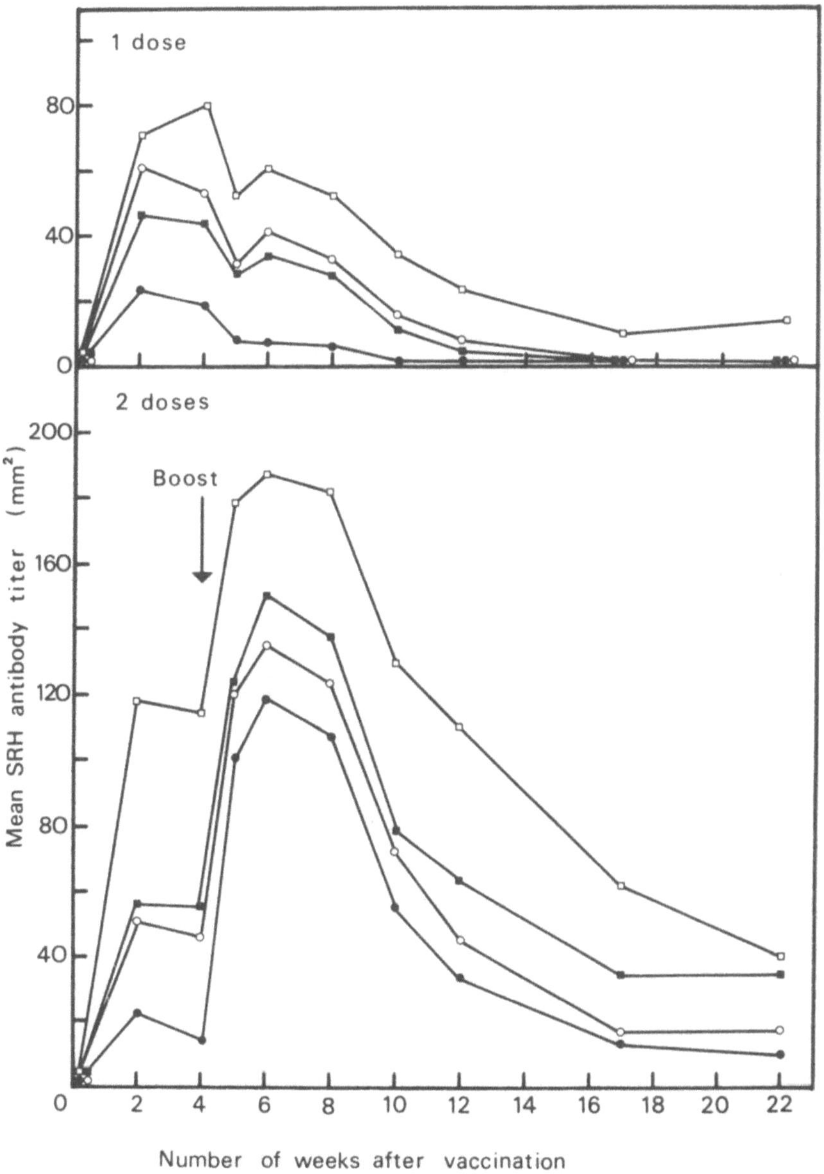

Figure 3. Postvaccination single-radial hemolysis (SRH) responses of seronegative yearling ponies to A/equine/Prague/56 virus. Ponies were vaccinated with either one or two doses (4 weeks apart) of β-propiolactone-inactivated bivalent aqueous vaccine containing A/equine/Miami/63 and A/equine/Prague/56 viruses. Responses to aqueous vaccines containing 5 (●), 15 (○), and 50 (■) μg hemagglutinin (HA) per dose and adjuvanted vaccine containing 15 μg HA per dose (□) are illustrated.

Table 1. Protection of Vaccinated Ponies against Challenge with A/Equine/Newmarket/79 Virus

Pony group		Temperature responses[b]		Virus excretion		SRH antibody responses to New/79	
Description	Prechallenge New/79 SRH antibody (mm^2)	No. of significant responses (%)	Mean duration (days)	No. of excreting virus (%)	Mean duration (days)	No. of significant responses (%)[c]	Mean titer (mm^2)
Unvaccinated	<4[a]	6/6 (100)[d]	3.0	6/6 (100)	5.0	6/6 (100)	62
Vaccinated[e]	<4	22/22 (92)	3.5	23/24 (96)	3.3	24/24 (100)	140
Vaccinated[e]	≥4	10/16 (62)	2.1	14/16 (87)	2.1	14/16 (87)	108

[a] Minimum detectable single-radial hemolysis (SRH) zone size: 4 mm^2.
[b] Significant temperature response is an increase in temperature to 38.9°C or above.
[c] Significant antibody response is an increase of 50% in SRH zone area.
[d] Number of ponies responding/number of ponies in group.
[e] Aqueous vaccine.

The results of the challenge (Table 1) demonstrated that protection correlated with prechallenge antibody levels. These studies indicate that equine influenza vaccine potency measured by SRD is relevant to the protective efficacy of vaccines.

IV. AVIAN INFLUENZA VACCINES

Avian influenza is a recurrent problem in domestic turkeys, particularly in North America (Lang *et al.*, 1968) but causes only sporadic outbreaks of disease in chickens. The majority of influenza viruses isolated from domestic poultry are of low virulence when examined experimentally; however, highly pathogenic viruses periodically cause severe outbreaks of disease with accompanying high economic losses (Alexander, 1985). There has been limited use of inactivated influenza vaccine in the turkey-raising areas of Minnesota (Poss *et al.*, 1982); experimentally, poultry influenza vaccines have been shown to stimulate protective antibody against pathogenic influenza viruses in both turkeys and chickens (Allan *et al.*, 1971; Brugh *et al.*, 1979). The recent lethal influenza outbreak in domestic poultry in Pennsylvania, which was caused by an H5N2 virus, A/chick/Penn/1370/83 (Bean *et al.*, 1985), provided an ideal opportunity for assessing standardization and efficacy of avian influenza vaccines.

In previous studies, vaccine potency was assessed by hemagglutination tests or by serologic tests in animals (Price, 1982), both of which are highly variable tests. In the H5N2 vaccine study (Wood *et al.*, 1985), five β-propiolactone-inactivated vaccines prepared from a duck H5N2 virus, A/duck/New York/189/82, were standardized by SRD to contain graded amounts of HA, ranging from 0.03 to 2.7 μg of HA per dose (Table 2). After addition of Freund's complete adjuvant (FCA), the vaccines were used to immunize seronegative chickens. One vaccine does stimulated good HI antibody responses related to

Table 2. Protection of Vaccinated Chickens against Challenge with A/Chick/Penn/83 Virus[a]

Vaccine HA[a] concn. (μg/dose)	Postvaccination HI antibody GMT[b]	Postchallenge responses: Virus isolation				
		Tracheal	Rectal	Disease signs	HI antibody[c]: GMT	No. of responses[d]
0.03	<10	5/5[e]	1/5	1/5	1470	5/5
0.1	<10	5/5	2/5	5/5 (2 dead)	508	3/3
0.3	92	1/5	0/5	0/5	3880	5/5
0.9	139	0/5	0/5	0/5	807	3/5
2.7	1470	0/5	0/5	0/5	1280	0/5

[a] GMT, geometric mean titer; HA, hemagglutinin; HI, hemagglutination-inhibition.
[b] Vaccine with Freund's complete adjuvant.
[c] Twenty-one days after vaccination.
[d] Fourteen days after challenge.
[e] Significant response in chickens with titers of <10 is taken as ≥40. For other chickens, a significant response is a four-fold increase in titer.
[f] Number of chickens responding/number in group.

vaccine HA concentration. Twenty-one days after vaccination, 70% of chickens receiving $\geq$ 0.3 μg HA possessed HI antibody titers $\geq$ 1:40. After challenge with lethal A/chick/Penn/1370/83 virus, all unvaccinated control chickens showed severe disease signs and died, whereas chickens vaccinated with 2.7 μg of HA were protected against infection and disease. This study demonstrates that vaccination against lethal avian influenza is effective and that SRD tests can be used to standardize vaccine potency.

V. INACTIVATED POLIO VACCINE

The *in vivo* potency tests of inactivated poliovaccine (IPV) in widespread use are based on stimulation of virus neutralizing antibody in guinea pigs, chickens (European Pharmacopeia), or monkeys (U.S. Pharmacopeia). Although IPVs are required to pass such tests, the tests are not sufficiently sensitive to detect small differences in potency (van Steenis *et al.*, 1981; Moynihan and Petersen, 1981) and are influenced by vaccine strain differences (Lapinleimu and Stenvick, 1981; Magrath *et al.*, 1986), and results from different laboratories are often not comparable (D. I. Magrath, unpublished data). Consequently, there is a need to develop alternative potency tests; the following is a description of *in vitro* D antigen potency tests based on SRD.

A. SRD Tests

Intact poliovirus particles are capable of diffusing through agarose gels and reacting with antisera to produce SRD zones (Fig. 4). The SRD reactions are poliovirus type specific, and the intensity of stained zones is greater for homologous than for heterologous antigens. Studies with ^{35}S-labeled poliovirus (Schild *et al.*, 1980) have demonstrated that SRD zones are produced only by D particles when the antiserum is against unfractionated virus. The poliovirus SRD test is capable of quantifying D antigen in concentrated virus preparations (Schild *et al.*, 1980); however, there is insufficient sensitivity for SRD potency tests of IPV, and more sensitive test systems are needed.

B. Zone-Enhancement Tests

As reported by Minor *et al.* (1980), SRD reaction zones can also be seen by autoradiography when ^{35}S-labeled poliovirus is added to SRD immunoplates. The use of labeled poliovirus in SRD permits the use of much lower antiserum concentrations (approximately 100-fold) than in conventional tests, which has the effect of increasing SRD sensitivity (Schild *et al.*, 1980). Studies were carried out to see what effect there was on the autoradiographic zones, when unlabeled homotypic poliovirus was added to labeled poliovirus D antigen (Fig. 5).

The mixtures were tested in immunoplates containing concentrations of serum suitable for use in direct SRD tests, so that stained and autoradiographic zones could be detected on the same immunoplate. Where stained and autoradiographic zones were of identical size, this indicated that the unlabeled antigen had enhanced the size of the radioactive zone produced by the labeled D antigen—a positive zone-enhancement (ZE) response. Where the size of an autoradiographic zone surrounding a well containing both

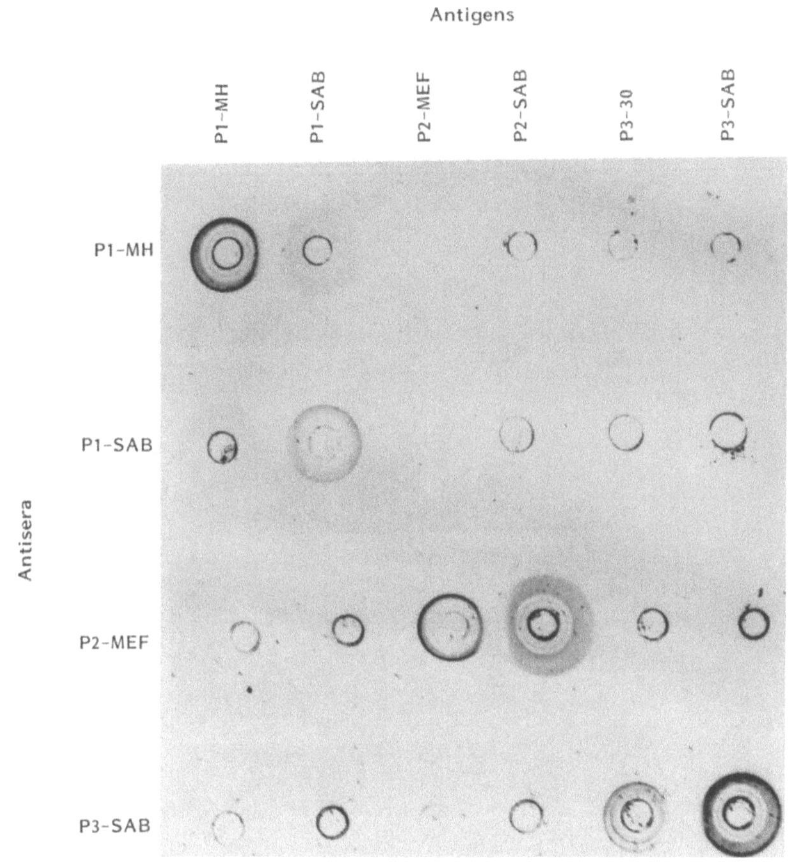

Figure 4. Specificity of direct single-radial diffusion (SRD) reactions of poliovirus antigens. Concentrates of poliovirus type 1 (Mahoney and Sabin strains), type 2 (MEF and Sabin), and type 3 (30 and Sabin) were added to wells in immunoplates containing immune rabbit sera to one of the following unfractionated viruses: poliovirus type 1, Mahoney (1.25 μl/ml gel); type 1, Sabin (6.5 μl/ml gel); type 2 MEF (14.3 μl/ml gel); and type 3, Sabin (11.1 μlp12l/ml gel). The reaction zones were stained with Coomassie blue.

reagents was the same as that produced by labeled virus alone, this indicated that co-migration of the two antigens did not occur; i.e., the unlabeled antigen had produced no ZE response.

It was concluded from these and other investigations that the conditions required for maximum ZE response are: (1) unlabeled virus antigenically homologous to the antiserum eg. P3-Sabin in Fig. 5, (2) labeled virus antigenically homologous to unlabeled virus. When neither of these conditions applies, the degree of enhancement varies according to the antigenic similarities of the three reagents. For example, in Fig. 5, P3-30-labeled D antigen was fully enhanced by the antigenically distinct P3-3/10 unlabeled virus, where the antiserum (anti-P3-Sabin) was heterologous to both antigens, yet unlabeled P3-Saukett antigen (IPV vaccine strain, antigenically distinct from P3-3/10 virus), and anti-P3-Sabin serum, gave only partial enhancement to the labeled virus. This illustrates the care

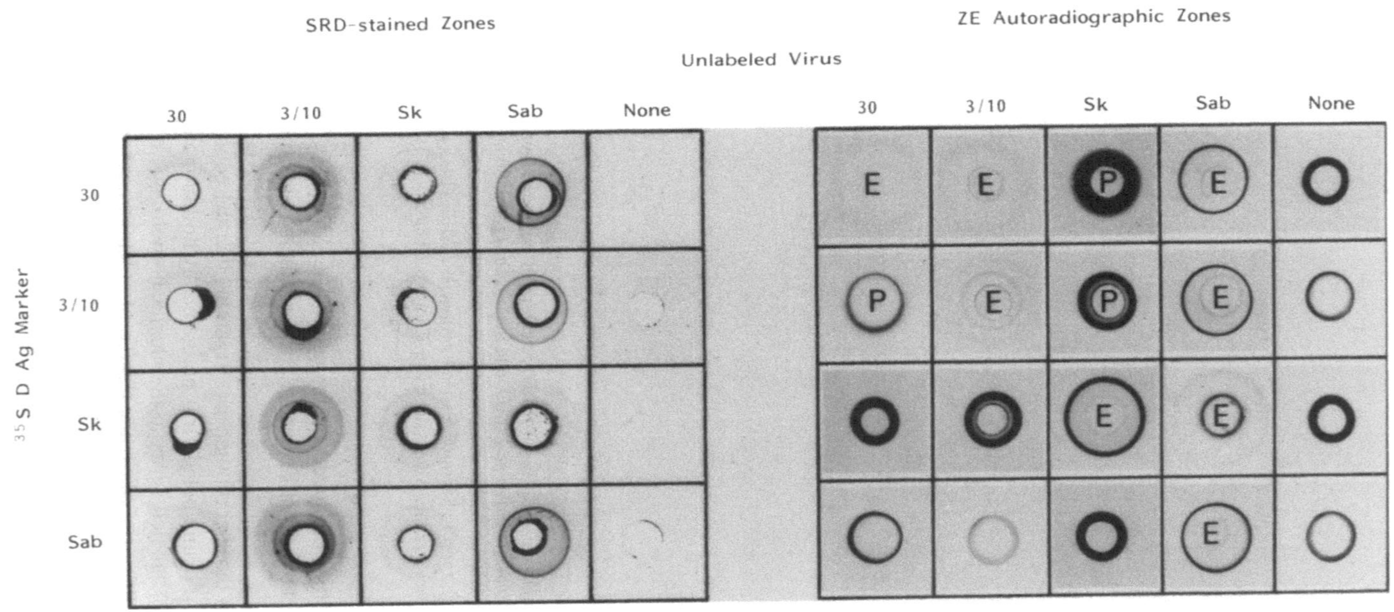

Figure 5. Specificity of poliovirus type 3 zone-enhancement (ZE) reactions using homologous and heterologous ^{35}S-labeled D antigen markers. Unlabeled concentrates of poliovirus type 3 strains 30, 3/10, Sabin, and Saukett were added to wells in immunoplates containing rabbit antiserum to unfractionated P3-Sabin virus. Sulfur-35-labeled D-antigen (P3-30, P3-3/10, P3-Sabin, or P3-Saukett) was added to the wells indicated and, after 2 days for antigen diffusion, the plates were stained by Coomassie blue, followed by autoradiography. The reactions demonstrated by staining and autoradiography of the same plate are shown. Maximum ZE (E) was seen when autoradiographic zones were the same size as stained zones; no enhancement was observed when autoradiographic zones were similar in size to those of label alone; and partial enhancement (P) was seen when autoradiographic zones were intermediate in size between stained zones and zones produced by label alone. The ^{35}S marker viruses did not produce stained zones due to insufficient antigen concentration.

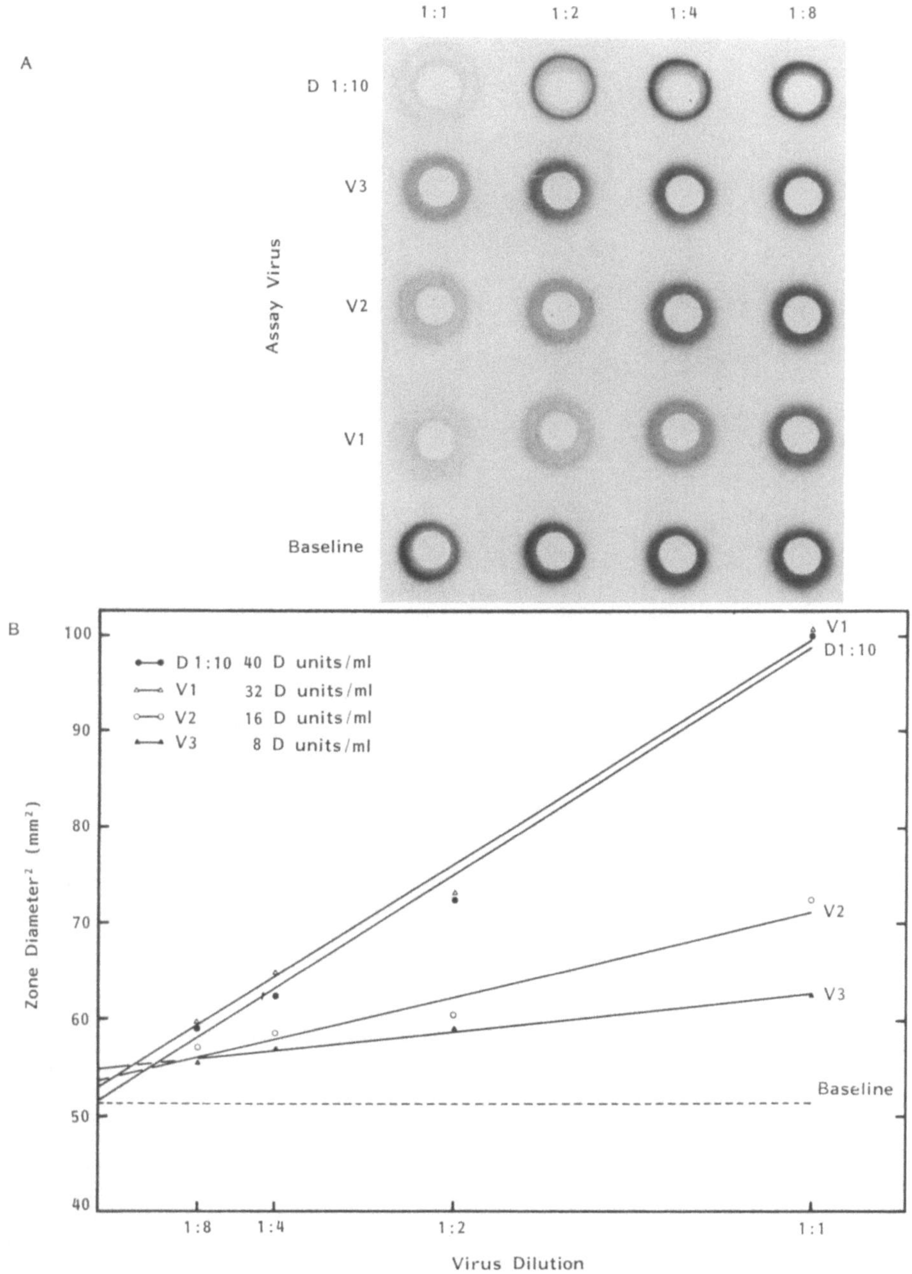

Figure 6. Zone-enhancement (ZE) dose–response study of inactivated polio vaccine (IPV) type 3 antigens. Serially diluted preparations (20-μl) (△) IPV1, 32 D units/ml, (○) IPV2, 16 D units/ml, (▲) IPV3, 8 D units/ml, and (●) a D antigen standard diluted 1 : 10, 40 D units/ml (Glaxo standard kindly provided by Dr. A. J. Beale) were added to wells in immunoplates containing rabbit antiserum to unfractionated P3-Sabin (0.003 μl/ml gel). This was followed by 10 μl of ^{35}S-labeled P3-Saukett D antigen marker. (A) Autoradiographic responses. (B) Resulting dose–response curves. Each type 3 preparation gave a linear dose response, with all responses meeting on the ordinate. A type 1 unlabeled virus produced no ZE response (baseline).

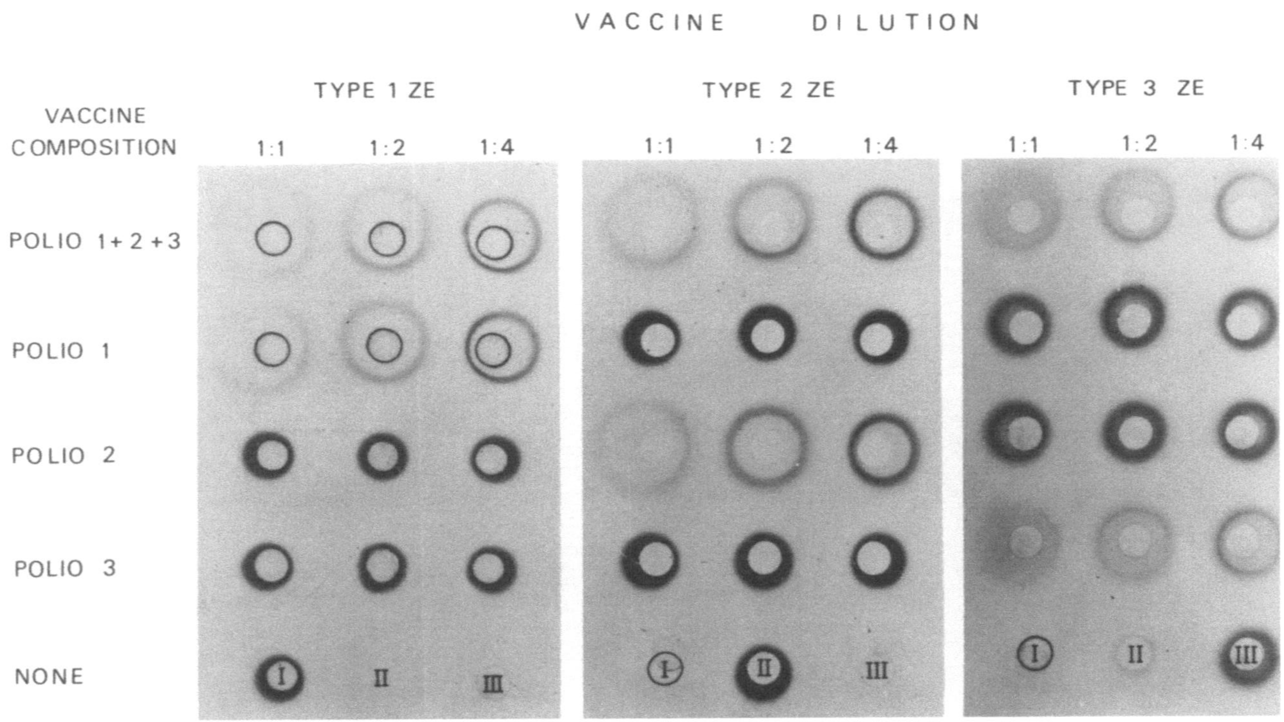

Figure 7. Specificity of types 1, 2, and 3 zone enhancement (ZE) for inactivated polio vaccine (IPV). Serial dilutions of trivalent IPV, monovalent type 1 IPV, type 2 IPV, and type 3 IPV were added to wells in type 1, 2, and 3 immunoplates. Immunoplates contained rabbit antisera to P1-Sabin virus (0.1 μl/ml gel), P2-Sabin virus (0.2 μl/ml gel), or P3-Sabin virus (0.01 μl/ml). The addition of vaccine was followed by ^{35}S-labeled D antigen (type 1, P1-Mahoney; type 2, P2-5616; type 3 P3-30), and the resulting autoradiographs are illustrated. Types 1, 2, and 3 label alone were added to the bottom row of wells of each immunoplate. The trivalent vaccine reacted in each assay, but monovalent vaccines reacted only in homologous ZE assays. Similarly, ^{35}S-labeled D antigen reacted only in homologous immunoplates.

necessary in the choice of labeled virus but nevertheless ensures that the ZE test is specific for poliovirus strain and for D antigen.

Figure 6 illustrates a type 3 ZE assay for serial dilutions of IPV in comparison with a D antigen standard antigen. Each antigen produced a linear dose response, and different dose–response curves converged to a common intercept on the d^2 ordinate. Thus, ZE potency assays can be analyzed in a similar manner to SRD assays. ZE assays have been shown to be 40–100-fold more sensitive than SRD assays (Schild *et al.*, 1980), and the limit of assay sensitivity is 10, 2, and 4 D-antigen units/ml DU/ml) (Beale and Mason, 1962) for vaccine types 1, 2, and 3, respectively. This is well within the limits required for testing conventional IPV containing approximately 40, 10, and 30 DU/ml for types 1, 2, and 3, respectively.

As the ZE assay is extremely specific antigenically, it is possible to quantify each of the three components separately in a trivalent IPV (Fig. 7). The trivalent vaccine reacted in type 1, 2, and 3 ZE assays, whereas the monovalent vaccines only reacted in the homologous ZE assays. In this and further experiments, no evidence of intertypic cross-reactivity has been discovered.

Alternative *in vitro* potency potency tests of IPV based on enzyme-linked immunosorbent assay (ELISA) have been described (van der Marel *et al.*, 1981). The tests are sensitive, and clinical trials have established that D antigen concentrations measured by ELISA correlate well with IPV immunogenicity in children (Salk *et al.*, 1981). The ZE test is less sensitive than ELISA, but there is evidence that some ELISA tests are not as specific as ZE tests (J. M. Wood, unpublished data). The reproducibility of ELISA tests for routine control purposes is unknown, whereas tests based on SRD have been in widespread use for influenza vaccines for several years. We have shown that the polio ZE test is as reproducible as SRD (coefficient of variation over 12 assays 6%) and would be a suitable assay system for IPV standardization.

VI. CONCLUSION

We have discussed the merits of assays based on SRD for potency testing of inactivated influenza vaccines and IPV. The SRD test is in international use for standardization of influenza vaccine potency and has virtually replaced potency tests based on hemagglutination or tests in animals. SRD tests are also applicable to equine and avian influenza vaccines, as they provide data relevant to clinical efficacy. The sensitivity and antigenic specificity of the SRD test are greatly increased by the use of radiolabeled markers in the poliovirus ZE test, yet the ZE test remains technically simple and reproducible. ZE provides an effective method for potency testing of IPV and may contribute toward improved vaccine efficacy and a reduction in the use of animals for vaccine testing.

ACKNOWLEDGMENTS

This study was supported in part by grant AI02649 from the National Institute of Allergy and Infectious Diseases, grant 12-16-9-080 from the U.S. Department of Agriculture, and the American Lebanese Syrian Associated Charities.

REFERENCES

Alexander, D. J. Orthomyxoviruses. In *Viral Infections of Vertebrates*, Vol. 3. Viral Infections of Birds (J. B. McFerran, ed.), Elsevier, Amsterdam (in press).

Allan, W. H., Madely, C. R., and Kendal, A. P. (1971). *J. Gen. Virol.* **12,** 79–84.

Beale, A. J., and Mason, P. J. (1962). *J. Hyg.* **60,** 113–120.

Bean, W. M., Kawaoka, Y., Wood, J. M., Pearson, J. E., and Webster, R. G. (1985). *J. Virol.* **54,** 151–160.

Brand, C. M., and Skehel, J. J. (1972). *Nature New Biol.* **238,** 145–147.

Brugh, M., Beard, C. W., and Stone, H. D. (1979). *Am. J. Vet. Res.* **40,** 165–169.

Dowdle, H. R., Downie, J. C., and Laver, W. G. (1974). *J. Virol.* **13,** 269–275.

Hinshaw, V. S., Naeve, C. W., Webster, R. G., Douglas, A., Skehel, J. J., and Bryans, J. (1983). *Bull. WHO* **61,** 153–158.

Hinshaw, V. S., and Webster, R. G. (1982). In *Basic and Applied Influenza Research* (A. S. Beare, ed.), CRC Press, Boca Raton, Florida.

Hobson, D., Beare, A. S., and Ward-Gardner, A. (1979). *Proceedings of the Symposium on Live Influenza Vaccines,* No. 73, Yugoslav Academy of Science and Arts.

Klingeborn, B., Rockborn, G., and Dinter, Z. (1980). *Vet. Rec.* **106,** 363–364.

Lang, G., Narayan, O., Rouse, B. I., Ferguson, A. E., and Connell, M. C. (1968). *Can. Vet. J.* **9,** 151–160.

Lapinleimu, K., and Stenvik, M. (1981). *Dev. Biol. Stand.* **47,** 109–112.

Magrath, D. I., Evans, D. M. A., Ferguson, M., Schild, G. C., Minor, P. D., Horaud, F., Crainic, R., Stenvik, M., and Hovi, T. (1986). *J. Gen. Virol.* **67,** 899–905.

Mayer, M. M., Rapp, H. J., Roizman, B., Klein, S. W., Cowan, K. M., Lukens, D., Schwerdt, C. E., Schaffer, F. L., and Charney, J. (1957). *J. Immunol.* **78,** 435–455.

Minor, P. D., Schild, G. C., Wood, J. M., and Dandawate, C. N. (1980). *J. Gen. Virol.* **51,** 147–156.

Moynihan, M., and Petersen, I. (1981). *Dev. Biol. Stand.* **47,** 129–133.

Mumford, J., Wood, J. M., Scott, A. M., Folkers, C., and Schild, G. C. (1983). *J. Hyg.* **90,** 385–395.

Nicholson, K. G., Tyrrell, D. A. J., Potter, C. W., Jennings, R., Clark, A., Schild, G. C., Wood, J. M., Yetts, R., Seagroatt, V., Huggins, A., and Anderson, S. G. (1979). *J. Biol. Stand.* **7,** 123–136.

Poss, P. E., Halverson, D. A., and Karunokaran, D. (1982). *Proceedings of First International Symposium on Avian Influenza, 1982,* Carter Comp. Corp., Richmond, Virginia, pp. 100–111.

Potter, C. W., Clark, A., Jennings, R., Schild, G. C., Wood, J. M., and McWilliam, P. K. A. (1980). *J. Biol. Stand.* **8,** 35–48.

Price, R. J. (1982). *Proceedings of the First International Symposium on Avian Influenza, 1981,* Carter Comp. Corp., Richmond, Virginia, pp. 178–179.

Salk, J., Stoeckel, P., van Wezel, A. L., Lapinleimu, K., and Van Steenis, G. (1982). *Ann. Clin. Res.* **14,** 204–212.

Schild, G. C., Wood, J. M., and Newman, R. W. (1975). *Bull. WHO* **52,** 223–230.

Schild, G. C., Wood, J. M., Minor, P. D., Dandawate, C. N., and Magrath, D. I. (1980). *J. Gen. Virol.* **51,** 157–170.

Simpson, D. (1973). *Symp. Ser. Immunobiol. Stand.* **20,** 326–331.

Van der Marel, P., Hazendonk, A. G., and van Wezel, A. L. (1981). *Dev. Biol. Stand.* **47,** 101–108.

Van Steenis, G., van Wezel, A. L., and Sekhuis, V. M. (1981). *Dev. Biol. Stand.* **47,** 119–128.

Wood, J. M., Schild, G. C., Newman, R. W., and Seagroatt, V. (1977*a*). *J. Biol. Stand.* **5,** 237–247.

Wood, J. M., Schild, G. C., Newman, R. W., and Seagroatt, V. (1977*b*). *Dev. Biol. Stand.* **39,** 193–200.

Wood, J. M., Seagroatt, V., Schild, G. C., Mayner, R. E., and Ennis, F. A. (1981). *J. Biol. Stand.* **9,** 317–330.

Wood, J. M., Schild, G. C., Folkers, C., Mumford, J., and Newman, R. W. (1983*a*). *J. Biol. Stand.* **11,** 133–136.

Wood, J. M., Mumford, J., Folkers, C., Scott, A. M., and Schild, G. C. (1983*b*). *J. Hyg.* **90,** 371–384.

Wood, J. M., Kawaoka, Y., Newberry, L. A., Bordwell, E., and Webster, R. G. (1985). *Avian Dis.* **29,** 867–872.

14

Rabies Vaccine Potency and Stability Testing

Applications of Single-Radial Immunodiffusion Methods

Morag Ferguson, Valerie Seagroatt, and G. C. Schild

I. INTRODUCTION

It is 100 years since Louis Pasteur carried out the first immunization of humans against rabies virus infection. Although great advances have been made in the development of more effective and purified vaccines, vaccines produced in adult or suckling animal nervous tissue are still in use in many countries. Cell culture systems now used for the production of human rabies vaccines include human diploid cell strain (HDCS), primary chick embryo fibroblasts, Vero cells, primary hamster, and dog or fetal bovine kidney cells.

Quality control is an important aspect of vaccine manufacture. Vaccines must be tested for sterility and effective inactivation and must be shown to be of adequate stability and potency. A mouse protection test, the National Institute of Health (NIH) test (Seligman, 1973), is currently recommended for potency testing of rabies vaccines (WHO, 1981). This test involves injecting groups of mice with dilutions of vaccine on days 0 and 7 followed by a challenge with a CVS strain of rabies virus on day 14. Although this test does give some indication of protective efficacy, it has several disadvantages: (1) it involves the use of large numbers of mice, (2) it is widely accepted that the potency estimates obtained in such tests are imprecise, and (3) it does not reflect the primary antigenicity of a vaccine, since it involves two injections of vaccine. In addition, the use of infectious virus is a potential hazard and can cause problems in countries in which special containment facilities are required for work with infectious rabies virus.

The major protective antigen of the rabies virion is the glycoprotein (Cox *et al.*, 1977). Several *in vitro* assays (Atanasiu *et al.*, 1982; van der Marel and van Wezel, 1981; Barth *et al.*, 1981) for rabies vaccine potency have been developed, but none has gained widespread acceptance. However, the single-radial immunodiffusion (SRD) method,

---

Morag Ferguson, Valerie Seagroatt, and G. C. Schild • National Institute for Biological Standards and Control, Potters Bar, Herts. EN6 3QG, England.

which has been used for the assay of the hemagglutinin content of inactivated influenza vaccines (Schild *et al.*, 1975; Wood *et al.*, 1977), has been applied to the assay of the glycoprotein of rabies vaccines produced in cell cultures (Ferguson and Schild, 1982).

This chapter describes studies demonstrating the reproducibility of SRD assays for rabies virus glycoprotein and the use of SRD in assessing the stability of HDCS rabies vaccines.

II. SINGLE-RADIAL IMMUNODIFFUSION TECHNIQUE

In SRD tests, virus preparations are treated with detergent to release the glycoprotein antigen from virus particles; dilutions of the treated virus are then added to wells in agarose gels in which antibody to purified glycoprotein is incorporated. Solubilized glycoprotein diffuses radially from wells and reacts with specific antibody to produce a zone of precipitation in the gel (Fig. 1). At equilibrium, the area of the diffusion zone has been found to be proportional to the amount of glycoprotein added, and linear dose–response curves are obtained (Fig. 2). Assay data can be analyzed by the slope ratio method (Finney, 1978) relating zone area to dose. Potency estimates obtained in SRD tests generally agree well with those obtained in NIH tests, although there is a tendency

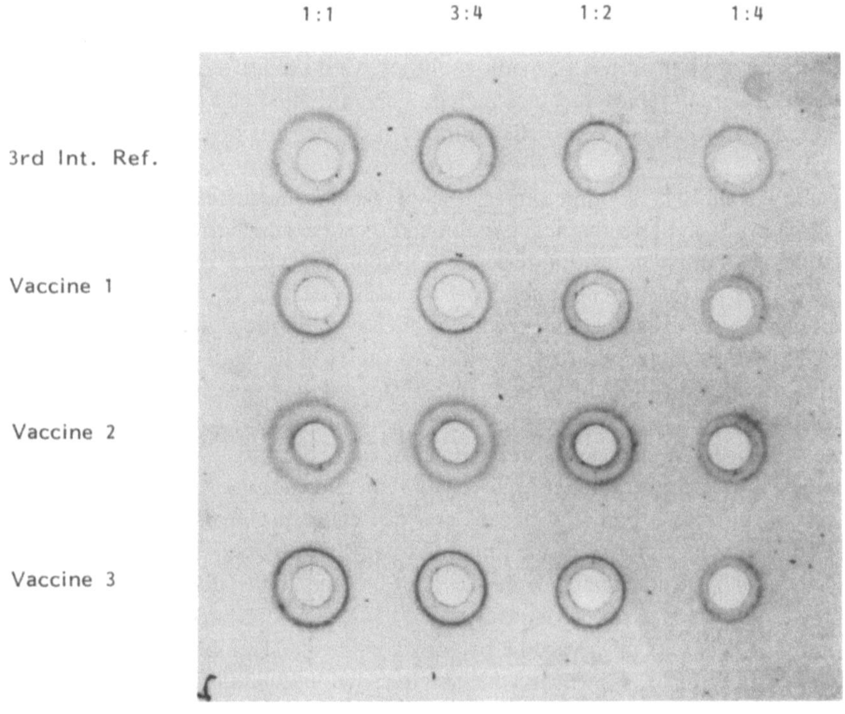

Figure 1. Zones produced in single-radioimmunodiffusion tests for the glycoprotein content of rabies vaccines.

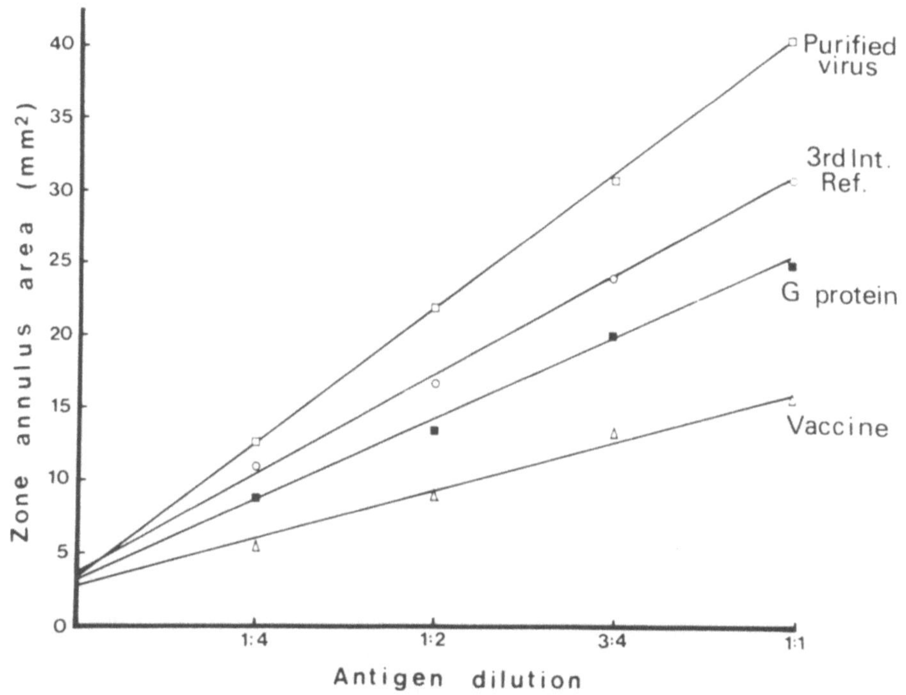

Figure 2. Relationship between zone annulus area and antigen dilution zone annulus area in single-radial immunodiffusion (SRD) assays of 3rd International Reference Preparation (IRP), purified virus, purified glycoprotein and a commercial vaccine.

for the SRD potencies to be slightly higher than those found in NIH tests (Table 1). However, potencies obtained in SRD tests are more precise than those obtained in NIH tests, as reflected by the wide confidence limits found in the latter (Table 1).

III. REPRODUCIBILITY OF SRD ASSAYS FOR RABIES VIRUS GLYCOPROTEIN

A collaborative study was conducted under the auspices of the World Health Organization (WHO) to examine the reproducibility of SRD assays for rabies virus glycoprotein (Ferguson *et al.*, 1984). Fourteen laboratories in seven countries were supplied with freeze-dried antigen and antiserum and were provided with a detailed protocol for the performance of the SRD tests. They were requested to test the 4th International Reference Preparation (4th IRP) (WHO 1983) and a working standard (CRV2). The 4th IRP and CRV2 were distributed and coded; duplicate samples of the 4th IRP were sent out with different codes. The Third International Reference Preparation for rabies vaccine (3rd IRP), which had been assigned a potency of 10 IU per ampoule (WHO, 1979), was used as a reference preparation in all tests. The identity of only the 3rd IRP was revealed to the

Table 1. Rabies Vaccines: Comparison of Potency Estimates and Widths of Confidence Limits Obtained in SRD and NIH Tests

	SRD tests		NIH tests	
Vaccine	Potency (IU per ampoule)	95% confidence interval	Potency (IU per ampoule)	95% confidence interval
V1	6.3	89–111	5.9	34–268
V2	5.7	89–109	7.9	25–390
V3	6.0	87–112	3.0	36–243
V4	13.2	78–119	9.2	34–304
V5	11.2	85–112	10.2	34–304
V6	9.1	87–111	4.9	33–302
V7	9.4	89–110	5.6	34–277
V8	7.7	86–112	4.1	41–234
V9	6.6	86–111	6.6	35–277
V10	3.8	87–113	2.3	26–296

[a] NIH, National Institutes of Health; SRD, single-radial immunodiffusion.

participants in the study. Some participants also assayed the vaccines by the NIH test.

Data from SRD assays were analyzed by the slope-ratio method and generally gave satisfactory results. The potency ratios of the coded duplicates of the 4th IRP were close to the expected value of unity. The potency estimates from the laboratories were found to agree well, as shown in the frequency distribution of potencies of the 4th IRP (Fig. 3). The interlaboratory variation, measured as the geometric coefficient of variation (GCV) (Kirkwood, 1979) of the laboratory mean potencies, ranged from 10 to 16%. The overall mean of potencies for each vaccine, taken as the geometric means of the laboratory mean estimates, is shown in Table 2.

Five laboratories also performed NIH tests, and the potency estimates were found to be homogeneous both within and between laboratories. The overall mean potencies with their 95% confidence intervals are compared with the values obtained from the SRD tests in Table 2. The potency estimates from the NIH and the SRD tests were in agreement, although the confidence intervals for the NIH estimates had much wider confidence intervals than did the SRD estimates.

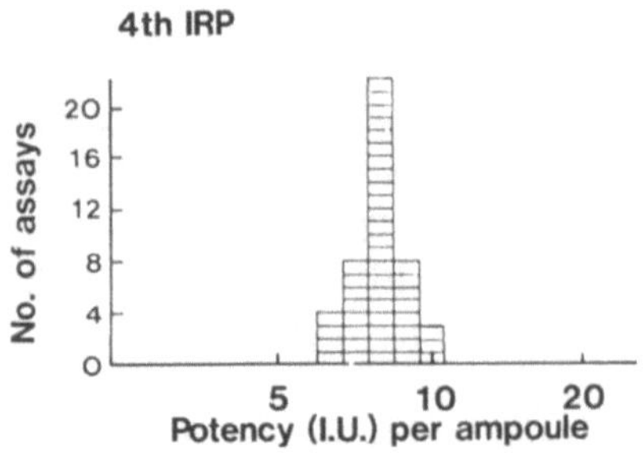

Figure 3. Frequency distribution of potency estimates of the 4th International Reference Preparation (IRP) preparation relative to the 3rd IRP obtained in single-radial immunodiffusion (SRD) assays in the collaborative study. Each box denotes the estimate from one assay.

Table 2. Comparison of Overall Potency Estimates in IU per Ampoule of the 4th International Reference Preparation (4th IRP) and the Working Standard (CRV2) Relative to the 3rd IRP Obtained from SRD Assays and from NIH Tests

		4th IRP		CRV2	
Assay method	No. of laboratories	Mean potency	95% confidence interval	Mean potency	95% confidence interval
SRD	13	7.8	7.4–8.3	8.8	8.1– 9.6
NIH test	5	4.9	3.1–7.6	7.3	4.9–11.9

IV. STABILITY OF HDCS RABIES VACCINES

Rabies vaccines are frequently used in countries with high ambient temperatures, where it is sometimes difficult to ensure storage of the vaccine at 4°C. The WHO therefore requires that rabies vaccines have adequate stability in an accelerated degradation test (WHO, 1981). The suggested test involves storage at 37°C for 4 weeks, after which the vaccine should retain the minimum required potency of 2.5 IU/dose in the NIH test.

A. NIH Tests

The stability of 38 batches of vaccine from a single manufacturer were stored for 4 weeks at 37°C and then assayed in a NIH test. The potency of these samples were expressed as percentages of the corresponding materials stored at 4°C. No significant losses in potencies were found, but the potency estimates, as would be expected, had wide confidence intervals. The frequency distribution of the potencies of the accelerated degraded samples is shown in Fig. 4. The wide variation in estimates from NIH tests ranged from less than 50% to more than 500%. As all vaccines were prepared in the same way, it can be assumed that these vaccines were homogeneous in their stability characteristics, and the estimates of relative potency were therefore combined. The combined value, calculated as the weighted geometric mean of the estimates of the individual vaccines, was 113% with a 95% confidence interval of 95–135%. Thus, there was no evidence of any loss in potency for the vaccines as a group. These results indicated that rabies

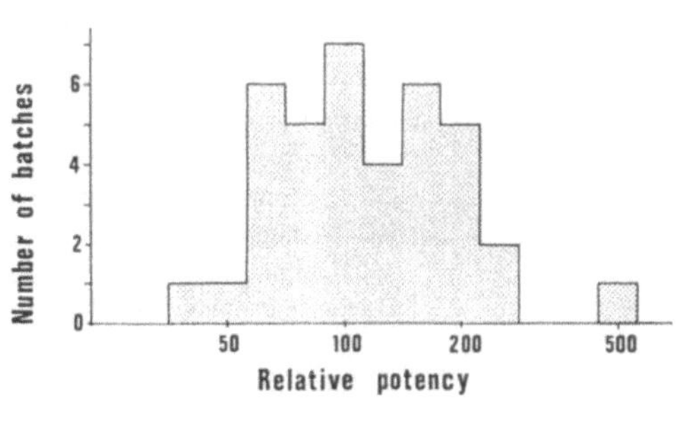

Figure 4. Frequency distribution of potency estimates from National Institutes of Health (NIH) tests of vaccines stored at 37°C for 4 weeks expressed as percentages of the potencies of vaccines stored at 4°C.

vaccines lose little, if any, potency after 4 weeks storage at 37°C and that the NIH test would be unlikely to detect in a single test any loss unless it was of the order of threefold or more.

B. SRD Tests

Twelve of the vaccines studied were also assayed against the 3rd IRP in SRD tests. The vaccines were reassayed after 2.5 years stored at 4°C; their potencies expressed as percentages of the original potencies are given in Table 3. No significant losses in potency were found. In fact, most of the potency estimates after 2.5 years storage were estimated to be somewhat higher than the original estimates, but the earlier results were obtained when the SRD technique was being developed and may not be as reliable as the later ones.

Table 3 also shows the estimates of potency of some of these samples after 4 weeks storage at 37°C as percentages of the vaccines stored at 4°C. These relative potencies are based on within-test comparisons and thus do not suffer from possible between-test variations, as did the potency estimates from SRD assays cited above. No significant losses in antigenic activity were found after storage at 37°C for 4 weeks.

Table 3. Potency Estimates Obtained in SRD Tests of HDCS Rabies Vaccines after Storage at 4°C for 2.5 Years or 37°C for 4 Weeks[a]

Vaccine	Relative potency (%)	
	2.5 years at 4°C[b]	4 weeks at 37°C[c]
1	111	97
2	115	99
3	108	83
4	105	91
5	119	104
6	114	93
7	130	ND
8	158	ND
9	115	ND
10	120	ND
11	98	ND
12	94	ND
13	103	ND
GM	114	95
GCV (%)	14	8

[a] GCV, geometric coefficient of variation; GM, geometric mean; HDCS, human diploid cell strain; ND, not done; SRD, single-radial immunodiffusion.

[b] Potencies expressed relative to values obtained 2.5 years previously.

[c] Samples of the vaccines stored at 37°C were tested concurrently with samples stored at 4°C.

V. DISCUSSION AND CONCLUSIONS

The results from the collaborative study showed that the SRD technique was simple to perform, even for those participants who were inexperienced with the technique. Close agreement was found between the laboratories in their estimation of potency by SRD tests. The GCVs of the laboratory mean potency estimates were about 12%, a magnitude similar to that found in a previous collaborative study on the assay of the hemagglutinin antigen (HA) of influenza vaccine by SRD (Wood *et al.*, 1981). However, the overall agreement between laboratories might not have been so close if they had not been provided with common assay reagents and followed the same assay procedure.

After 2.5 years, under usual storage conditions (i.e., 4°C), no losses in potencies were found in the rabies vaccines tested in SRD assays. Our results have shown that rabies vaccines generally lose little if any activity under these conditions. However, the imprecision of a NIH test is such that it is likely that only gross losses of activity (i.e., threefold or more) would be detected in a single test. By contrast, the SRD technique gives more precise estimates of relative potency than does the NIH test and would therefore be more likely to detect small losses in potency.

Although SRD assays of glycoprotein content do not provide direct evidence of immunogenicity, as does the NIH test, they do give precise estimates of the amount of the major protective antigen of rabies virus in a vaccine. The SRD technique gives a reproducible estimate of potency and is simple to perform. Its use would reduce the reliance on the imprecise *in vivo* test. In addition, SRD assays may be of value in the in-process control of vaccines by manufacturers in evaluating the consistency of batches of vaccine and in providing information on the stability of rabies vaccines.

REFERENCES

Atanasiu, P., Perrin, P., Delagneau, J. F., Versmisse, P., Chevallier, G., Reculard, P., Le Fur, R., Sureau, P. (1982). *J. Biol. Stand.* **10,** 289–296.

Barth, R., Gross-Albenhausen, E., Jaeger, O., and Milcke, L. (1981). *J. Biol. Stand.* **9,** 81–99.

Cox, J. H., Dietzschold, B., and Schneider, L. G. (1977). *Infection Immun.* **16,** 754–759.

Ferguson, M., and Schild, G. C. (1982). *J. Gen. Virol.* **59,** 197–201.

Ferguson, M., Seagroatt, V., and Schild, G. C. (1984). *J. Biol. Stand.* **12,** 283–294.

Finney, D. J. (1978). *Statistical Methods in Biological Assay,* 3rd ed., Griffin, London.

Kirkwood, T. B. (1979). *Biometrics* **35,** 908–909.

Schild, G. C., Wood, J. M., and Newman, R. W. (1975). *Bull. WHO* **52,** 223–231.

Seligman, E. B., Jr. (1973). In *Laboratory Techniques in Rabies,* 3rd ed. (M. M. Kaplan and H. Koprowski, eds.), WHO, Geneva.

van der Marel, P., and van Wezel, A. L. (1981). *Dev. Biol. Stand.* **50,** 267–275.

Wood, J. M., Schild, G. C., Newman, R. W., and Seagroatt, V. (1977). *J. Biol. Stand.* **5,** 237–247.

Wood, J. M., Seagroatt, V., Schild, G. C., Mayner, R. E., and Ennis, F. (1981). *J. Biol. Stand.* **9,** 317–330.

World Health Organization (1979). *Tech. Rep. Ser.* **638,** 15.

World Health Organization (1981). *Tech. Rep. Ser.* **658,** 54–95.

World Health Organization (1983). *Tech. Rep. Ser.* **700,** 15.

15

Hepatitis B Vaccine

Clinical Trials of New Vaccine Formulations and Dose Regimen

A. Goudeau, F. Dubois, J. Klein, A. Godefroy, J. Huchet, Y. Brossard, J. C. Soulié, C. Louq, and F. Tron

I. INTRODUCTION

The first results of active immunization against hepatitis B in humans were reported a little over 10 years ago (Maupas *et al.*, 1976). Since then, several serum-derived hepatitis B (HB) vaccines have been proved safe, highly immunogenic and efficacious in various populations at risk (Stevens *et al.*, 1984). However, the high cost of production has limited their use to high-risk volunteers in developed countries. Even there, it has been found difficult to expand immunization to all exposed individuals. In France, for instance, immunization of the estimated 1.5 million individuals at risk would cost around $100 million, i.e., 10-fold the cost of the entire rubella-prevention program. Endemic areas in which large-scale use of HB vaccines would be the most valuable are in Africa or Southeast Asia. Many countries in these areas simply cannot afford hepatitis B immunization unless prices of vaccines are significantly lowered. With currently available vaccines, immunization can be achieved by two means: *(1) reducing the number of doses for primary immunization, and (2) lowering the concentration of* HB_sAg *in the vaccine.*

A highly purified HB vaccine derived from HB_sAg^+ plasma was developed in France, underwent clinical trials in the fall of 1975, and was released for general use in 1981 under the name Hevac B (Pasteur-Vaccins, Paris). This vaccine was studied extensively in healthy adults (Crosnier *et al.*, 1981; Goudeau *et al.*, 1983), in children and newborn infants of endemic areas (Maupas *et al.*, 1981; Barin *et al.*, 1982), and in dialysis patients (Crosnier *et al.*, 1981; Goudeau *et al.*, 1983), in whom it was shown to be immunogenic and protective. The standard immunization protocol involves 3 monthly injections of an alum preparation of 5 μg HB_sAg and a booster dose after 1 year.

In a series of clinical trials, presented in this following chapter, we have investigated

A. Goudeau and F. Dubois • Institute of Virology, 37032 Tours, Cedex, France. *J. Klein and A. Godefroy* • Center of Blood Transfusion, 36000 Chateauroux, France. *J. Huchet, Y. Brossard, and J. C. Soulié* • Center of Perinatal Blood Biology, 75571 Paris, Cedex 12, France. *C. Louq and F. Tron* • Pasteur-Vaccins, 92430 Marnes La Coquette, France.

several means of reducing the cost of immunization with the French HB vaccine. Section I reports a study aiming at reducing the course of primary immunization to two injections instead of three, Section II presents a study of the dose response to HB vaccine preparations containing lower concentrations of HB_sAg (1.25 μg and 0.31 μg), and Section III discusses phase 3 clinical trials of a new HB vaccine formula containing 2 μg of HB_sAg that is being considered for licensure in France. Immunogenicity of this vaccine was studied in healthy adults and its efficacy assessed in newborn infants of HB_sAg-carrier mothers when given in association with anti-HB_s immunoglobulins.

II. MATERIALS AND METHODS

A. Laboratory Methods

HB_sAg, anti-HB_s, anti-HB_c, and HB_eAg were tested by enzyme immunoassays from Abbott (Auszyme, Ausab-EIA, Corzyme, and Hb_eAg-EIA). Quantitation of anti-HB_s in milli-international units (mIU/ml) was performed according to recommendations of the manufacturer with a cutoff value of 2.5 mIU/ml. Geometric mean titers (GMT) were calculated on responders only and expressed with their 95% confidence interval (95% CI).

B. Statistics

Chi-square and Student's *t*-test were used when appropriate.

III. PRIMARY IMMUNIZATION WITH TWO INJECTIONS INSTEAD OF THREE (STUDY I)

A. Studied Population and Protocol

The study consisted of 240 HBV seronegative subjects (227 females, 13 males). All volunteers were young adults (under 25 years of age) from schools of nurses, midwives, and laboratory technicians of the Tours University Hospital. Most were in the first year of their study, i.e., at low risk of hospital-acquired HBV infections.

Three groups were compared. In the first group, 79 subjects (73 females, 6 males) received the standard HB vaccine course of three injections. The second group comprised 134 subjects (129 females, 5 males) who received two injections of HB vaccine at a 2-month interval. The last group of 27 subjects (25 females, 2 males) received two injections of HB vaccine at a 4-month interval. Anti-HB_s seroconversion rates and titers were compared 1 month after the first and 1 month after the last dose of vaccine.

B. Results

The anti-HB_s response after the first injection was similar in all three groups (Table 1). However, striking differences were observed 1 month after completion of the primary immunization. The protocol with two injections at a 2-month interval performed poorly in

Table 1. Immunogenicity of Modified Protocols in Young Adults[a,b]

Protocol	N	Anti-HB_s positive after 1st dose	GMT (95% CI)	Anti-HB_s positive after last dose	GMT (95% CI)
1. 3 doses, 1 month apart	79	19 (24%)	27 (9–82)	77 (97%)	360 (256–506)
2. 2 doses, 2 months apart	134	34 (25%)	17 (10–29)	116 (87%)	150 (106–212)
3. 2 doses, 4 months apart	27	10 (37%)	20 (10–38)	26 (96%)	1514 (774–2961)

[a] Vaccine Hevac B, Pasteur-Vaccins, Paris. 1, lot 21; 2, lot 11; 3, lot 21.
[b] Geometric mean titers.

terms of seroconversion rate and anti-HB_s titers as compared with the standard three injections protocol: 87% versus 97% ($p < 10^{-2}$); GMT: 150 mIU/ml versus 360 mIU/ml ($p < 10^{-3}$). By contrast, the third option, two injections at a 4-month interval, yielded satisfactory results with even a GMT of responders (1514 mIU/ml) statistically higher than that observed after the standard three-dose course ($p < 10^{-3}$).

C. Discussion

Modified protocols of immunization reducing the number of injections to two instead of three yield a satisfactory anti-HB_s response in young adults, provided the interval between each dose is lengthened up to 4 months. Shorter intervals, such as 1 or 2 months, are less adequate, especially in older subjects (Goudeau *et al.*, 1984*a*). These adapted protocols represent a significant reduction of cost and could be applied to low-risk volunteers, such as students in medical or paramedical schools. Owing to the necessary delay between the two vaccinations, they are ill suited for immunization in high-risk settings.

Three clinical trials have been carried out with the same vaccine to assess the immunogenicity of a two-dose regimen in newborn infants or in children (Table 2). The

Table 2. Clinical Trials of Two-Dose Regimen for Hepatitis B Vaccine Immunization in Newborn Infants and Children

Investigators/ country	Populations studied	Group N	Protocol	Anti-HB_s seroconversion (%)
Yvonnet *et al.* (1984), Senegal	Children (mean age : 1 year)	1 (111)	3 × 5 μg, 1 month apart	95
		2 (72)	2 × 5 μg, 2 months apart	93
Piazza *et al.* (1985*a*,*b*), Italy	Newborns of HB_s^- mothers	1 (35)	2 × 5 μg, birth, 2 months	66
		2 (35)	3 × 5 μg, birth, 1, 2 months	97
	HBV^- children	3 (36)	2 × 5 μg, 2 months apart	92
	3 months	4 (42)	3 × 5 μg, 1 month apart	95
		5 (34)	2 × 5 μg, 2 months apart	97
	5 months	6 (38)	3 × 5 μg, 1 month apart	97
		7 (56)	2 × 5 μg, 2 months apart	96
	>1 year	8 (56)	3 × 5 μg, 1 month apart	98
Perrin *et al.* (1985), Burundi	Newborns	88	2 × 5 μg, birth, 2 months	96

results in newborn infants are discordant between two studies conducted in Southern Italy and Burundi and warrant further investigations before applying such protocol to neonates, especially in endemic areas. By contrast, results in older children (over 3 months of age) are satisfactory with seroconversion rate to anti-HB_s exceeding 90% in all three studies. These data are very encouraging for countries planning to undertake mass vaccination of their population. These countries could reserve the tight three-monthly dose protocol for high-risk individuals, such as children born to HB_sAg-carrier mothers, and could use cost-saving two-dose regimens for populations at lower risk.

IV. IMMUNIZATION WITH HB VACCINE PREPARATIONS CONTAINING REDUCED AMOUNT OF HB_sAG (STUDY II)

A. HB Vaccine

Three sub-lots of the commercial lot 08 of Pasteur-Vaccins HB vaccine were prepared. They contained, respectively, 5 μg (the standard dosage), 1.25 μg, and 0.31 μg of HB_sAg per dose, for a constant concentration of diluent and adjuvant (aluminum hydroxyide) equal to that of the licensed vaccine.

B. Studied Population and Protocol

The population to be immunized were students at the nursing, midwifery, and laboratory technician schools at the Tours University Hospital. The study consisted of 240 HBV seronegative volunteers, divided into three groups: 77 subjects of group A received the 5 μg vaccine, 83 subjects of group B received the 1.25-μg vaccine, and 80 subjects of group C received the 0.31-μg vaccine. Volunteers were young (mean age: 21.3, 22.0 and 21.8 years in groups A, B, and C, respectively) and almost all female (5, 1, and 1 male students in groups A, B, and C, respectively).

The protocol of immunization was that of the licensed vaccine, i.e., three injections 1 month apart and a booster injection after 1 year. The kinetics of the primary response were assessed by blood samples drawn 1 week before the second injection (P1), 1 week before the third injection (P2), and 5–6 weeks after the third injection (P3). Persistence of anti-HB_s was studied on blood samples drawn 34–37 weeks after the first injection (P4) and 50 weeks after the first injection (P5), i.e., 2 weeks prior to the booster injection. The anamnestic response to the booster was controlled 5–6 weeks post injection (P6). Completion of the follow-up evaluation was satisfactory, since 100% of vaccinees were controlled at P3, 96% at P4, 97% at P5, 97% received their booster injection, and 95% were checked on at P6.

C. Results

1. Primary Immunization

a. Anti-HB_s Response to Three Injections of Hepatitis B Vaccine. The kinetics of the anti-HB_s response in each group are given in Fig. 1. Overall seroconversion to anti-HB_s exceeded 95% after primary immunization in all three groups (76 of 77 in group A,

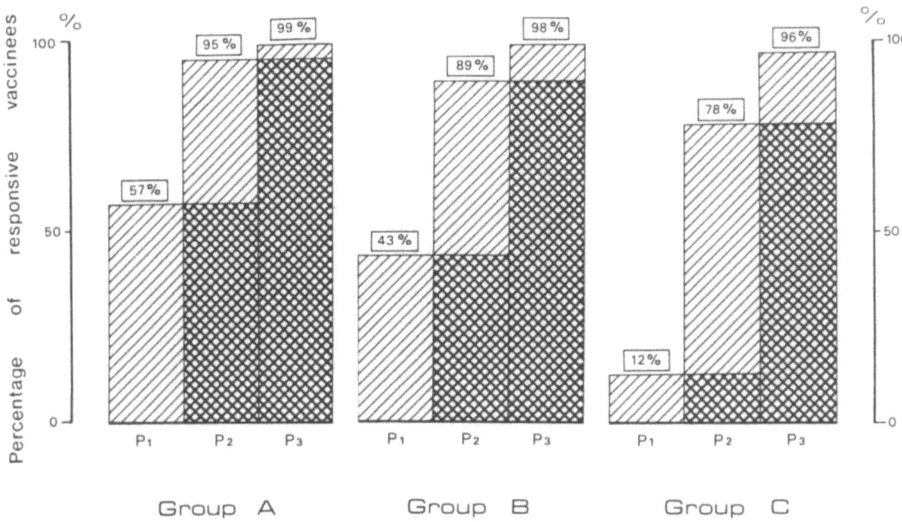

Figure 1. Kinetics of anti-HB_s response to successive injections of hepatitis B vaccine. Cumulative percentage of responders after the first (P1), second (P2), and third (P3) injection of hepatitis vaccine. Group A: 5 μg HB_sAg/dose; group B: 1.25 μg HB_sAg/dose; group C: 0.31 μg HB_s/Ag dose.

81 of 83 in group B, and 77 of 80 in group C). However, more volunteers of groups A (57%) and B (43%) responded to the first injection (anti-$HB_s{}^+$ at P1) as compared with the response in volunteers of group C (12%) ($p < 10^{-3}$). This difference has disappeared by P2.

Anti-HB_s GMT observed at P3 (Fig. 2) were significantly higher in group A: 875 mIU/ml than in group B: 361 mIU/ml ($p < 10^{-5}$) or group C: 119 mIU/ml ($p < 10^{-9}$); groups B and C were also significantly different ($p < 10^{-5}$). Also, 92% of volunteers in group A had titers above 100 mIU/ml as compared with 82% in group B and 58% in group C (A/B: not significant; B/C: $p < 10^{-3}$; A/C: $p < 10^{-6}$). Titers inferior or equal to 10 mIU/ml were observed in 3% of subjects in group A, in 4% of group B, and in 10% of group C (not significant).

b. Follow-up of Nonresponders. The one nonresponder in group A received a fourth injection of the 5-μg vaccine 23 weeks after the first injection and was found anti-$HB_s{}^+$ at P4 and P5 (45 mIU/ml). Of the two nonresponders in group B, one received a fourth injection of the 1.25-μg vaccine 18 weeks after the first and was anti-$HB_s{}^+$ at P4 and P5 (5 mIU/ml). The other did not receive a fourth injection and remained anti-$HB_s{}^-$. The three nonresponders in group C received a fourth injection of the 0.31-μg vaccine; one of them at 17 weeks and the other two at 20 weeks after the first injection. One patient remained anti-$HB_s{}^-$ and the other two seroconverted to anti-HB_s (3 mIU/ml and 38 mIU/ml at P5).

c. Follow-up of Responders. In group A, 73 of the 76 responders to three injections of vaccine were controlled at P4 and 74 at P5; all of them were still anti-$HB_s{}^+$ (Table 3). In group B, 77 of 81 responders were sampled at P4 and found to be anti-$HB_s{}^+$; 76 were controlled at P5, one had become anti-$HB_s{}^-$. In group C, 70 of 76 responders who were screened were still anti- $HB_s{}^+$ at P4, and 71 of 77 at P5 (92%). Of the six volunteers in

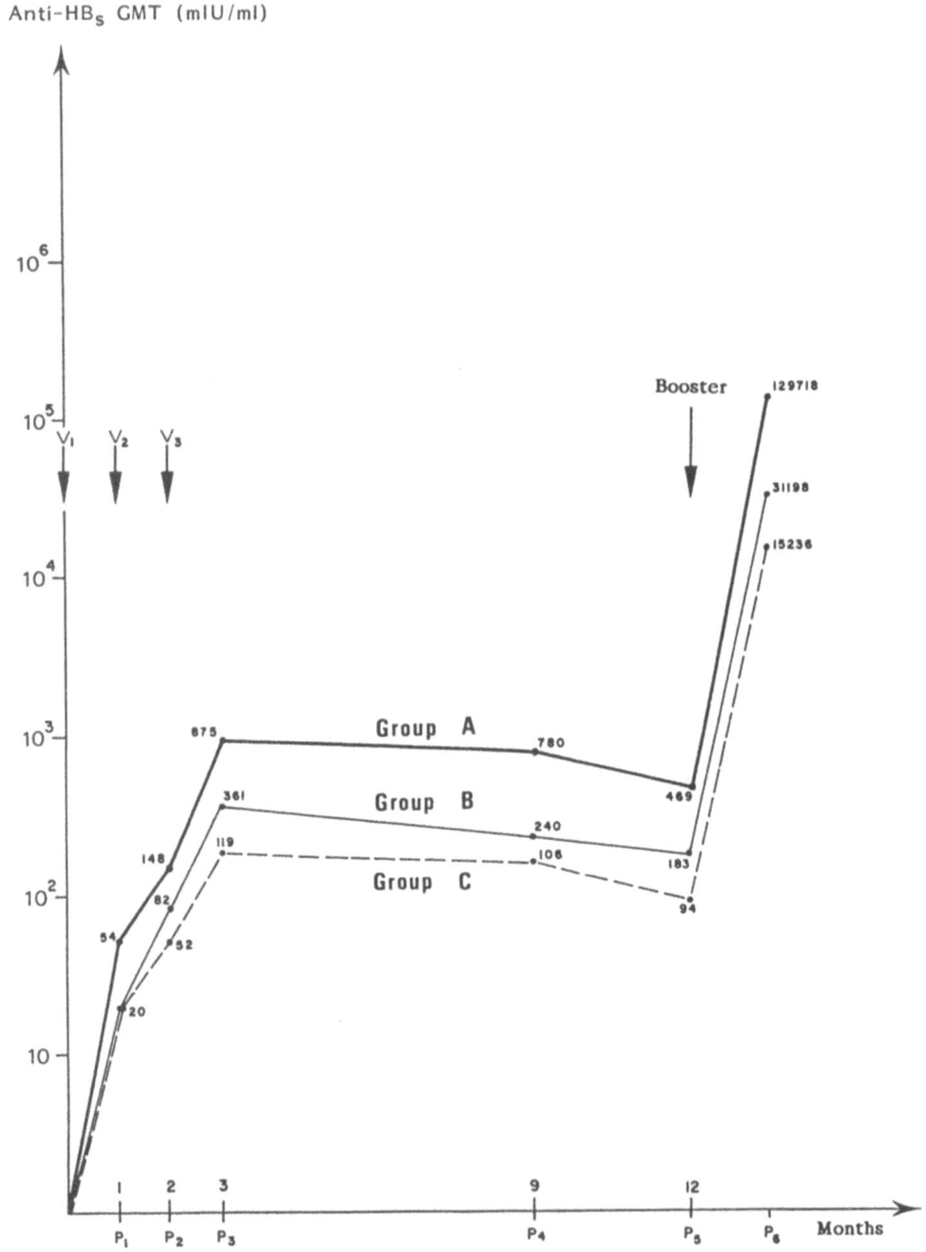

Figure 2. Evolution of anti-HB_s titer after three injections of hepatitis B vaccine (responders only). GMT, geometric mean titers.

Table 3. Follow-up of Responders to Three Injections of Hepatitis B Vaccine[a]

Anti-HB_s^+ at P3		Anti-HB_s^+		
Group	No.	at P4	at P5	at P6
A	76	73/73 (100%)	74/74 (100%)	72/72 (100%)
B	81	77/77 (100%)	75/76 (99%)	76/76 (100%)
C	77	70/76 (92%)	71/77 (92%)	79/79 (100%)

[a] Persistence of anti-HB_s during the first year.

group C who became anti-HB_s^- during the first year of follow-up, four had responded to the second injection and two to the third injection, and their anti-HB_s titers at P3 were lower (GMT:12 mIU/ml) than the mean for the whole of group C (GMT:119 mIU/ml), ($p < 10^{-3}$). Differences in GMT among groups A, B, and C observed at P3 remained significant at P5 (A/B: $p < 10^{-4}$; B/C: $p < 10^{-2}$; A/C: $p < 10^{-6}$). At this time, 2 weeks before the booster injection, 89% of group A had anti-HB_s titers above 100 mIU/ml, against 72% in group B and 51% in group C (A/B: $p < 10^{-2}$; B/C: $p < 10^{-2}$; A/C: $p < 10^{-6}$). Titers inferior or equal to 10 mIU/ml were observed in 1% of individuals in group A, 4% in group B, and 18% in group C (A/B: not significant; B/C: $p < 10^{-2}$; A/C: $p < 10^{-6}$) (Fig. 3).

2. Booster Injection

The booster injection was followed by a steep increase of anti-HB_s. GMT after the booster was 129,718 mIU/ml in group A, 31,198 in group B, and 15,236 in group C (A/B: $p < 10^{-5}$; B/C: $p < 10^{-2}$; A/C: $p < 10^{-6}$). No vaccinee of either group had a titer inferior to 100 mIU/ml after the booster injection. More than 90% of vaccinees in each group had an anti-HB_s above 10^3 mIU/ml; 58% of volunteers of group A had a titer above 10^5 mIU/ml after the booster versus 25% in group B and 11% in group C (Fig. 3). Twenty-five volunteers in group C were controlled 1 year after the booster injection, 24 (96%) were still anti-HB_s^+ with a GMT of 1516 mIU/ml, and one had become seronegative.

D. Discussion

The aim of this trial was to compare the immune response induced by HB vaccine containing low concentrations of HB_sAg. Two preparations derived from a commercial lot and adjusted to 1/4th (1.25 μg HB_sAg/dose, group B) and 1/16th (0.31 μg/HB_sAg dose, group C) of the HB_sAg concentration in the licensed vaccine were used. They were compared with a reference group (group A) receiving the standard 5-μg HB_sAg/dose vaccine.

The overall anti-HB_s seroconversion rate exceeded 90% in all three groups of vaccinees. Although decreasing concentrations were correlated with lower anti-HB_s titers, persistence of antibody was satisfactory as assessed by the results of group C, in which 92% of responders were still anti-HB_s^+ after a 1-year follow-up.

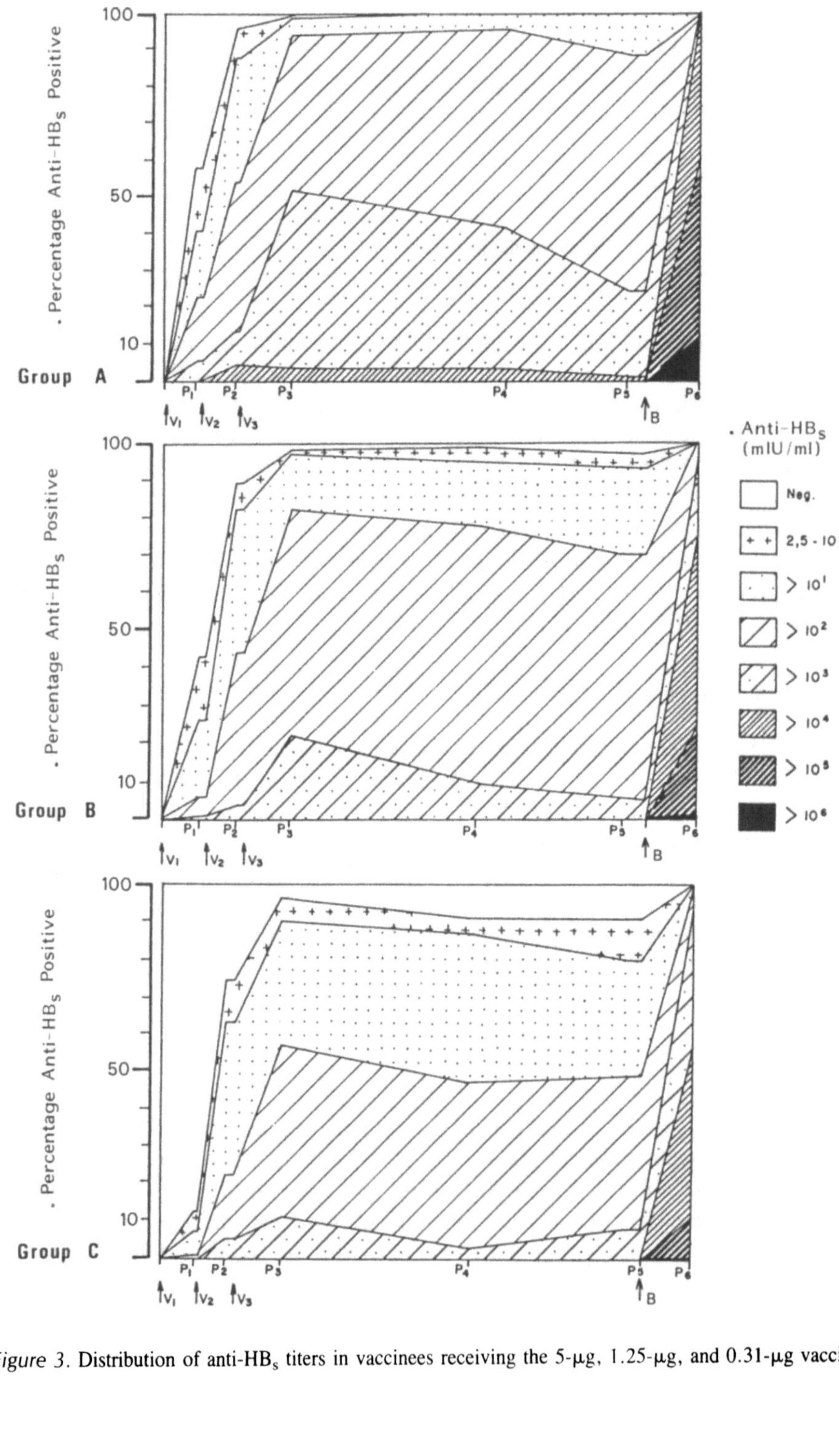

Figure 3. Distribution of anti-HB$_s$ titers in vaccinees receiving the 5-μg, 1.25-μg, and 0.31-μg vaccines.

After immunization, anti-HB_s decrease followed an exponential mode with a similar rate in all three groups (Fig. 2). Therefore, volunteers in group C (0.31 μg HB_sAg/dose) in whom the GMT was low at P3 were more likely to decrease to titers equal or inferior to 10 mIU/ml before completion of the first year. By contrast, volunteers of group B (1.25 μg HB_sAg/dose) developed sufficient anti-HB_s to sustain highly protective levels until administration of the booster injection. Boostering was very efficient, resulting in anti-HB_s titers above 10^3 mIU/ml in more than 90% of vaccinees, with GMT exceeding 10^5 mIU/ml in the three groups.

Our study provides evidence that the immune response to reduced concentrations of HB_sAg is satisfactory in young adults. Hepatitis B vaccine containing as low as 1.25 μg of HB_sAg/dose may be used without significant loss of immunogenicity when compared with the presently licensed 5-μg/dose vaccine. Reduction of HB_sAg concentration represents one means of stretching out the current vaccines that could lower the cost of production, hence their availability for wider use.

V. IMMUNIZATION WITH A NEW HB VACCINE FORMULA CONTAINING 2 μg OF HB_sAg (STUDY III)

A. HB Vaccine

A new vaccine preparation derived from commercial lot 21 of Pasteur-Vaccins HB vaccine was used in Study III. The preparation contained 2 μg of HB_sAg per dose but a total amount of adjuvant equal to that of the licensed vaccine.

B. Populations and Protocol

Two clinical trials were carried out in study III.

1. HBV-Seronegative Healthy Adults

The comparative trial involved 318 volunteers (201 females, 117 males); 134 (126 females, 8 males) were students from schools of nurses and midwifes of the Tours University Hospital, and 184 (75 females, 109 males) were healthy blood donors from the Chateauroux Blood Center. The population was randomized by age and sex into two groups: subjects of the first group (161 subjects) received the standard 5-μg HB_sAg vaccine, and subjects of the second group (157 subjects) received the modified 2-μg H B_sAg vaccine. The protocol of primary immunization for both vaccines consisted of three injections at 1-month intervals and a booster at 1 year. Vaccinations were administered subcutaneously in the arm or shoulder.

2. Newborn Infants of HB_sAg-Carrier Mothers

The study involved 110 neonates of HB_sAg-carrier mothers born in various maternity hospitals of Paris. With the informed consent of both parents, passive–active immunization of the infants was carried out with either the standard 5-μg (52 subjects) or the new

2-μg vaccine (58 subjects). The serovaccination combined intramuscular injection of 200 IU of anti-HB_s immunoglobulins (Centre National de Transfusion Sanguine, Paris) and subcutaneous injection of the first dose of HB vaccine within the first few days of life; 107 infants were immunized within the first 48 hr; two the third day, and one the fourth day of life. A second and third dose of HB vaccine was given, respectively, at 1 and 2 months of age.

C. Results

1. Healthy Adults

One month after the third injection, 86% (139 of 161) of vaccinees in the group receiving the 5-μg HB vaccine were found to be anti-HB_s^+ against 76% (119 of 157) of those receiving 2-μg HB vaccine ($p < 0.02$). The kinetics of anti-HB_s response was similar in both groups: 23 (14%) responding to one injection of the 5-μg vaccine and 20 (13%) to one injection of the 2-μg vaccine: 88 (55%) and 78 (50%) responding to two injections in the 5-μg and 2-μg groups, respectively.

In both arms of the study, the anti-HB_s response was age and sex dependent, with female volunteers responding better than males, and young volunteers aged 18–35 years responding better than older ones (Table 4). Overall, the anti-HB_s response of females was satisfactory with both HB vaccines: 97% (99 of 102) with the 5 μg vaccine and 90% (89 of 99) with the 2-μg vaccine (Table 5). However, a lower seroconversion rate, 70% (16 of 23) was observed in women over 36 years of age receiving the 2-μg vaccine. Seroconversion to anti-HB_s of male recipients was poor with both HB vaccine preparations and significantly different from that of female volunteers: 68% (40 of 59) in the 5-μg group and 52% (30 of 58) in the 2-μg group. In fact, the best results observed in males (in subgroup 18–35 years, receiving the 5-μg vaccine) were equivalent to the poorer results observed in females (in subgroup 36–65 years, receiving the 2-μg vaccine). The dif-

Table 4. Sex- and Age-Dependent Anti-HB_s Response to 5- and 2-μg Hepatitis B Vaccine Preparations

				Anti-HB_s positive			
HB vaccine	Sex	Age	No.	1 month after V1	1 month after V2	1 month after V3	GMT 1 month after V3 (95% CI)
5 μg	Females	18–35	79	20 (25%)	64 (81%)	78 (99%)	358 (256–500)
		36–65	23	1 (4%)	10 (43%)	21 (91%)	78 (42–144)
	Males	18–35	23	1 (4%)	7 (30%)	17 (74%)	46 (18–120)
		36–65	36	1 (3%)	7 (19%)	23 (64%)	37 (19–72)
2 μg	Females	18–35	76	17 (22%)	56 (74%)	73 (93%)	209 (171–258)
		36–65	23	0	7 (30%)	16 (70%)	46 (22–96)
	Males	18–35	24	0	7 (29%)	13 (54%)	66 (27–161)
		36–65	34	3 (9%)	8 (26%)	17 (50%)	42 (23–76)

Table 5. Sex-Dependent Response to the 5- and 2-μg Hepatitis Vaccines

HB vaccine	Sex	No.	Anti-HB_s positive: 1 month after V1	1 month after V2	1 month after V3		GMT 1 month after V3 (95% CI)	
5 μg	Females	102	21 (21%)	74 (73%)	99 (97%)	$p < 10^{-6}$	244 (178–335)	$p < 10^{-6}$
	Males	59	2 (3%)	14 (24%)	49 (68%)		41 (24–68)	
2 μg	Females	99	17 (17%)	63 (64%)	89 (90%)	$p < 10^{-6}$	160 (115–223)	$p < 10^{-4}$
	Males	58	3 (5%)	15 (26%)	30 (52%)		51 (32–81)	

ference between younger and older males and females was also obvious in terms of anti-HB_s titers (Fig. 4).

2. Newborn Infants of HB_sAg-Carrier Mothers

Preliminary data are available at 6–8 months of age for 76 children: 43 in the group receiving the 5-μg vaccine and 33 in the group receiving the 2-μg vaccine. The overall results are presented in Table 6. The anti-HB_s response of newborn infants to the 2-μg vaccine was significantly lower than that of children receiving the 5-μg vaccine, although the GMT of responders was similar for both groups (83 mIU/ml and 107 mIU/ml for the 5-μg and 2-μg groups, respectively).

Two failures were observed. The first child belonged to the 5-μg group and was born to one of the 11 HB_eAg^+ mothers included in the study. The other child had received the 2-μg vaccine and was born to a HB_eAg^- carrier mother. At 6–8 months, nine (12%) children were still HB_sAg and anti-HB_s^-, thus at risk of late HBV infection acquired from their carrier mother.

D. Discussion

Our initial studies with low-dose HB vaccines were carried out in young female volunteers (Goudeau *et al.*, 1984*b*). The results were found to be very satisfactory and prompted the preparation of a new HB vaccine formula containing 2 μg HB_sAg/dose instead of 5 μg/dose in the standard HB vaccine. Clinical trials of this vaccine showed that our first results could not be extended to other populations such as newborn infants or male adults. In adults, sex and age modulation of the anti-HB_s response played a significant role in the relatively disappointing results obtained with HB vaccine preparation containing only 2 μg of HB_sAg.

The low seroconversion rate in older individuals was not a surprise, since it had already been reported in hemodialysis patients (Goudeau *et al.*, 1983) and in healthy recipients over 60 years of age (Denis *et al.*, 1984). However, the use of a low-dose HB vaccine clearly amplified this phenomenon. This resulted in very low immune responses in subjects of the age group 36–65 receiving the 2-μg vaccine: 58% (33 of 57) as compared with corresponding recipients of the 5-μg vaccine: 75% (44 of 59).

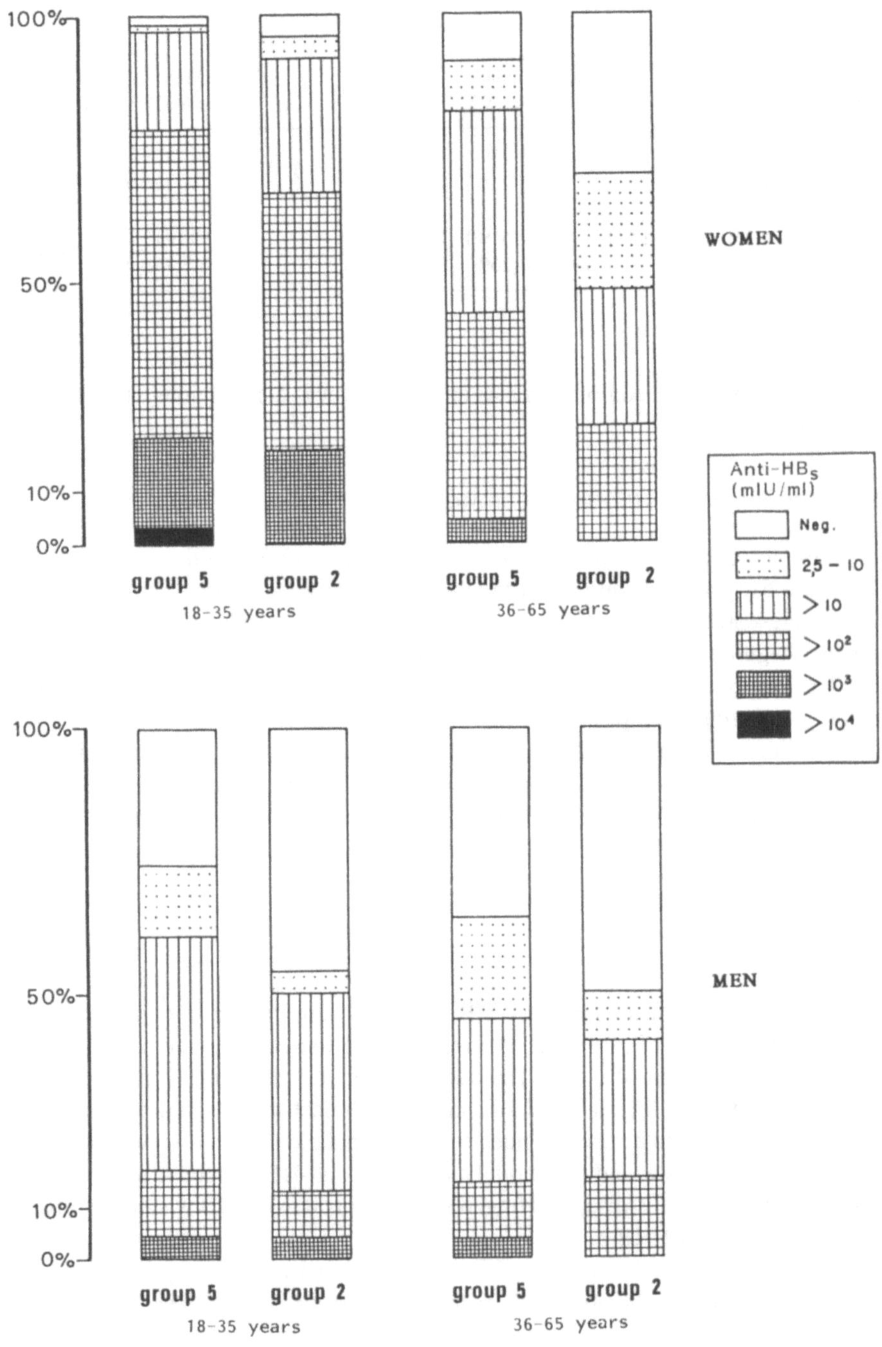

Figure 4. Distribution of anti-HB_s titers according to age and sex in vaccinees receiving the 5-μg or the 2-μg vaccines.

Table 6. Immune Response of Neonates to 5- and 2-μg Vaccines[a]

HB vaccine	N	Anti-HB_s^+/HB_sAg^-		Anti-HB_s^-/HB_sAg^-	Anti-HB_s^-/HB_sAg^+
5 μg	43	39 (91%)		3 (7%)	1 (2%)
			$p < 0.05$		
2 μg	33	26 (79%)		6 (18%)	1 (3%)

[a] HB_sAg and anti-HB_s status at 6–8 months of age.

The influence of sex on the immune response to HB vaccine had been overlooked in earlier studies carried out either in all male (Szmuness *et al.*, 1980) or in mostly female populations (Goudeau *et al.*, 1983). Better anti-HB_s seroconversion rates and higher titers among female volunteers were observed in both groups of our study. However, in women of the 36–65-age group receiving the 2-μg vaccine, the seroconversion rate of 70% (16 of 23) was clearly insufficient. The antibody response of men was poor in all subgroups, even in those receiving the standard 5-μg vaccine.

These disturbing data underline how hazardous it is to apply conclusions obtained after limited clinical trials to the population at large. They have also practical implication in the conduct and surveillance of vaccination programs. For instance, postimmunization control of anti-HB_s seroconversion and persistence seem highly advisable in male vaccinees, especially among those over 30 years of age.

Our clinical trial of the 2-μg vaccine in neonates of HB_sAg-carrier mothers is very limited both in numbers and in time of follow-up. However, our initial data indicate a significantly lower response of newborns to the 2-μg vaccine as compared with that observed after the 5-μg one. Further studies are clearly necessary before releasing this new formula for general use in very young infants, especially those exposed to a high risk of HBV infection.

VI. OVERALL CONCLUSIONS

We investigated two different approaches to reduce the cost of immunization against hepatitis B with a licensed serum-derived vaccine. A modified two-dose regimen may be used with the available vaccine, provided a convenient interval between injections (more than 2 months) is chosen. Such modification may prove very useful in endemic areas of Africa and Asia, where logistics problems and limited funds prohibit vaccination requiring multiple injections. It has been shown that the French HB vaccine could be used with most vaccines for children (Mazert *et al.*, 1983; Chiron *et al.*, 1984). Thus, a two-dose regimen for vaccination against HB may be easily inserted in the Extended Programmes of Immunization currently recommanded by WHO. Several clinical trials are under way to ascertain the feasibility and efficacy of such an approach.

Considering the disappointing results obtained with the 2-μg vaccine, it would seem premature to recommend such preparation for general use. Further studies in various populations including newborn infants, children, and immunocompromised patients are certainly warranted.

How reduction of HB_sAg concentrations or modified two-dose regimen will affect the cost of mass immunization significantly remains to be ascertained. One should also emphasize that all clinical data collected with the currently available serum-derived vaccine will be most valuable to design second-generation vaccines and refine convenient immunization protocols for their clinical use.

ACKNOWLEDGMENTS

We wish to thank Ms. Lacassagne, Ms. Leduc, Ms. Mignot, Ms. Raynaud, and Ms. Simon-Goguelat, members of the staff in schools where the immunization trials were carried out, for their sustained help, and all participants for their remarkable compliance with the protocols. We thank Ms. A. Bergeron for the artwork and Ms. D. Taranne and Ms. M. Soignier for skillful typing.

REFERENCES

Barin, F., Goudeau, A., Denis, F., Yvonnet, B., Chiron, J. P., Coursaget, P., and Diop Mar, I. (1982). *Lancet* **1,** 251–253.

Chiron, J. P., Coursaget, P., Yvonnet, B., Auger, F., Lee Quan, T., Barin, F., Denis, F., and Diop Mar, I. (1984). *Lancet* **1,** 623–624.

Crosnier, J., Jungers, P., Courouce, A. M., Laplanche, A., Benhamou, E., Degos, F., LaCour, B., Prunet, P., Cerisier, Y., and Guesry, P. (1981). *Lancet* **1,** 455–458; 797–800.

Goudeau, A., Dubois, F., Barin, F., Dubois, M. C., and Coursaget, P. (1983). *Dev. Biol. Stand.* **54,** 267–284.

Goudeau, A., Dubois, F., Jeussray, B., DuBois, M. C., Louq, C., Mazert, M. C., and Tron, F. (1984*a*). *Proceedings of the IASL–EASL Symposium, Bern.*

Goudeau, A., Dubois, F., Dubois, M. C., Louq, C., and Mazert, M. C. (1984*b*). *Lancet* **2,** 1091.

Maupas, P., Goudeau, A., Coursaget, P., Drucker, J., and Bagros, P. (1976). *Lancet* **1,** 1367–1370.

Maupas, P., Chiron, J. P., Barin, F., Coursaget, P., Goudeau, A., Perrin, J., Denis, F., and Diop Mar, I. (1981). *Lancet* **1,** 289–292.

Mazert, M. C., Chabanier, G., Adamowicz, P., and Gerfaux, G. (1983). *Dev. Biol. Stand.* **54,** 53–62.

Perrin, J., Ntareme, F., Coursaget, P., and Chiron, J. P. (1985). In *Virus-Associated Cancers in Africa* (A. O. Williams, G. T. O'Connor, and G. B. de The, C. A. Johnson, eds.), pp. 307–318, IARC, Lyons.

Piazza, M., Picciotto, L., Willari, R., Guadagnino, V., Orlando, R., Isabella, L., Macchia, V., Memoli, A. M., Vegnente, A., Borelli, A. M., Scarcella, A., Cascioli, C., Cirrilo, C., Coppola, P., Isabella, E., and Parisi, G. (1985*a*). *Lancet* **1,** 949–951.

Piazza, M., Picciotto, L., Villari, R., Guadagnino, V., Orlando, R., Macchia, V., Memoli, A. M., Vegnente, A., Guida, S., and Fusco, C. (1985*b*). *Lancet* **2,** 1120–1121.

Stevens, C. E., Taylor, P. E., Tong, M. J., Toy, P. T., and Vyas, G. N. (1984). In *Virale Hepatitis and Liver Disease* (G. N. Vyas, J. L. Dienstag, and J. H. Hoffnagle, eds.), pp. 275–291, Grune & Stratton, Orlando, Florida.

Szmuness, W., Stevens, C. E., Harley, E. J., Zang, E. A., Olesko, W. R., William, D. C., Sadovsky, R., Morrison, J. M., and Kellner, A. (1980). *N. Engl. J. Med.* **303,** 833–841.

Yvonnet, B., Coursaget, P., Petat, E. Tortey, E., Diouf, C., Barin, F., Denis, F., Diop Mar, I., and Chiron, J. P. (1984). *J. Med. Virol.* **14,** 137–139.

16

Secretory IgA Antibody Responses after Immunization with Inactivated and Live Poliovirus Vaccine

L. Mellander, B. Carlsson, L. Å. Hanson, F. Jalil, A. L. van Wezel, and S. Zaman

I. INTRODUCTION

Many defense factors are involved in protection of the mucous membranes, preventing adherence and penetration by potential pathogens (Hanson and Brandtzaeg, 1986, Walker, 1976). More than one-half of the lymphoid cells in the body are located close to the mucous membranes, where they produce secretory immunoglobulin A (SIgA), which is the main immunoglobulin in exocrine secretions.

SIgA is a dimer of IgA with a joining chain produced by plasma cells. This molecule is transported onto mucosal membranes by binding to a glycoprotein, the secretory component (SC), which is synthesized by epithelial cells. The addition of the SC to the IgA dimer molecule confers resistance to degradation by enzymes and pH variations, making it functional in the tough milieu on various mucosae (Newcomb *et al.*, 1968; Tomasi and Bienenstock, 1968; Hanson and Brandtzaeg, 1986). SIgA antibodies function on the mucosal membranes through neutralization of toxins and viruses and by inhibition of microbial adherence to epithelial surfaces (Williams and Gibbons, 1972; Taylor and Dimmock, 1985).

After exposure to antigen in the gut, SIgA is produced both locally and in exocrine secretions, from, e.g., the salivary and mammary glands through transport, homing, of committed B lymphocytes from the Peyer's patches to the exocrine glands (Goldblum *et al.*, 1975; Weisz-Carrington *et al.*, 1979). The SIgA antibodies in the breast milk are thus directed against antigens to which the mother and the newborn are exposed in the gut; in

---

L. Mellander, B. Carlsson, and L. Å. Hanson • Department of Clinical Immunology, Guldhedsgatan 10, S-413 46 Göteborg, Sweden; and Department of Pediatrics, East Hospital, S-9 41685 Göteborg, Sweden.
F. Jalil and S. Zaman • Department of Social and Preventive Pediatrics, King Edward Medical College, Lahore, Pakistan. *A. L. van Wezel*† • National Institute for Public Health and Environmental Hygiene, Bilthoven, The Netherlands.
†Deceased.

this way, the breast-fed infant receives passive mucosal protection against relevant antigens (Carlsson *et al.*, 1976). A similar transfer of IgA antibody production to exocrine glands and mucosal membranes results from homing to the mammary glands of lymphocytes after exposure of the bronchus-associated lymphoid tissue to antigen (Hanson and Brandtzaeg, 1986).

As the newborn has very low levels of SIgA antibodies in secretions, the passive protection provided by the breast milk SIgA is of great importance. In the small infant, secretory IgM antibodies also compensate for the relative lack of SIgA in secretions. The adult capacity for SIgA production is usually reached around 1 year of age (Mellander *et al.*, 1984), although intensive antigenic exposure to *Escherichia coli* O-antigens can induce increased levels of SIgA antibodies already at 1 month of age (Mellander *et al.*, 1985).

Bacterial and viral infections often give an antibody response in secretions as well as in serum (Waldman and Ganguly, 1974). A secretory antibody response can be seen after vaccinations, depending on which route of vaccination is used and whether the vaccine is live or inactivated. Using cholera toxoid in an animal model, subcutaneous priming and subsequent repeated oral boosting was found to give the best intestinal immunity (Pierce and Sack, 1977). In other studies, the most efficient mucosal IgA response has been attained by parenteral boosting after mucosal priming. Increased SIgA levels in breast milk were demonstrated against *Vibrio cholera* in naturally primed Pakistani women after parenteral vaccination, but this was not seen in nonexposed Swedish women (Svennerholm *et al.*, 1980). Infection with wild poliovirus and vaccination with live poliovirus vaccine (LPV) gave both a serum and a secretory antibody response, but a secretory antibody response was not found after vaccination with parenterally administered inactivated poliovirus vaccine (IPV) (Ogra *et al.*, 1968). A boosting of the SIgA response was found, however, in milk from Pakistani women after vaccination with IPV (Hanson *et al.*, 1984).

The secretory antibodies are an important part of the intestinal immunity that decreases dissemination of wild-type virus; the presence of such antibodies is therefore of epidemiologic significance (Sabin *et al.*, 1960; Nightingale, 1977). Still circulation of wild poliovirus has ceased even in countries that use IPV only (Böttiger, 1984). This study shows that IPV given in repeated doses can also produce an SIgA response.

II. DETERMINATION OF SIgA ANTIBODIES

The enzyme-linked immunosorbent assay (ELISA) was used for determination of antibodies against poliovirus type I antigen (Carlsson *et al.*, 1985). Poliovirus type I antigen, diluted 1 : 100 in phosphate-buffered saline (PBS) was incubated in microplates (polystyrene, Dynatech, Alexandra, Virginia) overnight for antigen coating. Saliva and milk samples were added to the washed plates in 10-fold dilutions in PBS containing 0.05% Tween 20 (Merck, Darmstadt, West Germany) and incubated at room temperature for 4 hr. Antisera specific against human IgA, IgM, IgG (Dakopatts AS, Copenhagen, Denmark) or SC (Seward Laboratories, London) were conjugated to alkaline phosphatase (Boehringer-Mannheim, Mannheim, West Germany) and used to detect antibodies belonging to the various immunoglobulin classes. The microplates were read in a Titertek

Multiskan (Flow Laboratories, Ayrshire, Scotland) at 405 nm after 100-min reaction time with the enzyme substrate, *p*-nitrophenylphosphate (Sigma Chemical Co., St. Louis, Missouri) in 1 M diethanolamine buffer containing 0.001 M Mg Cl_2, pH 9.8. Titers were given as means of duplicate determinations and represent the inverse of the interpolated dilutions, giving an absorbance of 0.2 above the background.

Immunoglobulin class-specific antibody titers against poliovirus antigens as determined by ELISA are not easily comparable with neutralization titers. A breast milk pool containing ELISA titers of 1 : 10 for IgG, 1 : 6400 for IgA, and 1 : 40 for IgM antibodies had a neutralization titer of 1 : 8. A standard serum that was tested in both systems with a neutralization titer of 1 : 400 had ELISA titers of 1 : 8000 for IgG, 1 : 2900 for IgA, and 1 : 935 for IgM antibodies. The specificity of the antigen–antibody reactions were tested in inhibition experiments using excess of homologous antigen and poliovirus type III antigen as well as a pool of somatic antigens from common *E. coli* strains. Only the homologous antigens could inhibit the antigen–antibody reactions in ELISA. Precautions were also taken to minimize the risk of cross-reactivity with other viruses via the C antigen. The poliovirus antigen was stored at −70°C on arrival and was absorbed with DEAE-Sephadex once more in connection with the coating procedure to remove C antigen (van Wezel and Hazendonk, 1979).

The live vaccine used was trivalent Oral-Virelon (kindly provided by Behringwerke AG, Marburg, West Germany) and given in three drops orally. Swedish inactivated trivalent poliovirus vaccine (National Bacteriological Laboratory, Solna, Sweden) was given as 1.0 ml SC. The Dutch vaccine produced at the Rijks Instituut voor de Volksgezondheid en Milieuhygiene (RIVM), Bilthoven, the Netherlands, was a trivalent inactivated vaccine containing 40 DU of type I, 8 DU of type II, and 32 DU of type III antigen and given in a dose of 1.0 ml S.C.

III. THE SIgA RESPONSE AFTER NATURAL EXPOSURE AND VACCINATION

A. In Children and Adults

The salivary immune responses were studied in children and adults in Pakistan and Sweden. Following natural exposure, the antibody response was studied in 81 Pakistani infants from birth to 6 months of age, both longitudinally up to 3 months of age and then cross-sectionally, as well as in 20 adults. In the same endemic area, 80 children were studied after vaccination with live or inactivated vaccines given in three or four doses at 2–7 months of age. Sixty-three Swedish children and adults with close to 100% coverage with inactivated vaccine at 8, 9, and 18 months and 5 years and no exposure to circulating poliovirus were studied as well.

The salivary SIgA antibodies to poliovirus type I antigen began to appear at 1 month of age in the unvaccinated Pakistani infants, reaching the levels obtained in unvaccinated Pakistani adults at 6 month of age. After completion of the vaccination at 8 months of age, the vaccinated children showed no difference in salivary antibody levels, whether in the group given IPV or that given LPV. All Swedish children lacked SIgA antibodies to polivorius type I antigen at the first vaccination, but after the fourth vaccination they all had such antibodies. The levels in Swedish adults who had been vaccinated as children

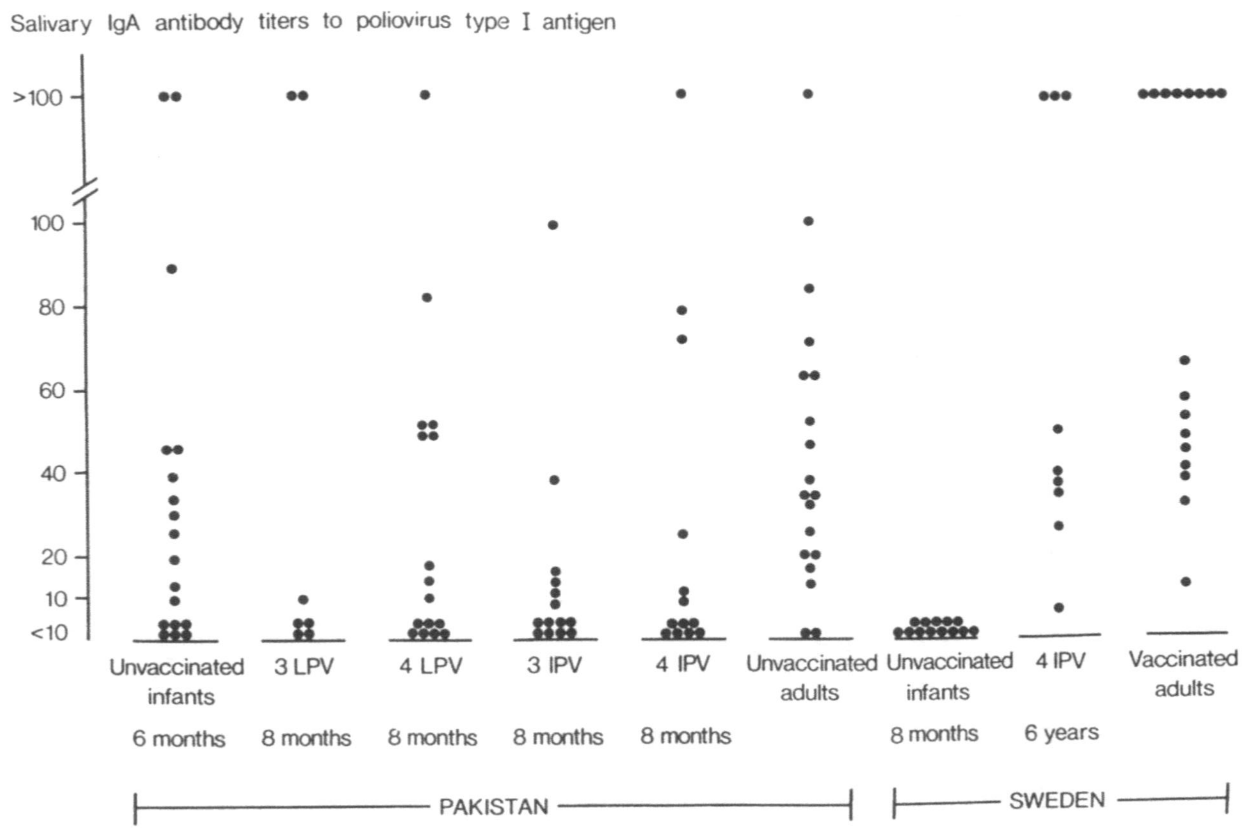

Figure 1. Salivary secretory immunoglobulin A (SIgA) antibodies against poliovirus type I antigen in groups of Pakistani and Swedish infants and adults. The vaccinated Pakistani infants were given either three or four doses of inactivated poliovirus vaccine (IPV) and live poliovirus vaccine (LPV), respectively, and tested at 8 months of age. Swedish infants were tested before the first vaccination at 8 months of age and after the fourth vaccination at 6 years; vaccinated adults had been vaccinated in childhood.

were in the same range. The Swedish adults had significantly higher levels than did the unvaccinated Pakistani adults (Carlsson *et al.*, 1985) (Fig. 1).

The disappearance of circulating poliovirus from countries using IPV is only partly explained by transudation of serum antibodies reducing virus excretion from the pharynx and affecting the mainly oral spread of the virus (Marine *et al.*, 1962). Our finding of SIgA antibodies to poliovirus type I antigen after repeated vaccinations with IPV may help explain the disappearance of circulating poliovirus in populations only vaccinated with such vaccine. In the endemic situation, the LPV and IPV poliovirus vaccines gave the same SIgA response, when given in repeated doses. Inhibition experiments support the specificity of the antibodies, but a boosting of the SIgA response by cross-reacting antigens cannot be completely ruled out.

B. In Lactating Women

Three groups of unvaccinated Pakistani mothers, 121 in total, were vaccinated with the three different vaccine preparations, LPV, and Swedish or Dutch IPV, in one or two doses at different intervals beginning at least 2 weeks after parturition. The milk SIgA antibodies against poliovirus type I decreased after the first dose of LPV (Fig. 2). When Swedish IPV was given as a second dose to the same individuals, the milk IgA antibodies increased significantly, independent of the time interval between the two doses, 2 or 4 weeks. These results are in agreement with our earlier findings (Hanson *et al.*, 1984).

When Swedish IPV only was given to Pakistani women, an increase was seen in SIgA milk antibodies in all the groups given either a single dose or two doses at 2- or 4-week intervals. The increase was not very long-lasting, and the antibody titers fell to prevaccination levels within 6–8 weeks. A more marked SIgA antibody increase was seen when the Dutch IPV was used. Within the first weeks after vaccination, the antibody levels were increased up to 50-fold, but the response was again rather short-lasting and dropped within 8 weeks to starting levels. The prevaccination levels were regularly quite high in all groups, probably due to the circulating wild poliovirus in the population. During the vaccination period, a control group of 11 women was followed, and their milk antibody levels showed only minor fluctuations over the entire study period.

Our investigation indicates that it is possible to affect the SIgA polio antibody levels in milk using both LPV and IPV. The LPV seemed to decrease the preexisting antibody levels quite regularly, whereas IPV given as a booster dose 2 weeks later could induce a response (Fig. 2). The responsible mechanism could be that the oral vaccine induces a temporary suppressor effect that can be overcome by the parenteral dose.

IV. CONCLUSION

Inactivated as well as live poliovirus vaccines produced a mucosal SIgA response. In the endemic situation and after repeated vaccinations in children, there was no difference in the SIgA antibody levels attained in the saliva. Even in a country without circulating poliovirus, IPV in repeated doses gave a mucosal SIgA response.

In breast milk, IPV was demonstrated effective in inducing a SIgA response, especially in boosting the response in mothers previously exposed to wild poliovirus. The

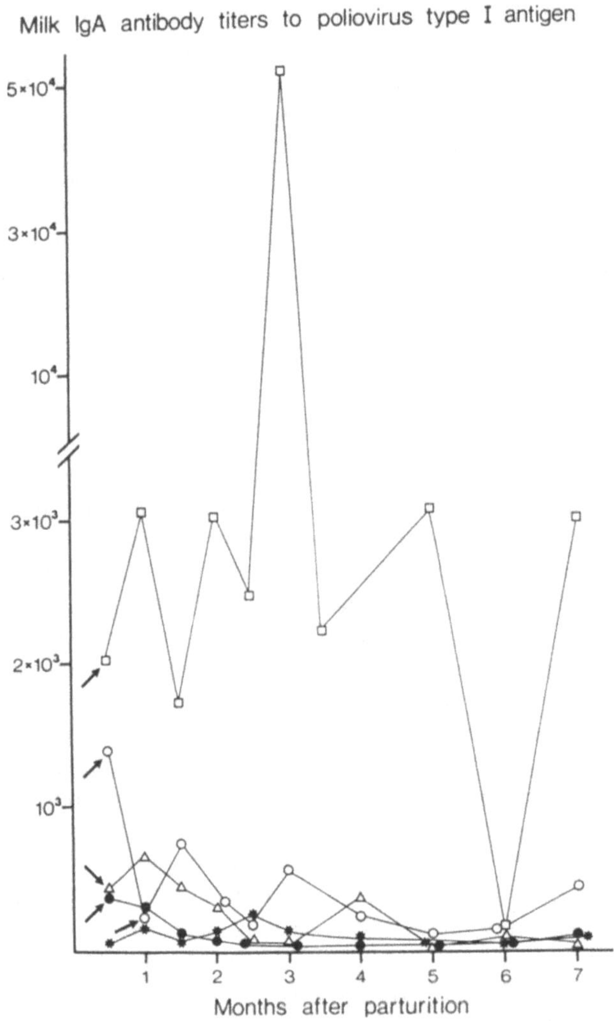

Figure 2. Secretory (SIgA) antibodies against poliovirus type I antigen in milk from Pakistani mothers given live poliovirus vaccine (LPV) (●), Swedish inactivated poliovirus vaccine (IPV) (△), or Dutch IPV (□) at least 2 weeks after parturition. One group was first given LPV and a second dose with Swedish IPV after 2 weeks (○). Asterisk (*) indicates unvaccinated controls. (↑) indicates vaccination.

more antigen-rich Dutch IPV yielded the most striking increase. The breast milk antibodies against poliovirus antigen may be of value especially in developing countries, since poliomyelitis is common in the youngest age groups, but the protective value of the milk SIgA antibodies is yet to be evaluated. It is obvious that the development of a new more potent antigen-rich vaccine for parenteral use is of great interest because it provides mucosal SIgA antibodies for both mothers and infants, in addition to providing serum antibodies.

ACKNOWLEDGMENTS

These studies were supported by grants from the Swedish Agency for Research Cooperation with Developing Countries, the Swedish Medical Research Council (No. 215), the Ellen, Walter and Lennart Hesselman Foundation for Research, the First of May Flower Annual Campaign for Children's Health, and the Swedish National Bank Foundation.

REFERENCES

Böttiger, M. (1984). *Rev. Infect. Dis.* **6,** S548–551.

Carlsson, B., Gothefors, L., Ahlstedt, S., Hanson, L. Å., and Winberg, J. (1976). *Acta Paediatr. Scand.* **65,** 216–224.

Carlsson, B., Zaman, S., Mellander, L., Jalil, F., and Hanson, L. Å. (1985). *J. Infect. Dis.* **152,** 1238–1244.

Goldblum, R. M., Ahlstedt, S., Carlsson, B., Hanson, L. Å., Jodal, U., Lidin-Janson, G., and Sohl-Åkerlund, A. (1975). *Nature (Lond.)* **257,** 797–799.

Hanson, L. Å, and Brandtzaeg, P. In *Immunologic Disorders in Infants and Children* (E. R. Stiehm, ed.), 3rd ed., WB Saunders, Philadelphia (in press).

Hanson, L. Å, Carlsson, B., Jalil, F., Lindblad, B., Raza Kahn, S., and van Wezel, A. L. (1984). *Rev. Infect. Dis.* **6,** S356–360.

Mellander, L., Carlsson, B., and Hanson, L. Å. (1984). *J. Pediatr.* **104,** 564–568.

Mellander, L., Carlsson, B., Jalil, F., Söderström, T., and Hanson, LÅ. (1985). *J. Pediatr.* **107,** 430–433.

Newcomb, R. W., Normansell, D., and Stanworth, D. R. (1968). *J. Immunol.* **101,** 905–914.

Nightingale, E. O. (1977). *N. Engl. J. Med.* **297,** 249–253.

Ogra, P. L., Karzon, D. T., Righthand, F., and MacGillivray, M. (1968). *N. Engl. J. Med.* **279,** 893–900.

Pierce, N. F., and Sack, R. B. (1977). *J. Infect. Dis.* **136,** S113–117.

Sabin, A. B., Ramos-Alvarez, M., Alvarez-Amezquita, J., Pelon, D. F. W., Michaels, R. H., Spigland, I., Koch, M. A., Barnes, J. M., and Rhim, J. S. (1960). *JAMA* **173,** 1521–1526.

Svennerholm, A. M., Hanson, L.Å., Holmgren, J., Lindblad, B. S., Nilsson, B., and Quereshi, F. (1980). *Infect. Immun.* **30,** 427–430.

Taylor, H. P., and Dimmock, N. J. (1985). *J. Exp. Med.* **161,** 198–209.

Tomasi, T. B., and Bienenstock, J. (1968). In *Adv. Immunol.* **9,** 1–96.

van Wezel, A. L., and Hazendonk, A. G. (1979). *Intervirology* **11,** 2–8.

Waldman, R. H., and Ganguly, R. (1974). *J. Infect. Dis.* **130,** 419–440.

Walker, W. A. (1976). *Pediatrics* **57,** 901–916.

Weisz-Carrington, P., Roux, M. E., McWilliams, M., Phillips-Qaugliata, J. M., and Lamm, M. E. (1979). *J. Immunol.* **123,** 1705–1708.

Williams, R. C., and Gibbons, R. J. (1972). *Science* **177,** 697–699.

IV

Antiviral Chemotherapy

17

Nucleoside Analogues in the Chemotherapy of Viral Infections

Recent Developments

Erik De Clercq

I. INTRODUCTION

Various nucleoside analogues with promising antiviral activity were recently described. They can be divided into two classes: pyrimidine nucleoside analogues, and purine nucleoside analogues.

II. PYRIMIDINE NUCLEOSIDE ANALOGUES

A. Bromovinyldeoxyuridine

Bromovinyldeoxyuridine (BVDU) was originally described by De Clercq *et al.* (1979) (Fig. 1) as a potent antiherpetic agent and still ranks among the most potent and most selective inhibitors of herpes simplex virus type 1 (HSV-1) and varicella–zoster virus (VZV) reported to date. Its antiviral activity spectrum encompasses quite a wide variety of herpes viruses, including Epstein–Barr virus (EBV) and several herpes viruses of veterinary importance (Table 1). The *in vitro* and *in vivo* efficacy of BVDU has been demonstrated in various experimental model systems (De Clercq and Walker, 1984); a number of clinical studies point to its therapeutic value in the topical treatment of herpetic eye infections (Maudgal *et al.*, 1984, 1985*a*) and systemic (peroral) treatment of HSV-1 and VZV infections in immunosuppressed patients (Wildiers and De Clercq, 1984; Benoit *et al.*, 1985; Maudgal *et al.*, 1985*b*; De Clercq *et al.*, 1985*b*; Tricot *et al.*, 1986). Double-blind controlled clinical trials have been initiated in which BVDU is compared with acyclovir (ACV) (Zovirax) in the systemic (peroral versus intravenous) treatment of VZV infections in immunosuppressed patients. Since the spectrum of BVDU extends to EBV,

Erik De Clercq • Rega Institute for Medical Research, The Catholic University of Louvain, B-3000 Louvain, Belgium.

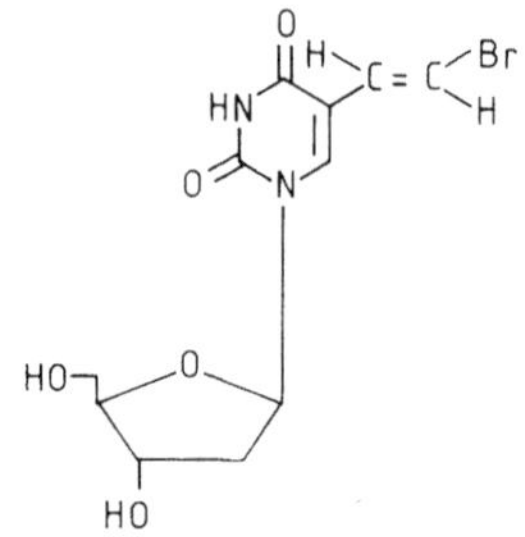

(*E*)-5-(2-Bromovinyl)-2′-deoxyuridine

Figure 1. Bromovinyldeoxyuridine (BVDU).

SHV-1, BHV-1 (Table 1), and bovine ulcerative mammilitis virus, BVDU should be further explored for its potential in the treatment of EBV-associated diseases (i.e., infectious mononucleosis, Burkitt lymphoma, nasopharyngeal carcinoma, and certain lymphosarcomas in immunosuppressed patients), SHV-1 infection (pseudorabies) in pigs, and infectious rhinotracheitis and ulcerative mammilitis in cows.

Bromovinyldeoxyuridine is an efficient substrate for pyrimidine nucleoside phosphorylases, i.e., 2′-deoxythymidine (dThd) phosphorylase (Desgranges *et al.*, 1983), which cleave the *N*-glycosidic linkage so as to release the free pyrimidine base bromovinyluracil (BVU). BVU has no antiviral activity by itself (De Clercq *et al.*, 1986*a*) but can be reconverted to BVDU through a pentosyl transfer reaction with any 2′-deoxynucleoside as the 2′-deoxyribosyl donor (Desgranges *et al.*, 1984). This regeneration process can be achieved repeatedly as long as BVU persists in the bloodstream (Desgranges *et al.*, 1985). In this sense, BVU can be considered a prodrug of BVDU. BVU also has therapeutic potential as an adjunct antitumor agent. Being itself not a substrate of dihydrothymine dehydrogenase (the enzyme that initiates the catabolic pathway of pyrimidines), BVU inhibits the degradation of 5-fluorouracil (5-FU) by this

Table 1. Antiviral Activity Spectrum of BVDU[a,b]

Herpes simplex virus type 1 (HSV-1)
Varicella–zoster virus (VZV)
Epstein–Barr virus (EBV)[c]
Suid herpes virus type 1 (SHV-1)
Bovid herpes virus type 1 (BHV-1)
Simian varicella virus (SVV)
Herpes virus platyrrhinae (HVP)
Multiple nucleocapsid polyhedrosis virus (MNPV)

[a] Minimum inhibitory concentration: ≤0.1 μg/ml.
[b] According to De Clercq (1984) and De Clercq and Walker (1984).
[c] See Lin *et al.* (1985).

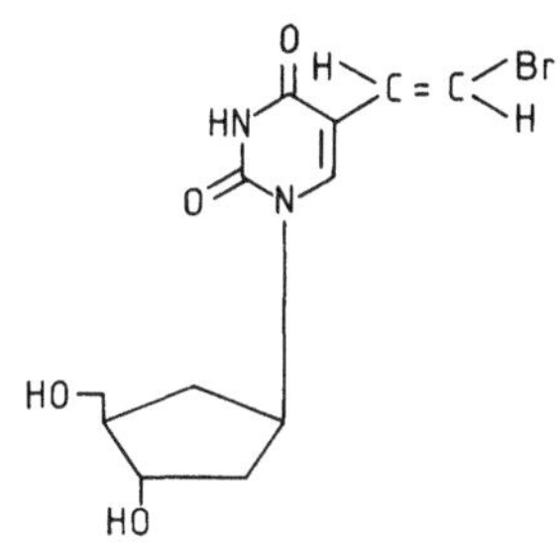

Carbocyclic (E)-5-(2-bromovinyl)-2'-deoxyuridine

Figure 2. Carbocyclic bromovinyldeoxyuridine (C-BVDU).

enzyme and, concomitantly, enhances the antitumor activity of 5-FU (Desgranges *et al.*, 1986b). The enhancing effect of BVU on the antitumor activity of 5-FU has been clearly demonstrated in mice bearing MOPC-315 plasmacytoma tumors (Ben-Efraim *et al.*, 1986).

B. Carbocyclic Bromovinyldeoxyuridine

Carbocylic bromovinyldeoxyuridine (C-BVDU) is completely resistant to degradation by dThd phosphorylase (De Clercq *et al.*, 1985*d*) (Fig. 2) and is as good, if not better, a substrate of the HSV-1 dThd kinase than BVDU (De Clercq *et al.*, 1985*a*). C-BVDU is not recognized as a substrate by cellular dThd kinase, which means that its phosphorylation (to the 5′-mono-, di-, and triphosphate) is restricted to the virus-infected cell (De Clercq *et al.*, 1985*c*). Like BVDU, C-BVDU is ultimately incorporated into DNA of the virus-infected cell (De Clercq *et al.*, 1985*c*); to what extent this incorporation alters the conformation and various biologic functions of the DNA is an interesting lead of research that is currently being pursued. Since it is not degraded to BVU, C-BVDU should have a greater bioavailability and, perhaps, greater *in vivo* efficacy than BVDU. Preliminary experiments with C-BVDU *in vivo,* however, did not seem to indicate that it was more effective than BVDU in the systemic or topical treatment of HSV-1 infections in mice (Herdewijn *et al.*, 1985).

C. Bromovinyluracil Arabinoside

Bromovinyluracil arabinoside (BVaraU), a closely related analogue of BVDU, is, like BVDU, an effective anti-HSV-1 and anti-VZV agent (Machida *et al.*, 1981, 1982) (Fig. 3). Its *in vivo* efficacy has been demonstrated in some animal models, i.e., HSV-1 encephalitis (Machida and Sakata, 1984; Reefschläger *et al.*, 1986) and simian varicella (Soike *et al.*, 1984). The anti-HSV-1 activity of BVaraU is more stringently dependent on the choice of the cell system than that of BVDU (De Clercq, 1982). A possible advantage of BVaraU over BVDU is that it is not recognized as substrate by dThd phosphorylase (De Clercq *et al.*, 1985*d*), suggesting that it may have a longer bioavailability *in vivo* than

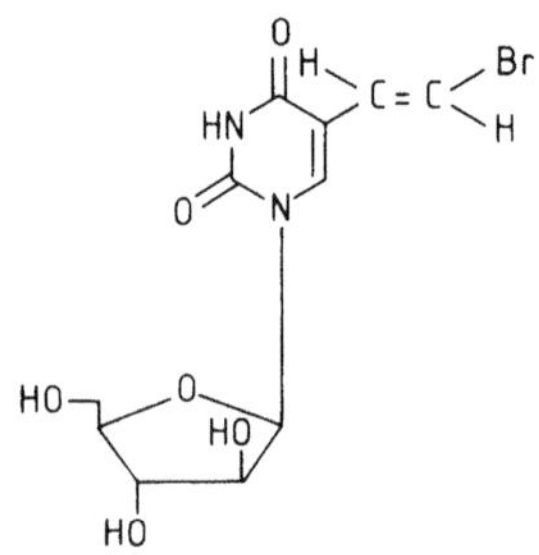

(E)-5-(2-Bromovinyl)-1-β-D-arabinofuranosyluracil

Figure 3. Bromovinyluracil arabinoside (BVaraU).

BVDU. Whether this longer bioavailability would have any therapeutic implications and how BVaraU is eventually metabolized and eliminated by the organism are important issues that remain to be resolved.

D. Other BVDU Derivatives

In addition to C-BVDU and BVaraU, various other structurally related analogues of BVDU have been described, including bromovinyldeoxycytidine (BVDC) (Fig. 4) and the 3′-amino and 3′-azido derivatives of BVDU (Busson *et al.*, 1981) BVDC is almost equally potent against HSV-1 as BVDU but, as its processing by the cells requires an additional deamination step at either the nucleoside or nucleotide level, BVDC gains some extra selectivity over that of BVDU (De Clercq *et al.*, 1982*a*). The 3′-azido and 3′-amino derivatives of BVDU are less potent anti-HSV-1 agents than BVDU but, like BVDU itself, they depend for their antiviral activity on a specific phosphorylation by the viral dThd kinase (De Clercq *et al.*, 1982*b*, 1983). Neither the 3′-azido nor the 3′-amino derivative of BVDU is a substrate for dThd phosphorylase (De Clercq *et al.*, 1985*d*). The *in vivo* efficacy of these derivatives, as well as that of BVDC, remains to be explored.

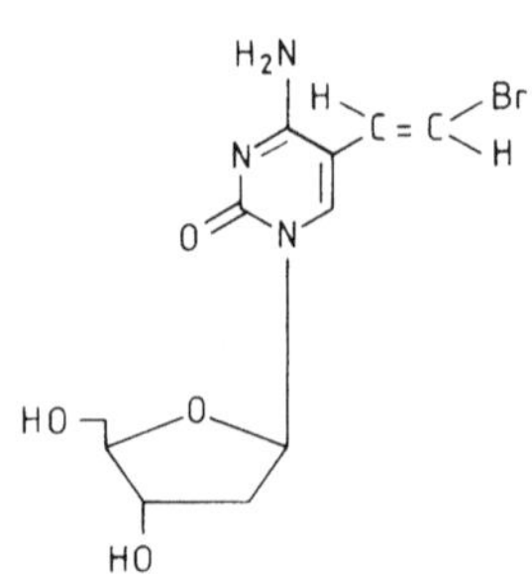

(E)-5-(2-Bromovinyl)-2′-deoxycytidine

Figure 4. Bromovinyldeoxycytidine (BVDC).

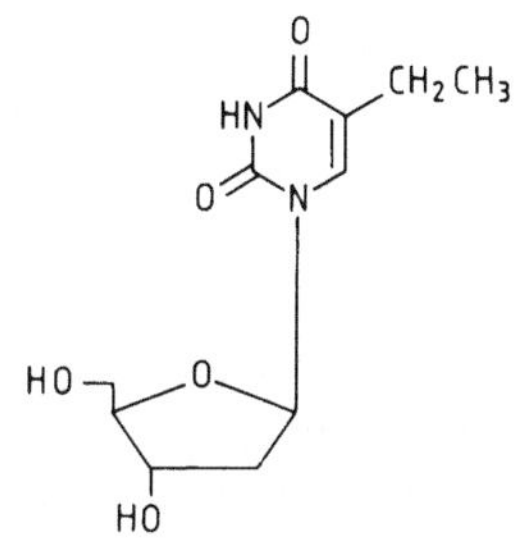

5-Ethyl-2'-deoxyuridine

Figure 5. Ethyldeoxyuridine (EDU).

E. Ethyldeoxyuridine

Ethyldeoxyuridine (EDU) has been known since 1967 (De Clercq and Shugar, 1975) (Fig. 5). Interest in the clinical potential of this drug has revived, since Schinazi *et al.* (1985) and Spruance *et al.* (1985) demonstrated its effectiveness in the topical treatment of HSV-1 and HSV-2 infections in mice and guinea pigs. Studies aimed at elucidating the mechanism of action of EDU were recently initiated in our laboratory; our preliminary findings indicate that [4-^{14}C]-EDU is phosphorylated in both mock- and HSV-infected cells, albeit to a larger extent in the virus-infected than mock-infected cells, and that [4-^{14}C]-EDU is finally incorporated into DNA of both mock- and HSV-infected cells, although, again, to a larger extent into DNA of virus-infected than mock-infected cells. Concomitantly with the incorporation of [4-^{14}C]-EDU into viral and cellular DNA of HSV-infected cells, there is a marked reduction in DNA synthesis and this reduction is more pronounced for viral DNA than for cellular DNA (R. Bernaerts and E. De Clercq, unpublished data, 1986).

F. Chloroethyldeoxyuridine

Chloroethyldeoxyuridine (CEDU) represents another dThd analogue recently synthesized by Griengl *et al.* (1985) (Fig. 6). CEDU is about 10 times less potent than BVDU

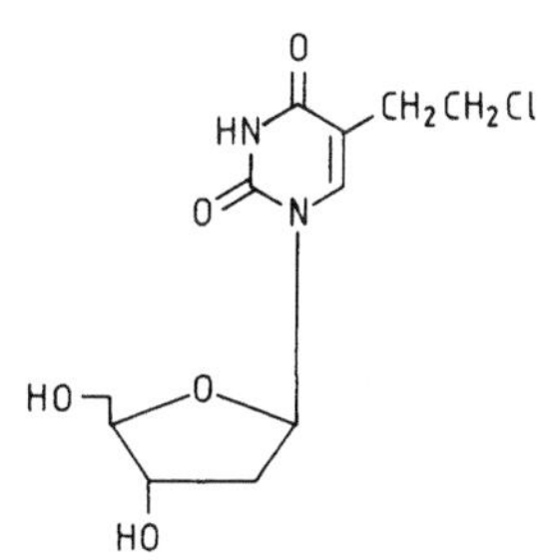

5-(2-Chloroethyl)-2'-deoxyuridine

Figure 6. Chloroethyldeoxyuridine (CEDU).

against HSV-1 *in vitro* and about equally active as BVDU in the topical treatment of cutaneous HSV-1 lesions in hairless mice; when given systemically, however, CEDU appears to be effective at a 5 to 15-fold lower dose than BVDU or ACV (De Clercq and Rosenwirth, 1985; Rosenwirth *et al.*, 1985). CEDU also holds great promise for the topical treatment of both epithelial and stromal HSV-1 keratitis and HSV-1 uveitis, readily penetrating the cornea (Maudgal and De Clercq, 1985; Maudgal *et al.*, 1986). It is not clear why CEDU would be more effective than BVDU upon systemic (peroral or intraperitoneal) administration. This difference in efficacy does not seem to be related to differences in phosphorolytic cleavage by pyrimidine nucleoside phosphorylases (i.e., dThd phosphorylase), since CEDU is as good a substrate for this enzyme as other dThd analogues, and it is as rapidly cleared from the bloodstream as is BVDU (Desgranges *et al.*, 1986*a*).

G. Azidothymidine

Azidothymidine (AZT) represents a 3′-modified thymidine derivative, first synthesized by Horwitz *et al.* (1964) and later by Lin and Prusoff (1978) and Colla *et al.* (1985) (Fig. 7). The compound has no activity against HSV-1 but is a rather potent inhibitor of cellular DNA synthesis, as demonstrated with both primary rabbit kidney and murine leukemia L1210 cells (De Clercq *et al.*, 1980). Mitsuya *et al.* (1985) recently showed that AZT is a potent and selective inhibitor of the replication of human T-cell lymphotropic virus type III/lymphadenopathy-associated virus (HTLV-III/LAV), the etiologic agent of acquired immunodeficiency syndrome (AIDS). It fully protects ATH8 cells, a T4 cell clone, against the cytopathic effect of HTLV-III/LAV at a concentration of about 1 μM, which is at least 50-fold lower than the concentration at which the growth of the uninfected cells is impaired. AZT is assumed to interact selectively with the HTLV-III/LAV reverse transcriptase following a nonselective phosphorylation by the host cell enzymes. Preliminary findings point to a partial immunologic reconstitution and clinical improvement in certain patients with AIDS or AIDS-related complex (ARC) following a 6-week course with AZT (Yarchoan *et al.*, 1986), but it remains to be seen whether these improvements will be sustained and whether AZT would ultimately affect the outcome of the disease.

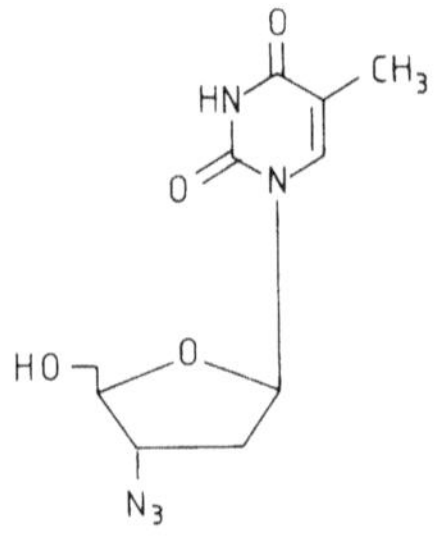

3′-Azido-2′,3′-dideoxythymidine

Figure 7. Azidothymidine (AZT).

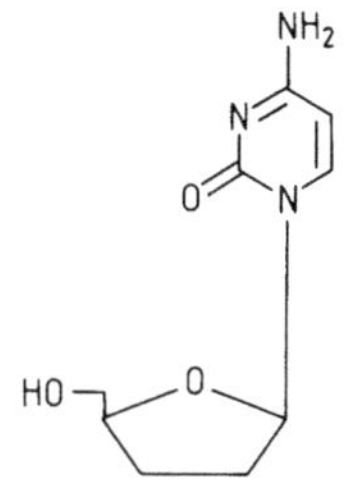

Figure 8. Dideoxycytidine (ddCyd).

H. Dideoxycytidine

The 2′,3′-dideoxynucleosides ddAdo, ddGuo, ddIno, ddCyd, and ddThd are all potent and selective inhibitors of HTLV-III/LAV replication, ddCyd (Fig. 8) being the most potent of the series (Mitsuya and Broder, 1986). At 0.5 μM, it confers complete protection of ATH8 cells against the cytopathogenicity of HTLV-III/LAV, while not being toxic for the host cells at concentrations up to 25 μM. The 2′,3′-dideoxynucleosides are assumed to be converted intracellularly to the corresponding 5′-triphosphates, upon which they may interact with the reverse transcriptase (Waqar *et al.*, 1984). This notion also holds for ddCyd (Cooney *et al.*, 1986). If ddCyd or any other 2′,3′-dideoxynucleoside were incorporated onto the 3′-end of the growing DNA chain, further elongation of the chain would be prevented. The reason(s) for the selectivity of ddCyd and the other 2′,3′-dideoxynucleosides as inhibitors of HTLV-III/LAV replication remain to be clarified. They may owe this selectivity to either a greater accessibility or a greater affinity (or both) of the 2′,3′-dideoxynucleoside 5′-triphosphates for the HTLV-III/LAV reverse transcriptase than for the cellular DNA polymerases. It is likely that in the near future other nucleoside analogues (or combinations thereof) may be found that are as effective, if not more effective, inhibitors of HTLV-III/LAV than AZT or ddCyd. It is noteworthy in this regard that, when combined with dThd, the protective effect of ddCyd against the cytopathogenicity of HTLV-III/LAV was strongly potentiated (J. Balzarini, H. Mitsuya, E. De Clercq, and S. Broder, unpublished data, 1986).

III. PURINE NUCLEOSIDE ANALOGUES

A. Dihydroxypropyladenine

Dihydroxypropyladenine (DHPA) represents the first inhibitor of S-adenosylhomocysteine (SAH) hydrolase—a key enzyme in transmethylation reactions involved in the maturation of viral mRNA (Votruba and Holý, 1980)—which was recognized for its broad-spectrum antiviral properties (De Clercq *et al.*, 1978; De Clercq and Holý, 1979) (Fig. 9). Its antiviral spectrum encompasses both DNA viruses (poxviruses: vaccinia) and RNA viruses [(−)RNA viruses: paramyxo- (parainfluenza, measles), rhabdo- (vesicular

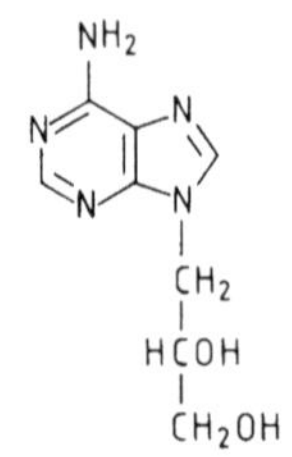

(S)-9-(2,3-Dihydroxypropyl)adenine

Figure 9. Dihydroxypropyladenine (DHPA).

stomatitis, rabies); (+)RNA viruses: retro- (Rous sarcoma) and (±)RNA viruses: reo (rota)] (Table 2). Also included in the activity spectrum of DHPA are some fish and plant viruses. However, relatively high concentrations are required for DHPA to achieve its antiviral activity.

B. Adeninylhydroxypropanoic Acid

Adeninylhydroxypropanoic acid (AHPA) is a more potent inhibitor of SAH hydrolase than DHPA (Holý *et al.*, 1985) (Fig. 10). Also, the alkyl esters of AHPA appear to be more potent antiviral agents than DHPA (De Clercq and Holý, 1985). Their activity spectrum appears to be similar to that of DHPA. It is postulated that the AHPA alkyl esters are as such taken up by the cells and hydrolyzed within the cell to release the parent compound, AHPA, which would then interact with the SAH hydrolase.

Table 2. Antiviral Activity Spectrum of DHPA[a–c]

Vaccinia virus
Parainfluenza virus
Measles virus
Rous sarcoma virus
Vesicular stomatitis virus
Rabies virus
Reovirus
Rotavirus
Infectious pancreatic necrosis virus
Tobacco mosaic virus
Potato virus X

[a] Minimum inhibitory concentration: ≤100 μg/ml.
[b] According to De Clercq *et al.* (1978, 1984); De Clercq and Holý (1979); Kara *et al.* (1979); Sodja and Holý (1980); Smee *et al.* (1982); Bussereau *et al.* (1983); Kitaoka *et al.* (1986); J. Bernstein (unpublished data, 1983); T. Kimura (unpublished data, 1986); G. De Fazio (unpublished data, 1986); B. Lerch (unpublished data, 1980).
[c] Also extends to other SAH hydrolase inhibitors, such as AHPA, C-cAdo, and neplanocin A.

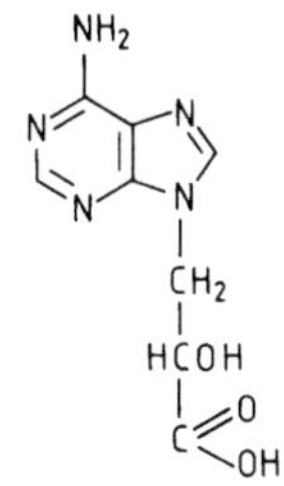

(*RS*)-3-Adenin-9-yl-2-hydroxypropanoic acid

Figure 10. Adeninylhydroxypropanoic acid (AHPA).

C. Carbocyclic Deazaadenosine

Carbocyclic deazaadenosine (C-cAdo), or the carbocyclic analogue of 3-deazaadenosine, is both a more potent inhibitor of SAH hydrolase (Montgomery *et al.*, 1982) and a more potent antiviral agent (De Clercq and Montgomery, 1983) than DHPA and AHPA (Fig. 11). Yet, its activity spectrum is remarkably similar to that of DHPA and AHPA (De Clercq *et al.*, 1984). C-cAdo has been found effective *in vivo* against experimental infections in mice with vesicular stomatitis virus (De Clercq and Montgomery, 1983) and vaccinia virus (De Clercq *et al.*, 1984). Its efficacy in the therapy of rabies and rotavirus infections is being examined.

D. Neplanocin A

Neplanocin A can be considered a cyclopentenyl derivative of adenine (Fig. 12). It was described by Borchardt *et al.* (1984) as a potent inhibitor of SAH hydrolase. Neplanocin A has a broad antiviral activity spectrum, similar to that of DHPA, AHPA, and C-cAdo (De Clercq, 1985), but is more potent than its congeners. The antiviral potency increases in the order DHPA < AHPA < C-cAdo < neplanocin A, as does their inhibitory potency for SAH hydrolase. In fact, a close correlation has been established between the antiviral potency of these compounds (i.e., against vesicular stomatitis virus) and their K_i/K_m values for bovine liver SAH hydrolase activity (measured in the direction of SAH

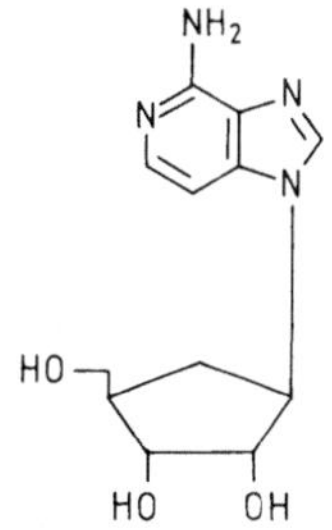

Carbocyclic 3-deazaadenosine

Figure 11. Carbocyclic deazaadenosine (C-cAdo).

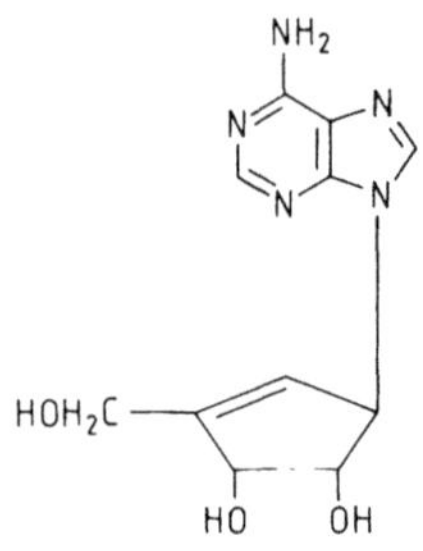

Neplanocin A

Figure 12. Neplanocin A.

synthesis) (De Clercq and Cools, 1985). The *in vivo* antiviral activity of neplanocin A remains the subject of further study. Although the compound is fairly toxic *in vivo* (De Clercq, 1985), marked activity against vaccinia virus-induced tail lesions can be achieved with neplanocin A given at nontoxic doses to mice (E. De Clercq, unpublished data, 1986).

E. Hydroxyphosphonylmethoxypropyladenine

Hydroxyphosphonylmethoxypropyladenine (HPMPA) represents a novel compound with a uniquely broad anti-DNA virus activity spectrum (Table 3; Fig. 13). It can be

Table 3. Antiviral Activity Spectrum of HPMPA[a]

Adenovirus (types 1–8)
Herpes simplex virus type 1 (HSV-1)
Herpes simplex virus type 2 (HSV-2)
Varicella–zoster virus (VZV)
Cytomegalovirus (CMV)
Suid herpes virus type 1 (SHV-1)
Bovid herpes virus type 1 (BHV-1)
Equid herpes virus type 1 (EHV-1)
Herpes virus platyrrhinae (HVP)
Seal herpes virus (SeHV)
African swine fever virus (ASFV)
TK$^-$ (dThd kinase-deficient) HSV-1
TK$^-$ (dThd kinase-deficient) VZV
PAA^r (phosphonoacetate-resistant) HSV-1
Vaccinia virus
Murine (Moloney) sarcoma virus

[a] Minimum inhibitory concentration: $\leq$1–2 μg/ml.
[b] According to De Clercq *et al.* (1986*b*); Maudgal *et al.* (1986*b*); Osterhaus *et al.* (1987); Baba *et al.* (1986); E. De Clercq (unpublished data, 1986); C. Gil-Fernandez (unpublished data, 1986).

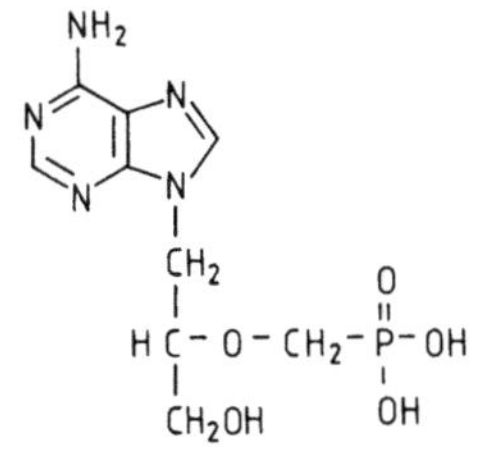

(S)-9-(3-hydroxy-2-phosphonylmethoxypropyl) adenine

Figure 13. Hydroxyphosphonylmethoxypropyladenine (HPMPA).

considered a derivative of DHPA in which the 2′-hydroxyl group has been replaced by a phosphonylmethoxy group. The resulting product has proved effective against all DNA viruses that have so far been looked at, including HSV-1, HSV-2, VZV, and dThd kinase-deficient (TK^-) HSV-1 and VZV mutants resistant to ACV, BVDU, EDU, and CEDU (De Clercq *et al.*, 1986*b*). The activity spectrum of HPMPA extends to adenovirus, cytomegalovirus (CMV) and various other herpes viruses (see Table 3) that are not sensitive to either ADV or BVDU, or both. In this sense, HPMPA may resemble 2′-nor-cGMP, the cyclic phosphate of 2′-nor-deoxyguanosine or 9-(1,3-dihydroxy-2-propoxymethyl)guanine, which is also endowed with broad-spectrum anti-DNA virus activities (Tolman *et al.*, 1985; Field *et al.*, 1986). HPMPA appears to be selective in its antiviral action, as it does not inhibit normal cell growth or metabolism up to concentrations that are 100–1000-fold higher than those required for inhibition of virus replication. The mechanism of action of HPMPA remains to be determined.

The *in vivo* antiviral efficacy of HPMPA has been demonstrated in several animal models, including cutaneous HSV-1 infection in hairless mice and HSV-1 keratitis in rabbits (De Clercq *et al.*, 1986*b*; Maudgal *et al.*, 1987*b*). In these models, HPMPA completely suppressed all manifestations of the infection, even when applied topically at a concentration as low as 0.1 or 0.2%. Particularly noteworthy was the effect seen with HPMPA in the treatment of keratitis caused by a TK^- HSV-1 mutant. The latter was isolated from an immunosuppressed patient who had become clinically resistant to ACV, following a single intravenous ACV treatment course for an orofacial HSV-1 infection (Vinckier *et al.*, 1986). The isolated HSV-1 strain was resistant to ACV and co-resistant to BVDU, EDU, and CEDU; it was identified as a TK^- mutant because it induced (in TK^- HeLa cells) only 1–5% of the TK activity induced by the wild-type TK^+ HSV-1 strains. This ACV^r, $BVDU^r$, EDU^r, $CEDU^r$, TK^- HSV-1 mutant proved highly pathogenic for the rabbit eye and, while neither BVDU nor CEDU (0.2% eyedrops) had any influence on the course of the infection, HPMPA (0.2% eyedrops), brought about an almost complete healing of the keratitis within 2–3 days (Maudgal *et al.*, 1987).

IV. CONCLUSION

The nucleoside analogues currently pursued for their potential as antiviral drugs are targeted at any of the following enzymes:

1. *Viral DNA polymerase,* via a specific phosphorylation by the *virus-encoded dThd kinase,* as shown for those compounds (BVDU, C-BVDU, BVaraU, BVDC, EDU, CEDU) that selectively inhibit the replication of HSV-1 and VZV
2. *Viral DNA polymerase,* without a specific phosphorylation by a virus-encoded dThd kinase, as may be postulated for those compounds (i.e., HPMPA) that are inhibitory to a broad spectrum of DNA viruses, including TK^- mutants of HSV and VZV
3. *SAH hydrolase* and possibly other enzymes involved in transmethylation reactions, which would account for the activity of the acyclic adenosine analogues (DHPA, AHPA) and carbocyclic adenosine analogues (C-cAdo, neplanocin A) against poxvirus, paramyxovirus, rhabdovirus, and reovirus
4. *Reverse transcriptase,* which seems to explain the inhibitory effects of 2′,3′-dideoxynucleoside analogues (AZT, ddCyd) on HTLV-III/LAV replication

ACKNOWLEDGMENTS

The experimental work described in this chapter was supported by grants from the Belgian Fonds voor Geneeskundig Wetenschappelijk Onderzoek (Project no. 3.0040.83) and the Belgian Geconcerteerde Onderzoeksacties (Project no. 85/90-79). I am most grateful to Christiane Callebaut for her dedicated editorial assistance.

REFERENCES

Baba, M., Mori, S., Shigeta, S., and De Clercq, E. (1987). *Antimicrob. Agents Chemother.* **31,** 337–339.
Ben-Efraim, S., Shoval, S., and De Clercq, E. (1986). *Brit. J. Cancer,* **54,** 847–852.
Benoit, Y., Laureys, G., Delbeke, M.-J., and De Clercq, E. (1985). *Eur. J. Pediatr.* **143,** 198–202.
Borchardt, R. T., Keller, B. T., and Patel-Thombre, U. (1984). *J. Biol. Chem.* **259,** 4353–4358.
Bussereau, F., Chermann, J. C., De Clercq, E., and Hannoun, C. (1983). *Ann. Virol. (Inst. Pasteur)* **134E,** 127–134.
Busson, R., Colla, L., Vanderhaeghe, H., and De Clercq, E. (1981). *Nucleic Acids Res. (Symposium Series No. 9)* **9,** 49–52.
Colla, L., Herdewijn, P., De Clercq, E., Balzarini, J., and Vanderhaeghe, H. (1985). *Eur. J. Med. Chem.* **20,** 295–301.
Cooney, D. A., Dalal, M., Mitsuya, H., McMahon, J. B., Nadkarni, M., Balzarini, J., Broder, S., and Johns, D. G. (1986). *Biochem. Pharmacol.* **35,** 2065–2068.
De Clercq, E. (1982). *Antimicrob. Agents Chemother.* **21,** 661–663.
De Clercq, E. (1984). *J. Antimicrob. Chemother.* **14** (Suppl. A), 85–95.
De Clercq, E. (1985). *Antimicrob. Agents Chemother.* **28,** 84–89.
De Clercq, E., and Cools, M. (1985). *Biochem. Biophys. Res. Commun.* **129,** 306–311.
De Clercq, E., and Holý, A. (1979). *J. Med. Chem.* **22,** 510–513.
De Clercq, E., and Holý, A. (1985). *J. Med. Chem.* **28,** 282–287.
De Clercq, E., and Montgomery, J. A. (1983). *Antiviral Res.* **3,** 17–24.
De Clercq, E., and Rosenwirth, B. (1985). *Antimicrob. Agents Chemother.* **28,** 246–251.
De Clercq, E., and Shugar, D. (1975). *Biochem. Pharmacol.* **24,** 1073–1078.
De Clercq, E., and Walker, R. T. (1984). *Pharmac. Ther.* **26,** 1–44.
De Clercq, E., Descamps, J., De Somer, P., and Holý, A. (1978). *Science* **200,** 563–565.
De Clercq, E., Descamps, J., De Somer, P., Barr, P. J., Jones, A. S., and Walker, R. T. (1979). *Proc. Natl. Acad. Sci. USA* **76,** 2947–2951.

De Clercq, E., Balzarini, J., Descamps, J., and Eckstein, F. (1980). *Biochem. Pharmacol.* **29,** 1849–1851.

De Clercq, E., Balzarini, J., Descamps, J., Huang, G.-F., Torrence, P. F., Bergstrom, D. E., Jones, A. S., Serafinowski, P., Verhelst, G., and Walker, R. T. (1982*a*). *Mol. Pharmacol.* **21,** 217–223.

De Clercq, E., Busson, R., Colla, L., Descamps, J., Balzarini, J., and Vanderhaeghe, H. (1982*b*). In *Current Chemotherapy and Immunotherapy* (P. Periti and G. G. Grassi, eds.), Vol. II, pp. 1062–1064, American Society for Microbiology, Washington, D. C.

De Clercq, E., Descamps, J., Balzarini, J., Fukui, T., and Allaudeen, H. S. (1983). *Biochem. J.* **211,** 439–445.

De Clercq, E., Bergstrom, D. E., Holý, A., and Montgomery, J. A. (1984). *Antiviral Res.* **4,** 119–133.

De Clercq, E., Balzarini, J., Bernaerts, R., Herdewijn, P., and Verbruggen, A. (1985*a*). *Biochem. Biophys. Res. Commun.* **126,** 397–403.

De Clercq, E., Benoit, Y., Laureys, G., and Delbeke, M. J. (1985*b*). In *Herpes Viruses and Virus Chemotherapy* (R. Kono, ed.), pp. 49–52, Elsevier Science Publishers, Amsterdam.

De Clercq, E., Bernaerts, R., Balzarini, J., Herdewijn, P., and Verbruggen, A. (1985*c*). *J. Biol. Chem.* **260,** 10621–10628.

De Clercq, E., Desgranges, C., Herdewijn, P., Sim, I. S., and Walker, R. T. (1985*d*). In *Proceedings of the Eighth International Symposium on Medicinal Chemistry* (R. Dahlbom, and J. L. G. Nilsson, eds.), Vol. 1, pp. 198–210, Swedish Pharmaceutical Press, Stockholm.

De Clercq, E., Desgranges, C., Herdewijn, P., Sim, I. S., Jones, A. S., McLean, M. J., and Walker, R. T. (1986*a*). *J. Med. Chem.* **29,** 213–217.

De Clercq, E., Holý, A., Rosenberg, I., Sakuma, T., Balzarini, J., and Maudgal, P. C. (1986*b*). *Nature (London)* **323,** 464–467.

Desgranges, C., Razaka, G., Rabaud, M., Bricaud, H., Balzarini, J., and De Clercq, E. (1983). *Biochem. Pharmacol.* **32,** 3583–3590.

Desgranges, C., Razaka, G., Drouillet, F., Bricaud, H., Herdewijn, P., and De Clercq, E. (1984). *Nucleic Acids Res.* **12,** 2081–2090.

Desgranges, C., Razaka, G., Bricaud, H., and De Clercq, E. (1985). *Biochem. Pharmacol.* **34,** 405–406.

Desgranges, C., De Clercq, E., Razaka, G., Drouillet, F., Belloc, I., and Bricaud, H. (1986*a*). *Biochem. Pharmacol.* **35,** 1647–1653.

Desgranges, C., Razaka, G., De Clercq, E., Herdewijn, P., Balzarini, J., Drouillet, F., and Bricaud, H. (1986*b*). *Cancer Res.* **46,** 1094–1101.

Field, A. K., Davies, M. E. M., DeWitt, C. M., Perry, H. C., Schofield, T. L., Karkas, J. D., Germershausen, J., Wagner, A. F., Cantone, C. L., MacCoss, M., and Tolman, R. L. (1986). *Antiviral Res.* **6,** 329–341.

Griengl, H., Bodenteich, M., Hayden, W., Wanek, E., Streicher, W., Stütz, P., Bachmaer, H., Ghazzouli, I., and Rosenwirth, B. (1985). *J. Med. Chem.* **28,** 1679–1684.

Herdewijn, P., De Clercq, E., Balzarini, J., and Vanderhaeghe, H. (1985). *J. Med. Chem.* **28,** 550–555.

Holý, A., Votruba, I., and De Clercq, E. (1984). *Collect. Czechoslov. Chem. Commun.* **50,** 262–279.

Horwitz, J. P., Chua, J., and Noel, M. (1964). *J. Org. Chem.* **29,** 2076–2078.

Kára, J., Vácha, P., and Holý, A. (1979). *FEBS Lett.* **107,** 187–192.

Kitaoka, S., Konno, T., and De Clercq, E. (1986). *Antiviral Res.* **6,** 57–65.

Lin, J.-C., Smith, M. C., Choi, E. I., De Clercq, E., Verbruggen, A., and Pagano, J. S. (1985). *Antiviral Res. (Suppl. No. 1)* **1,** 121–126.

Lin, T.-S., and Prusoff, W. H. (1978). *J. Med. Chem.* **21,** 109–112.

Machida, H., Kuninaka, A., and Yoshino, H. (1982). *Antimicrob. Agents Chemother.* **21,** 358–361.

Machida, H., and Sakata, S. (1984). *Antiviral Res.* **4,** 135–141.

Machida, H., Sakata, S., Kuninaka, A., and Yoshino, H. (1981). *Antimicrob. Agents Chemother.* **20,** 47–52.

Maudgal, P. C., and De Clercq, E. (1985). *Arch. Ophthalmol.* **103,** 1393–1397.

Maudgal, P. C., De Clercq, E., Bernaerts, R., Dieltiens, M., Breemersch, M., and Van Eeckhoutte, L. (1986*a*). *Invest. Ophthalmol. Vis. Sci.* **27,** 1453–1458.

Maudgal, P. C., De Clercq, E., and Huyghe, P. (1987). *Invest. Opthalmol. Vis. Sci.* **28,** 243–248.

Maudgal, P. C., De Clercq, E., and Missotten, L. (1984). *Antiviral Res.* **4,** 281–291.

Maudgal, P. C., Dieltiens, M., De Clercq, E., and Missotten, L. (1985*a*). In *Herpetic Eye Diseases* (P. C. Maudgal and L. Missotten, eds.), pp. 247–256, W. Junk Publishers, Dordrecht.

Maudgal, P. C., Dieltiens, M., De Clercq, E., and Missotten, L. (1985*b*). In *Herpetic Eye Diseases* (P. C. Maudgal and L. Missotten, eds.), pp. 403–411, W. Junk Publishers, Dordrecht.

Mitsuya, H., and Broder, S. (1986). *Proc. Natl. Acad. Sci. USA* **83,** 1911–1915.

Mitsuya, H., Weinhold, K. J., Furman, P. A., St. Clair, M. H., Nusinoff Lehrman, S., Gallo, R. C., Bolognesi, D., Barry, D. W., and Broder, S. (1985). *Proc. Natl. Acad. Sci. USA* **82,** 7096–7100.

Montgomery, J. A., Clayton, S. J., Thomas, H. J., Shannon, W. M., Arnett, G., Bodner, A. J., Kim, I.-K., Cantoni, G. L., and Chiang, P. K. (1982). *J. Med. Chem.* **25,** 626–629.

Osterhaus, A. D. M. E., Groen, J., and De Clercq, E. (1987). *Antiviral Res.* **7,** 221–226.

Schinazi, R. F., Scott, R. T., Peters, J., Rice, V., and Nahmias, A. J. (1985). *Antimicrob. Agents Chemother.* **28,** 552–560.

Smee, D. F., Sidwell, R. W., Clark, S. M., Barnett, B. B., and Spendlove, R. S. (1982). *Antimicrob. Agents Chemother.* **21,** 66–73.

Sodja, I., and Holý, A. (1980). *Acta Virol.* **24,** 317–324.

Soike, K. F., Baskin, G., Cantrell, C., and Gerone, P. (1984). *Antiviral Res.* **4,** 245–257.

Spruance, S. L., Freeman, D. J., and Sheth, N. V. (1985). *Antimicrob. Agents Chemother.* **28,** 103–106.

Reefschläger, J., Wutzler, P., Thiel, K.-D., and Herrmann, G. (1986). *Antiviral Res.* **6,** 83–93.

Rosenwirth, B., Griengl, H., Wanek, E., and De Clercq, E. (1985). *Antiviral Res. (Suppl. No. 1)* **1,** 21–28.

Tolman, R. L., Field, A. K., Karkas, J. D., Wagner, A. F., Germershausen, J., Crumpacker, C., and Scolnick, E. M. (1985). *Biochem. Biophys. Res. Commun.* **128,** 1329–1335.

Tricot, G., De Clercq, E., Boogaerts, M. A., and Verwilghen, R. L. (1986). *J. Med. Virol.* **18,** 11–20.

Vinckier, F., Boogaerts, M., Declerck, D., and De Clercq, E. (1986). J. Oral Maxillofacial Surg. (in press).

Votruba, I., and Holý, A. (1980). *Collect. Czechoslov. Chem. Commun.* **45,** 3039–3044.

Waqar, M. A., Evans, M. J., Manly, K. F., Hughes, R. G., and Huberman, J. A. (1984). *J. Cell. Physiol.* **121,** 402–408.

Wildiers, J., and De Clercq, E. (1984). *Eur. J. Cancer Clin. Oncol.* **20,** 471–476.

Yarchoan, R., Klecker, R. W., Weinhold, K. J., Markham, P. D., Lyerly, H. K., Durack, D. T., Gelmann, E., Nusinoff Lehrman, S., Blum, R. M., Barry, D. W., Shearer, G. M., Fischl, M. A., Mitsuya, H., Gallo, R. C., Collins, J. M., Bolognesi, D. P., Myers, C. E., and Broder, S. (1986). *Lancet* **1,** 575–580.

18

New Analogues to Acyclovir with Increased Potency to Herpes Viruses

A. Paul Fiddian

I. INTRODUCTION

Acyclovir is highly active *in vitro* against herpes simplex virus types 1 and 2 (HSV-1 and HSV-2, respectively), moderately active against varicella–zoster virus (VZV) and Epstein–Barr virus (EBV), and relatively inactive against cytomegalovirus (CMV). The full extent of the clinical value of acyclovir is still being defined, but we already know much about its efficacy and tolerance in a wide range of herpes virus infections. This has previously been discussed in a number of reviews, including Fiddian (1984), Fiddian *et al.* (1984), Rees and Fiddian (1985), and most recently Fiddian and Grant (1985).

Although still at an early stage, the developments of analogues to acyclovir have been reported. Since the primary aim of such work should be to improve on the usefulness of acyclovir, it is important to consider the progress and prospects of this against the background of the full potential of acyclovir.

II. HERPES SIMPLEX VIRUS

Briefly, intravenous, oral, dermal, and ophthalmic formulations of acyclovir have variously been shown to be effective in systemic, mucocutanous, and ocular manifestations of HSV infection. For example, acyclovir is the most potent agent available in the treatment of herpes simplex encephalitis (Whitley *et al.*, 1986), is the only drug of proven clinical value in the treatment of both immunocompromised (Meyers, 1985) and normal (Rees and Fiddian, 1985) patients with HSV infections of the skin and mucous membranes, and constitutes first-line treatment in HSV eye infections (Nicholson, 1984). In addition, there is extensive evidence to support the use of acyclovir for prophylaxis of HSV infections in the immunocompromised host and for prevention of recurrences of HSV infection in normal patients (Fiddian and Grant, 1985). These data further attest to the excellent tolerance of acyclovir even over prolonged periods of time.

A. Paul Fiddian • Clinical Antiviral Therapy Section, Wellcome Research Laboratories, Beckenham, Kent, BR3 3BS, England.

The development of analogues to acyclovir with improved potency to HSV is therefore of secondary importance to the need for consideration of the other herpes viruses. However, it is worthwhile to dwell for a moment on those aspects of the use of acyclovir in HSV infections that have been raised as possible limitations (Table 1). As can be seen, many of the supposed limitations of therapy are in fact erroneous or inherent to the virus and the natural history of the infections caused by it. The continuing development of acyclovir and its prodrug, 6-deoxyacyclovir (BW A515U), may lead to improvements in the ease of management of HSV infections, but these are of lesser importance than the need for effective therapy for other herpes virus infections. Therefore, this prodrug is discussed in more detail in the following section. A number of other analogues to acyclovir have been reported as well; the major analogues are listed in Table 2. Although BW B759U has similar *in vitro* activity to acyclovir against HSV, Collins and Oliver (1985) reported *in vivo* efficacy in two mouse models of systemic HSV infection to be greater with BW B759U. However, significant toxicities associated with BW B759U are likely to preclude its routine use in HSV infections.

Table 1. Possible Limitations to the Use of Acyclovir in HSV Infections

Aspect	Comments
No effect on latency	Prevention may be impracticable because of need for very early initiation of therapy. Established latency is almost certainly not influenced by antiviral agents.
Modest efficacy	
Recurrent disease	A factor of the natural history of recurrent infections, it may be improved by patient education and earlier therapy. Greatest efficacy is shown with suppressive treatment.
Dermal therapy	As above, but improved drug delivery is also important (e.g., acyclovir cream).
Insolubility	
Dermal therapy	It is used with penetration enhancers (e.g., propylene glycol, in acyclovir cream.)
Ocular therapy	Apply as an ointment or consider more soluble analogues (e.g., 2′-0-glycylacyclovir).
Intravenous use	Adjustment of the pH can make administration easier (but see below).
Renal toxicity	Caused by crystallization in renal tubules, this problem can be avoided by giving the drug as an intravenous infusion and ensuring adequate hydration.
Pharmacokinetics	
GI absorption	At 20%, it is adequate for most mucocutaneous HSV infections but is greatly increased as BW A515U. Severe HSV infections require intravenous dosing.
Half-life	At 3 hr, it requires frequent oral administration of standard treatment dose. It may be helped by using higher doses or BW A515U.
Long-term safety	Accumulating data indicate that long-term usage is well tolerated.
Resistance	*In vivo* evidence does not warrant limitation of drug usage in normal patients. No advantage for acyclovir analogues is likely.
Cost	Necessitated by drug-development costs, newer analogues are unlikely to be cheaper.

Table 2. Analogues to Acyclovir with Activity against HSV[a]

Analogue	*In vitro* activity (compared with ACV)	Comment
Prodrugs		
2,6-diamino-9-[2-hydroxyethoxy-methyl]-9H-purine (BW A134U)	As ACV	Insufficient advantage to justify selection over BW A515U
2-amino-9-[2-hydroxyethoxymethyl]-9H-purine (6-deoxyacyclovir, BW A515U)	As ACV	Ideal oral prodrug
2′-0-glycylacyclovir	As ACV	Possible potential as eyedrops
Analogues		
9-[2-hydroxy-1-(hydroxymethyl)eth-oxymethyl]guanine (BW B759U, DHPG)	As ACV	Too toxic for use in HSV infections
(R)-9-(3,4-dihydroxybutyl)guanine (DHBG)	10× less active	Uncertain potential

[a] ACV, acyclovir; HSV, herpes simplex virus.

III. VARICELLA–ZOSTER VIRUS

Despite being less active against VZV, acyclovir has been shown to be effective in VZV infections. Intravenous acyclovir at a dose of 5 mg/kg q8h is of benefit to immunocompetent patients with herpes zoster (Peterslund *et al.*, 1981; McGill *et al.*, 1983). For immunocompromised patients a higher dose has also been successfully employed, i.e., 10 mg/kg q8h (or 500 mg/m^2 for children). Prober *et al.* (1982) reported benefit in immunocompromised children with varicella and Balfour *et al.* (1983) in immunocompromised patients with herpes zoster. More recently, Shepp *et al.* (1986) demonstrated the superiority of acyclovir over vidarabine in such cases and concluded that intravenous acyclovir is the treatment of choice.

Since an outpatient therapy is desirable for the management of less severe VZV infections, oral acyclovir has been studied extensively in immunocompetent patients with herpes zoster. Initial trials employed doses of 400 mg five times daily, twice the recommended dose for the treatment of HSV infections in normal patients. Results from these studies indicate only limited benefits at this dose (McKendrick *et al.*, 1984; Balfour, 1986). It was therefore decided that even higher doses (800 mg five times daily) should be evaluated. Further work by McKendrick (1985) and others (unpublished data) has confirmed that this dose is more effective but is also well tolerated; thus, patients with herpes zoster may be managed at home in the future. Initial experience in immunocompromised children with varicella reported by Novelli *et al.* (1984) also suggests that oral acyclovir may be effective in these cases. However, until this has been confirmed, immunocompromised patients at particular risk from VZV infections may be better treated with intravenous acyclovir.

The demonstration by McKendrick (1985) that higher doses of oral acyclovir are

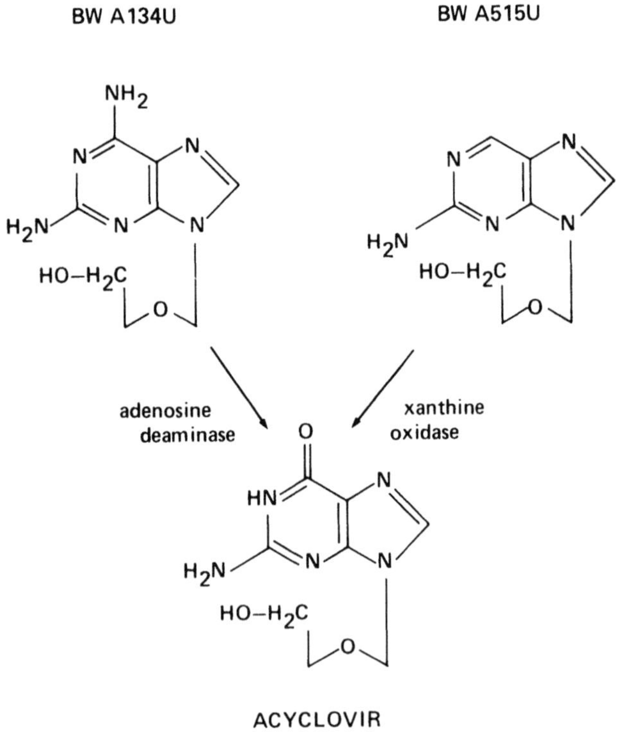

Figure 1. Conversion of prodrugs BW A134U and BW A515U to acyclovir.

required for herpes zoster than are recommended for HSV infections lends greater support for the development of prodrugs of acyclovir that might deliver higher plasma levels of active drug following oral administration. An additional benefit that might be achieved by such therapies is the convenience of less frequent dosing. The first prodrug to be seriously considered was BW A134U (Table 2). Despite increased absorption compared with oral acyclovir and subsequent conversion by the enzyme adenosine deaminase (Fig. 1), the consequent plasma levels of acyclovir in humans are disappointing (Table 3). This finding as well as concerns about its potential safety in humans (King and Fiddian, 1986) have diverted attention to another prodrug candidate.

BW A515U (Table 2) is converted to acyclovir by xanthine oxidase (Fig. 1) (Krenitsky *et al.*, 1984), an enzyme that is abundant in humans. Phase I studies employing

Table 3. Pharmacokinetics of Acyclovir following Oral Administration of Single 400-mg Doses

	Acyclovir	BW A134U	BW A515U
AUC (0–12) μM/hr	11.7 (±4.7)	30.6 (±13.8)	72.5 (±12.7)
C_{max} (μM)	2.8 (±1.2)	7.5 (±5.0)	25.8 (±6.0)
T_{max} (hr)	1.2 (±0.6)	1.0 (±0.4)	0.6 (±0.3)

Table 4. Pharmacokinetics of Acyclovir following Oral Administration of Multiple Doses

	Acyclovir[a]	BW A515U[b]
AUC (0–6) μM/hr	31.0 (±7.9)	71.7 (±18.9)
C_{max} (μM)	7.7 (±2.3)	23.7 (±5.3)
T_{max} (hr)	1.7 (±1.0)	1.3 (±0.5)

[a] Given as 800 mg q6h.
[b] Given as 250 mg q6h.

single doses up to 400 mg (Whiteman *et al.*, 1984) (Table 3) and multiple doses of 250 mg q6h (Selby *et al.*, 1984*a*) (Table 4) have been reported. Absorption and subsequent conversion to acyclovir are almost complete, making this prodrug an ideal candidate for further development. Phase II trials are currently in progress, but results will not be available for some time. Apart from BW A515U, the other analogues listed in Table 2 are of no interest in relationship to VZV infections.

IV. EPSTEIN–BARR VIRUS

Acyclovir was reported by Pagano *et al.* (1983) to have an ED_{50} of 0.3 μM against EBV replication *in vitro* (Table 5) despite an apparent lack of EBV-specified thymidine kinase. This activity is believed to be due to a greater affinity of acyclovir triphosphate for the viral DNA polymerase (Colby *et al.*, 1981), although Allaudeen *et al.* (1982) questioned whether this is the primary mechanism of action. Acyclovir suppresses but does not cure the EBV infection in virus-producing lines (Pagano and Datta, 1982), and clinical experience indicates that *in vivo* EBV is less sensitive than HSV. Indeed, it would appear that EBV is more similar to VZV in sensitivity to acyclovir, suggesting that earlier reports by Colby *et al.* (1980) indicating that an ED_{50} of around 6 μM may be more relevant to the clinical situation.

In a pilot study, Pagano *et al.* (1983) reported that high-dose intravenous acyclovir (500 mg/m^2 q8h) interrupted virus excretion in the throat but had minimal clinical bene-

Table 5. In Vitro Activity of Acyclovir and BW B759U against the Herpes Viruses

Virus type	Acyclovir ED_{50} (μM)	BW B759U ED_{50} (μM)	Comparative activity[a]
Herpes simplex virus type 1 (HSV-1)	0.1	0.1	1
Herpes simplex virus type 2 (HSV-2)	0.3	0.3	1
Epstein–Barr virus (EBV)	0.3	0.05	6
Varicella–zoster virus (VZV)	3.0	3.0	1
Cytomegalovirus (CMV)	45.0	1.5	30

[a] ED_{50} acyclovir ÷ ED_{50} BW B759U.

fits in patients with infectious mononucleosis. Andersson *et al.* (1986) confirmed these findings using a similar intravenous dose of 10 mg/kg q8h for 7 days. Acyclovir significantly, but reversibly, inhibited oropharyngeal shedding of EBV and was associated with a significantly more rapid overall symptomatic recovery compared with placebo. However, individual clinical symptoms and laboratory parameters were not significantly influenced by acyclovir in this study. Since an orally administered therapy would facilitate outpatient management of this disease, it is rational to evaluate both high-dose oral acyclovir and the prodrug of acyclovir, BW A515U, perhaps given for longer periods of time.

The only other analogue of interest in relationship to EBV is BW B759U. This was reported by Lin *et al.* (1984) to have a sixfold greater activity against EBV than acyclovir as measured by its ED_{50} 0.05 μM (Table 5). But perhaps more importantly the inhibitory effect is more prolonged after removal of the drug. Together with a low *in vitro* cytotoxicity, this suggested BW B759U as a promising candidate for clinical trials. Unfortunately, toxicity in animal studies and in early clinical experience noted by King and Fiddian (1986) is likely to preclude its use in humans for all but the most serious EBV infections. Nevertheless, it is of great interest to elucidate fully the mechanism of action of this drug, since it may lead to the development of other effective but potentially safer agents.

V. CYTOMEGALOVIRUS

Most cytomegalovirus (CMV) strains are relatively resistant to acyclovir *in vitro,* as generally borne out by a lack of apparent effect reported in the majority of clinical papers. An early study by Balfour *et al.* (1982) suggested some benefits in immunocompromised patients, especially renal transplant recipients, with CMV infection. Subsequent experience has not confirmed these observations, although a number of anecdotal reports claim at least partial efficacy in organ transplant patients, reviewed by Fiddian and Grant (1985). The most severe and potentially fatal CMV infections occur in bone marrow transplant recipients; acyclovir appears ineffective in these patients (Meyers, 1985).

Since the efficacy of acyclovir against other herpes viruses is most readily apparent when the drug is given prophylactically, it was considered worthwhile to evaluate this form of therapy. In a controlled trial, Gluckman *et al.* (1983) surprisingly found low-dose oral acyclovir prophylaxis effective in preventing CMV disease after marrow transplantation. This result is most likely aberrant, since others have reported both high-dose oral acyclovir (Lundgren *et al.*, 1985; Selby *et al.*, 1984*b*) and low-dose intravenous acyclovir (Hann *et al.*, 1983; Saral *et al.*, 1981) ineffective in this situation. It remains to be determined whether high-dose intravenous acyclovir might have any value in the prevention of CMV disease in susceptible patients.

More recently, attention has been directed toward an analogue of acyclovir, BW B759U, with greater *in vitro* activity against CMV (Table V). BW B759U differs from acyclovir only by the addition of a 3′-carbon and hydroxyl group attached to the acyclic side chain. Although this modification is not drastic, it confers a greater spectrum of *in vitro* activity against the herpes viruses. The altered structure is also associated with a greater propensity to cause toxicity particularly in gonadal function, myelosuppression, and mutagenicity. Nevertheless, the risk–benefit ratio may be sufficiently favorable in immunocompromised patients with severe and/or life-threatening CMV infections to warrant selected use in such target populations.

In particular, bone marrow transplant recipients and patients with acquired immunodeficiency syndrome (AIDS) are subject to a high risk of morbidity and/or mortality from CMV disease, including pneumonitis, retinitis, and gastrointestinal (GI) disease. Several groups of such patients are currently being evaluated for evidence of acceptable tolerance and efficacy. Although no definitive data are yet available, preliminary results are encouraging. Felsenstein *et al.* (1985), Back *et al.* (1985), and Masur *et al.* (1986) reported benefits in AIDS patients with CMV retinitis as evidenced by healing of retinal lesions and resolution of viraemia. However, relapse is common following cessation of treatment, so such patients may need to receive maintenance therapy for prolonged periods. Shepp *et al.* (1985) also demonstrated an antiviral effect in marrow transplant patients with CMV pneumonia, but survival was not influenced. This finding suggests that even a potent agent such as BW B759U may have to be employed prior to the development of pneumonitis if outcome is to be favorably modified.

VI. CONCLUSIONS

Acyclovir is highly potent against herpes simplex virus (HSV) and has been shown to be well tolerated and effective in a wide range of HSV infections. There are no reported analogues of acyclovir with greater potency that are likely to be developed for use in humans. A prodrug of acyclovir, BW A515U, which offers greater gastrointestinal bioavailability of acyclovir may prove a suitable alternative to intravenous acyclovir for some severe HSV infections (excluding herpes encephalitis and neonatal herpes).

Although less active against varicella–zoster virus (VZV), acyclovir is effective intravenously, or in higher oral doses, for the treatment of VZV infections. By attaining higher plasma levels of acyclovir, BW A515U may provide a more convenient outpatient regimen for herpes zoster. No other analogues have been identified that offer a greater potential for VZV. The situation is similar in relationship to Epstein–Barr virus (EBV), except that clinical benefit has not been clearly established for acyclovir despite evidence of a marked antiviral effect. Furthermore, BW B759U has a more prolonged and potent effect *in vitro* against EBV than acyclovir but is limited by its toxicity.

For cytomegalovirus (CMV), there is a definite need for a more potent agent, and BW B759U has been identified as such. Although there are concerns regarding the toxicity of this analogue, it seems that its use in managing certain serious CMV infections may be justified. Nevertheless, further work is still required before it can be recommended even for this. Identification of the mode of action of BW B759U against CMV and EBV may also prove useful for the development of further analogues.

REFERENCES

Allaudeen, H. S., Descamps, J., and Sehgal, R. K. (1982). *Antiviral Res.* **2,** 123–133.

Andersson, J., Britton, S., Ernberg, I., Andersson, U., Henle, W., Sköldenberg, B., and Tisell, A. (1986). *J. Infect. Dis.* **153,** 283–290.

Bach, M. C., Bagwell, S. P., Knapp, N. P., Davis, K. M., and Hedstrom, P. S. (1985). *Ann. Intern. Med.* **103,** 381–382.

Balfour, H. H., Jr. (1986). In *Clinical Advances in Antiviral Therapy* (P. Lietman, S. Ohtani, and A. P. Fiddian, eds.), pp. 27–33, University of Tokyo Press, Tokyo, Japan.

Balfour, H. H., Jr., Bean, B., Mitchell, C. D., Sachs, G. W., Boen, J. R., and Edelman, C. K. (1982). *Am. J. Med.* **73**(1a), 241–248.

Balfour, H. H., Jr., Bean, B., Laskin, O. L., Ambinder, R. F., Meyers, J. D., Wade, J. C., Zaia, J. A., Aeppli, D., Kirk, L. E., Segreti, A. C., and Keeney, R. E. (1983). *N. Engl. J. Med.* **308,** 1448–1453.

Colby, B. M., Shaw, J. E., Elion, G. B., and Pagano, J. S. (1980). *J. Virol.* **34,** 560–568.

Colby, B. M., Furman, P. A., Shaw, J. E., Elion, G. B., and Pagano, J. S. (1981). *J. Virol.* **38,** 606–611.

Collins, P., and Oliver, N. M. (1985). *Antiviral Res.* **5,** 145–156.

Felsenstein, D., D'Amico, D. J., Hirsch, M. S., Neumeyer, D. A., Cederberg, D. M., de Miranda, P., and Schooley, R. T. (1985). *Ann. Intern. Med.* **103,** 377–380.

Fiddian, A. P. (1984). In *Applied Virology* (E. Kurstak, ed.), pp. 103–120, Academic, Orlando, Florida.

Fiddian, A. P., and Grant, D. M. (1985). *Abs. Hyg. Commun. Dis.* **60,** R1–R22.

Fiddian, A. P., Brigden, D., Yeo, J. M., and Hickmott, E. (1984). *Antiviral Res.* **4,** 99–117.

Gluckman, E., Lotsberg, J., Devergie, A., Zhao, X. M., Melo, R., Gomez-Morales, M., Nebout, T., Mazeron, M. C., and Perol, Y. (1983). *Lancet* **2,** 706–708.

Hann, I. M., Prentice, H. G., Blacklock, H. A., Ross, M. G. R., Brigden, D., Rosling, A. E., Burke, C., Crawford, D. H., Brumfitt, W., and Hoffbrand, A. V. (1983). *Br. Med. J.* **2,** 384–388.

King, D. H., and Fiddian, A. P. (1986). In *Clinical Advances in Antiviral Therapy* (P. Lietman, S. Ohtani, and A. P. Fiddian, eds.), pp. 65–71, University of Tokyo Press, Tokyo, Japan.

Krenitsky, T. A., Hall, W. W., de Miranda, P., Beauchamp, L. M., Schaeffer, H. J., and Whiteman, P. D. (1984). *Proc. Natl. Acad. Sci. USA* **81,** 3209–3213.

Lin, J-C., Smith, M. C., and Pagano, J. S. (1984). *J. Virol.* **50,** 50–55.

Lundgren, G., Wilczek, H., Lönnqvist, B., Lindholm, A., Wahren, B., and Ringden, O. (1985). *Scand. J. Infect. Dis. (Suppl)* **47,** 137–144.

McGill, J., MacDonald, D. R., Fall, C., McKendrick, G. D. W., and Copplestone, A. (1983). *J. Infect. Dis.* **6,** 157–161.

McKendrick, M. W. (1985). *Scand. J. Infect. Dis. (Suppl)* **47,** 76–79.

McKendrick, M. W., Care, C., Burke, C., Hickmott, E., and McKendrick, G. D. W. (1984). *J. Antimicrob. Chemother.* **14,** 661–665.

Masur, H., Lane, C., Palestine, A., Smith, P. D., Manischewitz, J., Stevens, G., Fujikawa, L., Macher, A. M., Nussenblatt, R., Baird, B., McGill, M., Wittek, A., Quinnan, G. V., Parrillo, J. E., Rook, A. H., Eron, L. J., Poretz, D. M., Goldenberg, R. I., Fauci, A. S., and Gelmann, E. P. (1986). *Ann. Intern. Med.* **104,** 41–44.

Meyers, J. D. (1985). *Scand. J. Infect. Dis. (Suppl.)* **47,** 128–136.

Nicholson, K. G. (1984). *Lancet* **2,** 617–621.

Novelli, V. M., Marshall, W. C., Yeo, J., and McKendrick, G. D. (1984). *J. Infect. Dis.* **149,** 478.

Pagano, J. S., and Datta, A. K. (1982). *Am. J. Med.* **73**(1A), 18–26.

Pagano, J. S., Sixbey, J. W., and Lin, J.-C. (1983). *J. Antimicrob. Chemother.* **12**(B), 113–121.

Peterslund, N. A., Seyer-Hansen, K., Ipsen, J., Esmann, V., Schonheyder, H., and Juhl, H. (1981). *Lancet* **2,** 827–830.

Prober, C. G., Kirk, L. E., and Keeney, R. E. (1982). *J. Pediatr.* **101,** 622–625.

Rees, P. J., and Fiddian, A. P. (1985). *J. Drug Ther. Res.* **10,** 637–640.

Saral, R., Burns, W. H., Laskin, O. L., Santos, G. W., and Lietman, P. S. (1981). *N. Engl. J. Med.* **305,** 63–67.

Selby, P., Powles, R. L., Blake, S., Stolle, K., Mbidde, E. K., McElwain, T. J., Hickmott, E., Whiteman, P. D., and Fiddian, A. P. (1984*a*). *Lancet* **2,** 1428–1430.

Selby, P., Powles, R., Stolle, K., and Stern, H. (1984*b*). *Br. Med. J.* **2,** 253.

Shepp, D. H., Dandliker, P. S., de Miranda, P., Burnette, T. C., Cederberg, D. M., Kirk, L. E., and Meyers, J. D. (1985). *Ann. Intern. Med.* **103,** 368–373.

Shepp, D. H., Dandliker, P. S., and Meyers, J. D. (1986). *N. Engl. J. Med.* **314,** 208–212.

Whiteman, P. D., Bye, A., Fowle, A. S. E., Jeal, S., Land, G., and Posner, J. (1984). *Eur. J. Clin. Pharmacol.* **27,** 471–475.

Whitley, R. J., Alford, C. A., Hirsch, M. S., Schooley, R. T., Luby, J. P., Aoki, F. Y., Hanley, D., Nahmias, A. J., and Soong, S.-J. (1986). *N. Engl. J. Med.* **314,** 144–149.

19

Carrier-Mediated Antiviral Therapy

M. Kende, D. J. Gangemi, W. Lange, D. A. Eppstein, J. Kreuter, and P. G. Canonico

I. INTRODUCTION

Drug-delivery systems that appear to be suitable for antiviral compounds can be grouped in three major categories (Table 1). Accordingly, when a drug is encapsulated in a carrier or attached to macromolecules, endocytosis is the only mode of entry. Two types of endocytosis, phagocytosis and pinocytosis, constitute the physiologic basis of the delivery systems. In phagocytosis, particulate materials are transported in large intracellular vesicles. The drug is encapsulated in an insoluble carrier engulfed by phagocytic cells and is released after enzymatic breakdown of the carrier. In pinocytosis, soluble materials are transported in small vesicles. The linkage between drug and soluble carrier is stable in the plasma but is susceptible to hydrolysis by lysosomal enzymes. The third is an in-between type. The substance, the drug or the biological is encapsulated in a polymeric or liposomal carrier that, because of its location or size, is not taken up by the cells. Because of diffusion of the substance and bioerosion of the carrier, the biological or the drug is constantly released from the carrier into the circulation and eventually enters the cell via active or passive transport.

II. DRUG TRANSPORT VIA ENDOCYTIC PHAGOCYTOSIS

A. Treatment of Rift Valley Fever Virus Infection with Liposome-Encapsulated Ribavirin

The best-known example of drug transport via phagocytosis involves liposomes, which are spontaneously formed phospholipid vesicles.

Three types of liposomes can be produced based on the method of preparation:

M. Kende and P. G. Cononico • United States Army Medical Research Institute for Infectious Diseases, Fort Detrick, Frederick, Maryland 21701. *D. J. Gangemi* • School of Medicine, University of South Carolina, Columbia, South Carolina 29425 *W. Lange* • Robert Koch Institute, Federal Health Office, D-1000 West Berlin, Germany *D. A. Eppstein* • Institute of Bio-Organic Chemistry, Syntex Research, Palo Alto, California 94304. *J. Kreuter* • Institute of Pharmaceutical Technology, Johann Wolfgang Goethe University, D-6000 Frankfurt, Federal Republic of Germany

Table 1. Systems for Targeted and/or Prolonged Delivery of Antiviral Compounds

Drugs encapsulated or attached to macromolecular carrier enter cells via:		
Endocytic phagocytosis	Endocytic pinocytosis	Active or passive transport
Particulate materials transported in large intracellular vesicles	*Soluble* materials transported in small vesicles	Spherical, circulating, or localized reservoir
Drug encapsulated in an insoluble carrier engulfed by phagocytic cells	Linkage between drug and soluble carrier stable in the plasma but is cleaved in the cell by enzyme(s)	Release of the incorporated substance controlled by diffusion and erosion
Drug released after enzymatic breakdown of the carrier		Free drug enters the cell by active or passive transport

1. Small unilamellar vesicles (SUV) with a diameter of 30 μm are prepared by sonication. SUV of 30–110 μm are formed when an ethanol solution of the lipid is injected into an aqueous phase solution. A third method of producing SUV is to pass multilamellar vesicles (MLV) through a French press with a resulting size range of 30–50 μm.
2. Large unilamellar vesicles (LUV) are formed with reverse-phase evaporation and with detergents. The molar ratio of the lipid/detergent determines the size of the liposomes, which ranges from 50 to 10,000 μm.
3. Multilamellar vesicles with a size range of 400–7000 μm are prepared by forming a lipid film in a pear-shaped flask by using a rotary evaporator. The drug-encapsulation efficacy is best with LUV and MLV but is also dependent on the drug.

The size of the liposome is believed to have an impact on the uptake by cells in different organs. Inclusion of a targeting ligand further promotes targeting into specific cells. Liposomal encapsulation is frequently used to deliver drugs to macrophages infected with fungi, bacteria, and parasites (Alving, 1983; Alving *et al.*, 1978; Desiderio and Campbell, 1983; Graybill *et al.*, 1982; Lopez-Berestein *et al.*, 1983; New *et al.*, 1981; Tremblay *et al.*, 1984). The best result was obtained with an antileishmanial drug: treatment of experimental leishmanias required about 700 times less liposome-encapsulated drugs than did free drugs (Alving, 1983; Alving *et al.*, 1978).

Encapsulation of ribavirin into MLV increased the amount of drug in various organs. When compared to treatment with free drug, four times more drug was delivered into the liver, 25 times more to the spleen, and 10 times more into the lung (Kende *et al.*, 1985) (Fig. 1). Figure 2 shows the increased therapeutic efficacy of liposome-encapsulated ribavirin against Rift Valley Fever virus (RVFV) infection in mice.

B. Treatment of Herpes Simplex Virus Type 1 Infection with Liposome-Encapsulated Ribavirin or Muramyl Tripeptide

A study at the University of South Carolina examined the therapeutic advantage of liposomal encapsulation of ribavirin and muramyl tripeptide-phosphatidyl ethanolamine

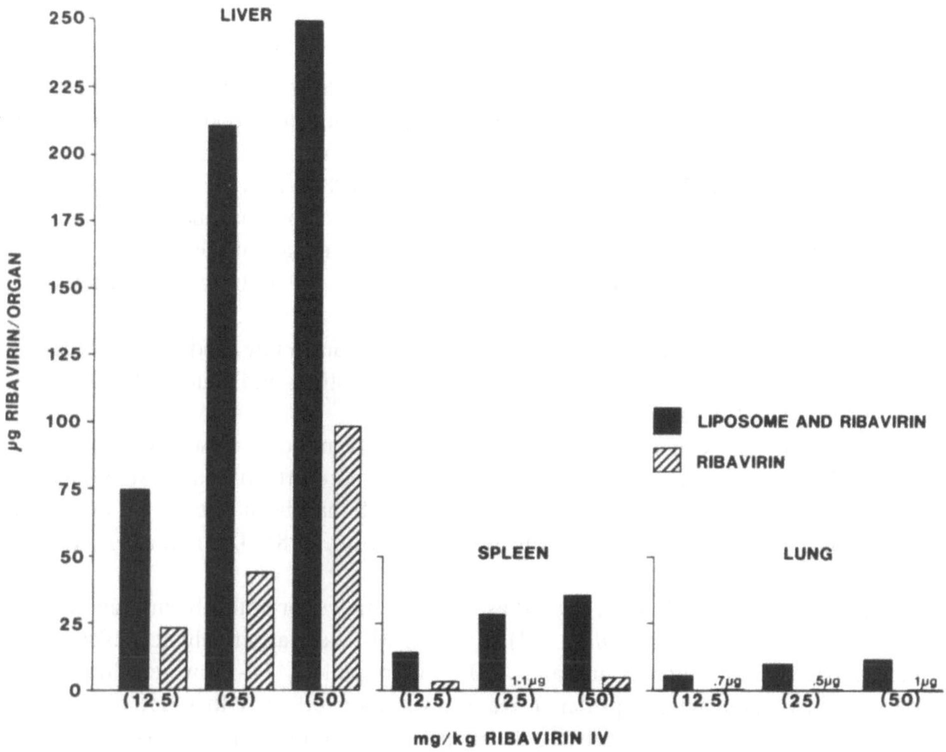

Figure 1. Distribution of [^{14}C]ribavirin, with or without liposomes in mouse tissue 1 hr after IV injection.

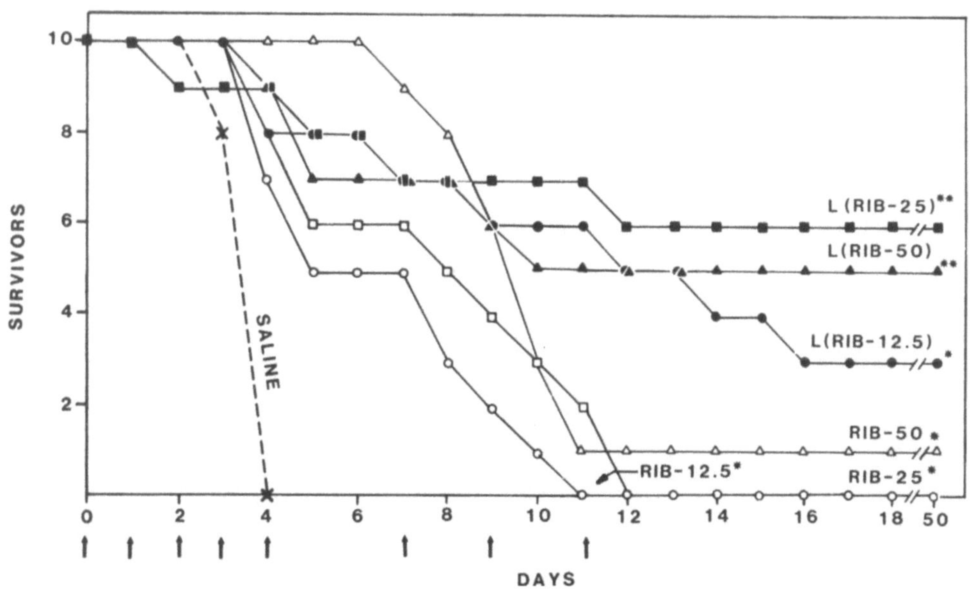

Figure 2. Efficacy of liposome-encapsulated ribavirin (L-RIB) against Rift Valley fever virus infection (day 0) in Swiss Webster mice ($N = 10$). Arrows indicate the days of treatment. Open symbols: 12.5, 25, and 50 mg/kg ribavirin (RIB). Filled symbols: 12.5, 25, and 50 mg/kg L(-RIB). *$p > 0.001$: not significant; **$p < 0.01$: significant in comparison with placebo treatment. The p value of all L-RIB-treated mice versus RIB-treated mice was 0.0001.

(MTP-PE) in the treatment of viral diseases involving the lung, liver, or central nervous system (CNS) as the primary sites of viral infection. Both compounds appear to be potentially useful chemotherapeutic agents for the treatment of human viral diseases (Sidwell *et al.*, 1973; Fidler *et al.*, 1982; Gisler *et al.*, 1982); however, the intravenous dosages required to maintain effective drug levels in infected organs, such as the lung or liver, may adversely affect the normally safe therapeutic indices of these drugs.

The murine models of HSV-1 infection were selected for study for several reasons:

1. Infection of weanling mice is followed by a high death rate, and the pathology is similar to that observed in human disease (Nachtigal and Caulfield, 1984; De Clercq and Luczak, 1976).
2. Ribavirin has been shown to inhibit the replication of both viruses (Sidwell *et al.*, 1973; Durr *et al.*, 1975) but is considerably more effective against influenza.
3. Activation of tissue macrophages by MTP-PE has been shown by others to enhance resistance to HSV-2 infection (Koff *et al.*, 1985; Dietrich *et al.*, 1983).

Negatively charged MLV were used as carrier vehicles for both ribavirin and MTP-PE (Kende *et al.*, 1985; Koff *et al.*, 1984). By using MLV as carrier vehicles, encapsulation efficiencies of 20% for ribavirin and >90% for lipophilic MTP-PE were achieved. In efficacy studies, liposome-encapsulated ribavirin was more effective than free ribavirin in enhancing survival (Fig. 3); moreover, the ribavirin dosage used in liposome carriers (2 mg per mouse) was only one-fifth that used in free form (10 mg per mouse).

The effect provided by liposomal encapsulation on the antiviral efficacy of MTP-PE was examined in three models of HSV-1 disease in which different organs serve as

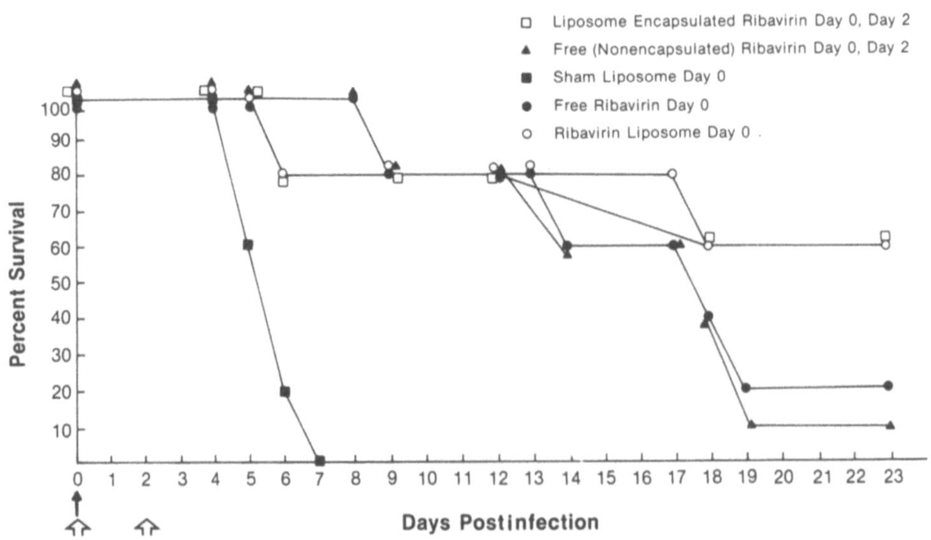

Figure 3. Ribavirin treatment of influenza virus-infected mice. Mice ($N = 10$) were inoculated intranasally with 10 LD_{50} of influenza virus and treated intravenously with either free (10 mg per mouse) or liposome-encapsulated ribavirin (3 mg per mouse) on days 0 and 2.

Table 2. Effect of MTP-PE Treatment on the Survival of Mice Infected Intranasally on Day 0 with HSV-1

Treatment groups	Treatment schedule[b]	Survivors Day 8	Survivors Day 23	% Survival Day 23	Mean survival time
Uninoculated controls	—	0/20	0/20	0	5.10
Sham liposome	−3, 0, 2	7/20	3/20	15	7.35
Free MTP-PE[a]	−3, 0, 2	10/20	9/20	45[c]	8.09[c]
Free MTP-PE[a]	0, 2	9/20	6/20	30[a]	8.75[c]
Liposome-encapsulated MTP-PE[a]	−3, 0, 2	18/20	16/20	80[c]	9.50[c]
Liposome-encapsulated MTP-PE[a]	0, 2	14/20	8/20	40[c]	8.25[c]

[a] 100 μg per mouse.
[b] Days pre- and postinfection, IV drug administration.
[c] $p < 0.001$ compared with uninoculated control group.

primary sites for viral replication: (1) intranasal instillation resulted in interstitial pneumonia and adrenal involvement, (2) footpad inoculation resulted in viral passage through the sciatic nerve to the spinal cord and brain, and (3) intravenous (IV) inoculation resulted in a generalized infection where the liver appeared to be the primary site of viral replication; but the lungs, spleen, adrenals, and CNS were also involved late in the course of infection.

Table 2 illustrates the effect of IV MTP-PE on the survival of mice injected intranasally with HSV-1. A small but significant enhancement of survival was observed when either free or liposome-encapsulated MTP-PE was administered on the day of inoculation and 2 days postinfection. More dramatic protection was observed when an additional dose of MTP-PE was administered 3 days prior to infection. Moreover, liposome-encapsulated MTP-PE was superior to free MTP-PE.

The enhanced survival observed after MTP-PE treatment was correlated with a reduction in viral titers in both the lung and adrenal glands (Fig. 4). MTP-PE-treated survivors had elevated neutralizing antibody titers and were resistant to viral rechallenge 21 days after the initial infection (Table 3).

By contrast, mice were not protected from the IV administration of virus, which leads to a more generalized infection. Neither liposome-encapsulated nor free MTP-PE, given several days following infection, protected mice from IV HSV-1 challenge (data not shown). Experiments were performed in which intranasal and IV routes of MTP-PE administration were combined. Figure 5 illustrates that combined routes of drug administration were highly effective in protecting mice from lethal IV HSV-1 challenge. The therapeutic activity of intranasally administered MTP-PE on a pure CNS disease induced by footpad inoculation of HSV-1 is illustrated in Fig. 6: free MTP-PE given intranasally was more effective than liposome-encapsulated MTP-PE. In addition, mice receiving free MTP-PE had lower viral titers in their spinal cords (Fig. 7).

In combination studies, single doses of MTP-PE on day 3, or ribavirin on day 0, had little effect on survival; however, when given together, 80% of the animals were protected (Table 4). These data suggest that when a virus infects an organ that is part of the reticuloendothelial system (RES), liposomes can be used to enhance the therapeutic index

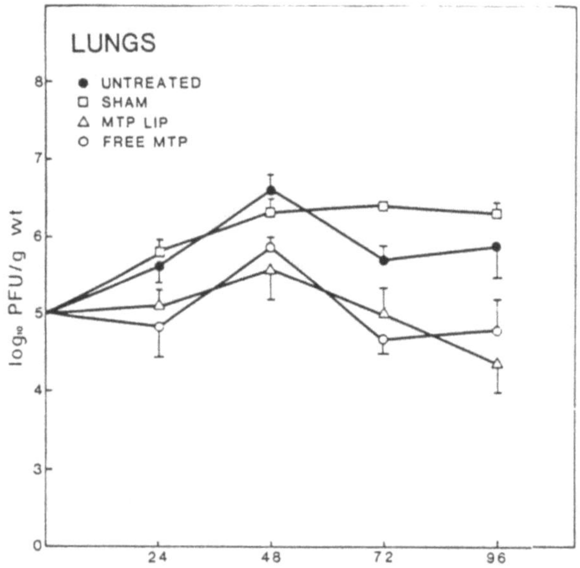

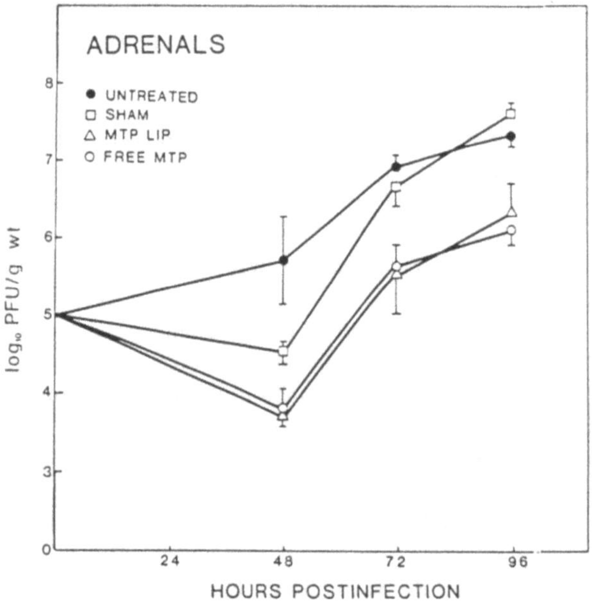

Figure 4. Inhibition of virus replication. Herpes simplex virus type 1 (HSV-1) titers in lungs and adrenals following MTP-PE treatment; 4- to 5-week-old mice were inoculated intravenously with 0.2 ml of either sham, liposome-free, or liposome-encapsulated MTP-PE (100 μg/day) prior to, on the day of, and 3 days following a challenge with 10 LD_{50} of HSV-1. Virus in cell-free extracts of lungs and adrenals was titrated on Vero cells ($N = 3$).

Table 3. Antibody Response of Mice to HSV-1 Infection in MTP-PE-Treated Survivors[a]

Titer	Number of animals with serum-neutralizing titers[b]	
	Free MTP	MTP-PE LIP
<40	3	0
40–100	4	2
100–200	3	3
500–1000	2	8
1000–2500	5	6
Mean titer	489	840

[a] MTP-PE, muramyl tripeptide-phosphatidyl ethanolamine; LIP, liposome.
[b] $p = 0.09$ by Wilcoxon's ranking test, when free MTP-PE and MTP-PE LIP groups were compared.

of selected antiviral agents. This enhancement probably results from macrophage trapping and localization. By contrast, the enhanced resistance observed in the HSV-1 encephalitis model appears to be due to the ease with which free MTP-PE can cross the blood–brain barrier (BBB) following intranasal inoculation. Nonetheless, liposomes may provide a unique means by which drug combinations (e.g., immunostimulants and antivirals) can be

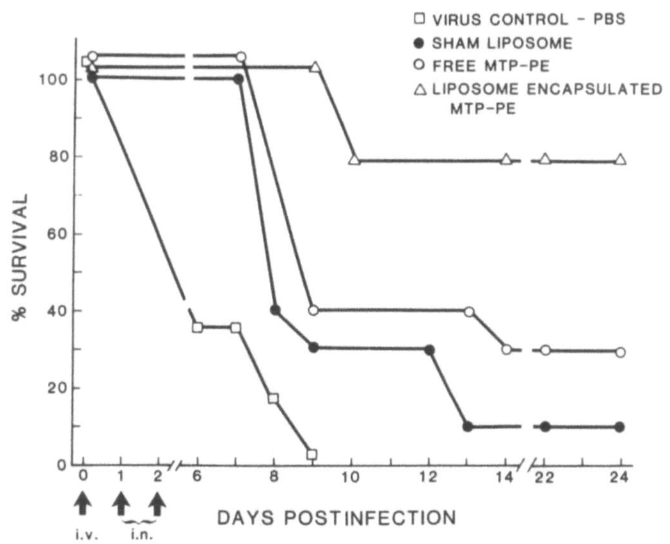

Figure 5. Therapeutic value of MTP-PE in disseminated herpes simplex virus type 1 (HSV-1) infections. Mice were infected IV with 10^4 PFU of the virus and treated on day 0 with 100 μg (IV) of either liposome-encapsulated or free MTP-PE. This was followed by the intranasal administration of 100 μg free MTP-PE on days 1 and 2 postinfection.

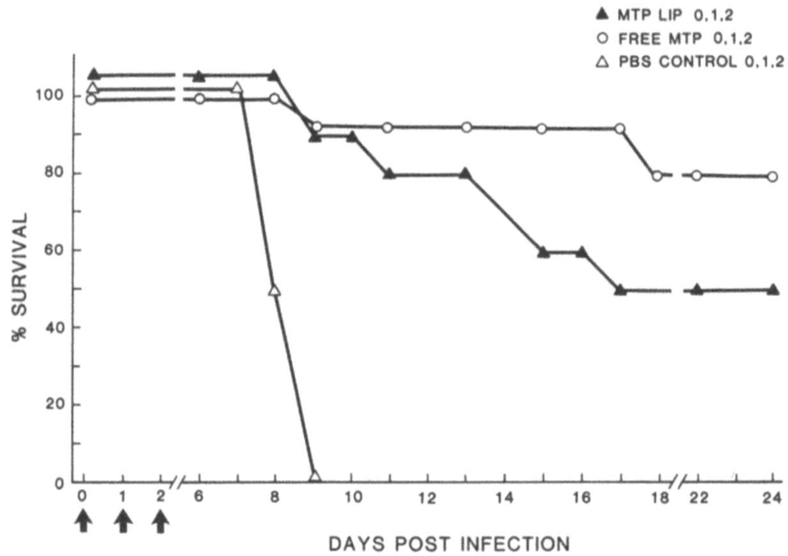

Figure 6. MTP-PE therapy of herpes simplex virus type 1 (HSV-1)-induced encephalitis; 4-week-old mice were inoculated via footpad with 5×10^5 PFU of HSV-1. Either free or liposome-encapsulated MTP-PE (100 μg per mouse) was administered on days 0, 1, and 2 postinfection.

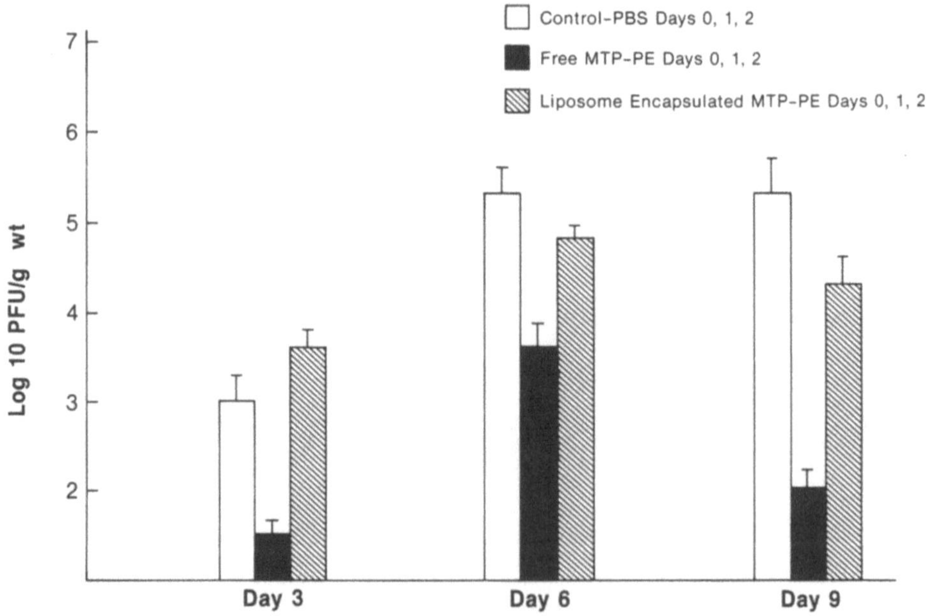

Figure 7. Herpes simplex virus type 1 (HSV-1) viral titers in spinal cords of MTP-PE-treated mice; 4-week-old mice were treated as described in Fig. 5. Spinal cords were removed on days indicated and homogenized, and virus was titrated on Vero cells.

Table 4. Combined Ribavirin/MTP-PE Therapy of HSV-1 Pneumonitis

Treatment groups[a]	Treatment schedule[b]	Survivors Day 10	Survivors Day 23	% Survival day 23	Mean survival times[c]
Uninoculated control	—	0/20	0/20	0	7.4
Sham liposome	−3 or 0	0/10	0/10	0	7.6
L-MTP-PE alone	−3	2/10	2/10	20	7.8
L-RIB alone	0	1/10	0/10	0	7.7
L-MTP-PE + L-RIB	−3 and 0	8/10	8/10	80[d]	7.5

[a] MTP-PE (100 μg per mouse); ribavirin (3 mg per mouse).
[b] Days pre- and postinfection, IV drug administration.
[c] Of animals that died.
[d] $p < 0.001$ compared with control group.

combined in the same carrier vehicle and delivered to specific sites of viral replication in a diseased host.

III. DRUG TRANSPORT VIA ENDOCYTIC PINOCYTOSIS

Targeting of Polymer Matrix–Drug Complex into Specific Cells

Hydroxypropylmethacrylamide (HPMA) polymer represents the second type of drug-delivery system (Kopacek *et al.*, 1985), in which the carrier-drug complex enters the cells via pinocytosis. The matrix of the carrier is HPMA, and the drug (*p*-nitroaniline) is attached to the matrix via an enzymatically degradable amino acid spacer.

Two important points must be studied. First, in the blood (serum), the bond between the drug and the carrier should be stable. It was found that the release of *p*-nitroaniline in plasma and serum is negligible with respect to the rate of cleavage by lysosomal enzymes (Kopacek *et al.*, 1985). Second, inside the cell, however, the bond between the drug and the carrier should be cleaved by lysosomal enzymes. These workers showed that the time-dependent cleavage of *p*-nitroaniline from the polymeric substrate was catalyzed by rat liver tritosomes and cathepsin H. Cathepsin B and L did not cleave the drug.

To avoid accumulation of nondegraded polymers, low-molecular-weight poly-HPMA chains can be prepared with cross-linking. Such polymers can be cleaved by chymotrypsin, trypsin, papains, and cathepsin B and L.

To achieve selective targeting of polymer-bound drugs, it is necessary to incorporate a specific residue, such as a glycolipid that interacts with membrane receptors unique to specific cell types. For instance, hepatocytes have membrane receptors that recognize galactose residues, which permit specific targeting of the drug carrier into liver hepatocytes.

Distribution studies in rats with ^{125}I-labeled HPMA water-soluble cross-linked copolymer showed that 60 min after injection, galactosamine-containing HPMA polymer assumes a high (63%) content in the liver, while without targeting, only 6% of labeled polymer is found in the liver. This system has not yet been deployed to deliver antiviral compounds, but it has all the features necessary for drug targeting into specific cells.

IV. DRUG UPTAKE BY ACTIVE OR PASSIVE TRANSPORT

For reservoir type of prolonged and controlled delivery of a substance, localized liposomes, circulating (systemic) polylactic acid, or polylactic/polyglycolic acid copolymer microcapsules have been used. The drug release is controlled by a combination of diffusion of the compound and erosion of carrier.

A. Localized Sustained Release of Interferon by Liposomes

Interferons α and β (IFN_{α} and IFN_{β}), produced both by recombinant DNA technology as well as purified from natural sources, have been shown to be efficacious in treating certain cancers and viral diseases. Studies with IFN_{γ} have more recently been undertaken; thus their clinical value is not yet as well defined. The treatment schedules usually involve multiple injections of the interferon over a period of several weeks. By using such treatment regimens, the doses of the interferons needed to obtain efficacy can result in toxic side effects. For all these reasons, methods of increasing the ease of administration as well as the therapeutic ratio of interferons are warranted.

Sustained-release liposomal–interferon preparations were employed for systemic and/or localized treatment. IFN_{β} or IFNγ was incorporated with high efficacy (100% for IFN_{β}, 50–100% for IFN_{γ}) into lyophilized multilamellar liposomes prepared by hydrating lyophilized lipids with an aqueous solution of the interferon (Eppstein *et al.*, 1985; Eppstein, 1986*a,b*; Eppstein and Felgner, 1987).

The presence of serum and cells induces the leakage of the interferon from the liposome preparations, with the most solid vesicles (i.e., those made with phospholipids having saturated acyl chins) being the least leaky. Accordingly, liposome compositions can be prepared to control the rate of release of the interferon to different degrees after subcutaneous (SC) or intramuscular (IM) injection. Formulations of recombinant human $IFN\beta_{ser17}$ ($rHuIFN_{\beta ser17}$, obtained from Tirton Biosciences Inc., Alameda, California) in MLV consisting of diarachidoyl-phosphatidylcholine–dipalmitoylphosphatidylglycerol (DAPC–DPPG), 9 : 1, were able to retain the interferon at the site of IM or SC injection in mice. This resulted in a slow interferon release from the vicinity of the injection site over 1–2 weeks. Therefore, interferon-containing liposomes will not be phagocytized by macrophages, which would result in the enzymatic digestion of interferon. Thus, 50% of the interferon was still retained after 2 days, 15% remained after 6 days, and 5% remained after 9 days. By contrast, free interferon was gone after 1 day (Eppstein, 1986*a*; Eppstein and Felgner, 1987). When saturated phospholipids of progressively shorter acyl chain length (C_{18}, C_{16}, and C_{14}) were employed, the duration of retention of the interferon was progressively reduced (Faulkner, Schryver, and Eppstein, unpublished results).

The efficacy of liposomal formulations of $rHuIFN_{\beta ser17}$ was studied in a primate model of varicella–zoster virus (VZV). African green monkeys were infected systemically with simian VZV, and 24 hr later, interferon treatments were initiated. Initial results indicated that free $rHuIFN_{\beta ser17}$, given IM twice daily at 10^6 U/kg/dose for 10 days (2×10^7 U/kg total dose), showed very good efficacy in suppressing systemic VZV infection. The treated monkeys had minimal viremia and rash, and they did not die. Placebo-treated controls showed extensive viremia and rash and died 10–11 days postinfection. However, the same dosage of interferon was marginally effective when given in only two or three injections (on days 1 and 6 or on days 1, 4, and 7 postinfection) instead of the 20 twice-daily injections. By contrast, if this same dosage of interferon was given on days 1 and 6, but in a liposomal formulation, distearoylphosphatidylcholine (DSPC)–DPPG (9 : 1), 40 μmoles lipid/kg per dose, intermediate efficacy was obtained; death was prevented and viremia and rash were reduced, although not to the extent obtained with the 20 doses of free interferon (Soike *et al.*, in press). Formation of neutralizing antibodies to the human IFN_{β} was not obtained in monkeys 1–4 weeks after injection of the second dose of liposomal interferon. These initial results suggest that liposomal interferon formulations can be obtained that will enhance the efficacy of an injection of interferon, most likely by providing a slow release of the drug over several days.

B. Polymeric Microcapsule Carriers for Drugs and Viral Vaccine

Persistent levels of the contraceptive norethisterone hormone were obtained in baboon serum with polylactate microcapsules. The amount of norethisterone is related to size of the capsules. Polylactide–polyglycolide microcapsules containing norethisterone were used in humans to maintain serum levels for 6 months (Beck *et al.*, 1980).

In diabetic rats that received SC implants of insulin containing ethylene vinyl acetate polymer, the blood glucose level was normal for 4 weeks, in contrast to the high glucose level in those rats that received placebo-containing polymer (Creque *et al.*, 1980). Sustained release with ethylene vinyl acetate polymer pellets as carrier and immunologic adjuvant of bovine serum albumin (BSA) has also been described (Langer, 1981). The immune response evoked in rats with a single injection of BSA–polymer was comparable to or better than two injections of BSA alone.

The development of more specific and purer vaccines very often leads to a decrease in the antibody response, hence protection. This effect was observed some years ago during the development of split vaccines (Kreuter and Speiser, 1976*a*; Kreuter *et al.*, 1976) but may be more pronounced during the development of vaccines produced by genetic engineering. This problem may be overcome by the use of adjuvants.

Because of their side effects, very few adjuvants can be used in human vaccines. Besides the classic aluminum adjuvants, particulate polymeric carriers, the so-called nanoparticles, hold promise for use as adjuvants. Nanoparticles are solid colloidal particles ranging in size from 10 to 1000 nm (1 μm). They consist of macromolecular materials in which the active principle (antigen or drug) is dissolved, entrapped, or encapsulated or to which the active principle is adsorbed or attached (Kreuter, 1983). These materials are biodegradable (Grislain *et al.*, 1983; Kreuter *et al.*, 1983). The tissue response 1 year after IM injection of poly(methylmethacrylate) nanoparticles was very mild, comparable to that of a fluid vaccine (Kreuter *et al.*, 1976).

These particles may be produced either by x-ray-induced polymerization in the case of methyl methacrylate (Kreuter and Speiser, 1976*a*; Kreuter *et al.*, 1976) or by base-catalyzed polymerization in the case of the cyanocrylates (Couvreur *et al.*, 1979). Polymerization in the presence of the antigen led to a partial incorporation of the antigen into the particles, while polymerization in absence of the antigen led to the adsorption of the antigen onto the particle surface (Kreuter and Speiser, 1976*b*). γ-Irradiation was chosen as one of the polymerization methods because it does not destroy the antigenicity of a number of antigens (Kreuter and Zehnder, 1978).

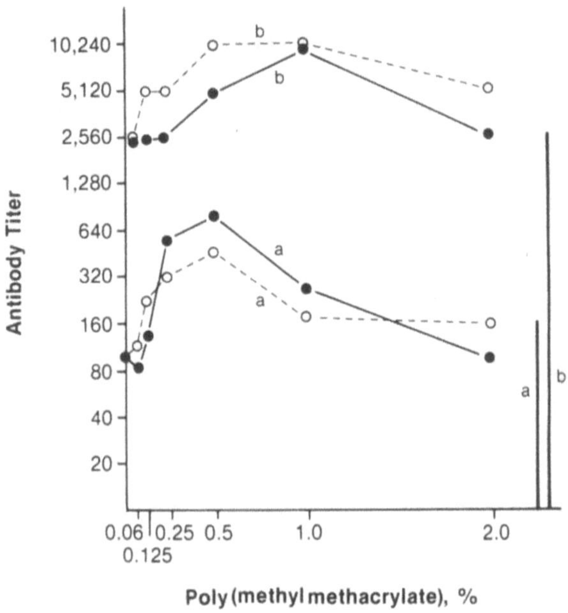

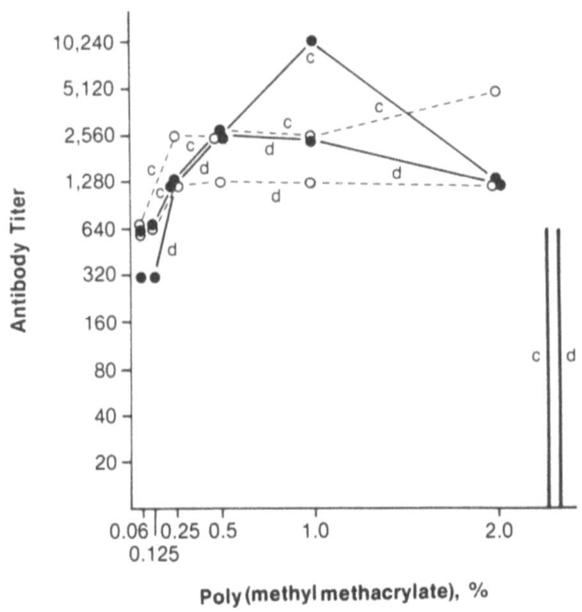

Figure 8. Influence of different contents of poly(methylmethacrylate) on the antibody response of mice after 28 (a), 36 (b), 64 (c), and 92 (d) days. Boosting occurred after 28 days with the same vaccine as was used in the primary vaccination. (–●–) Incorporation into poly(methyl methacrylate). (–○–) Adsorption onto poly(methylmethacrylate) (|) Adsorption onto 0.2% aluminum hydroxide.

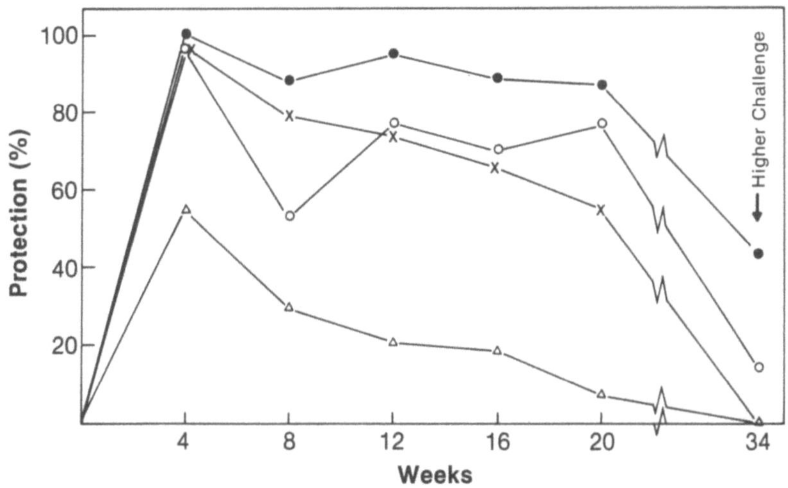

Figure 9. Protection against infection after challenge with 50 LD_{50} of mouse-adapted influenza virus. Higher challenge = 250 LD_{50}. The vaccines contained the following adjuvants. (●) Incorporation into 0.5% poly(methylmethacrylate). (○) Adsorption onto 0.5% poly(methylmethacrylate). (x) Adsorption onto 0.2% aluminium hydroxide. (△) Fluid vaccine.

The vaccines with nanoparticles as adjuvants were tested by determination of the antibody response in guinea pigs (Langer, 1981; Kreuter and Speiser, 1976*a*) and mice (Langer, 1981; Kreuter and Speiser, 1976*a*; Kreuter and Zehnder, 1978) as well as by measurement of the protection of mice against infection with an LD_{50} of mouse-adapted live influenza virus (Kreuter and Liehl, 1978, 1981).

Poly(methylmethacrylate) nanoparticles obtained an optimal adjuvant effect at a concentration of 0.5%. This optimum was considerably more pronounced with incorporated influenza antigen than with the adsorbed product (Fig. 8). At optimal concentrations, this adjuvant yielded a better antibody response (Kreuter and Speiser, 1976*a*; Kreuter *et al.*, 1976; Kreuter and Liehl, 1981) and better protection (Kreuter and Liehl, 1981) than did aluminum hydroxide. This improvement in adjuvant effect was especially pronounced after extended times (Fig. 9). The use of the polymeric adjuvants did improve the storage stability of the vaccines at higher temperatures (Fig. 10).

C. Controlled Systemic Release of Interferon with Biodegradable Polymer

To achieve controlled release of an interferon over a longer time frame than was possible with liposomal formulations, biodegradable polymers of poly(*d,l*-lactide-co-glycolide) (PLGA) were employed. Such polymers have been used in the development of controlled-release systems for small peptide hormones, such as luteinizing hormone-releasing hormone (LH-RH) (Sanders *et al.*, 1984), but their application to larger more labile polypeptides requires quite different formulation techniques in order to avoid protein denaturation.

It was possible to prepare $rHuIFN_{\beta ser17}$ in PLGA polymer matrices with 100% incorporation of interferon and full retention of biologic activity by using novel formula-

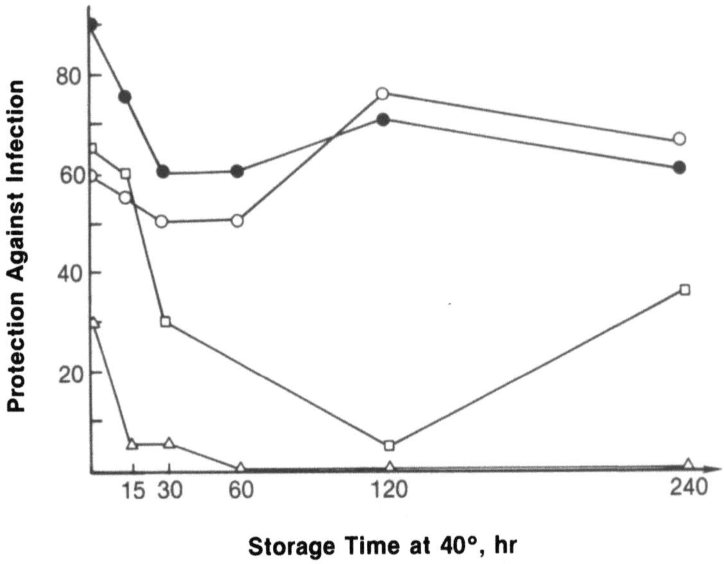

Figure 10. Stability of vaccines against heat inactivatiion, showing protection of mice against morbidity after immunization with vaccines that were stored at 40°C for different time periods. The vaccines contained the following adjuvants. (●) Incorporation into 0.5% poly(methyl methacrylate). (○) Adsorption onto 0.5% poly(methyl methacrylate). (x) Adsorption onto 0.2% aluminum hydroxide. (△) Fluid vaccine without adjuvants. The mice were challenged with 50 times the LD_{50} value of homologous mouse-adapted virus.

tion techniques (Schryver, van der Pas, and Eppstein, manuscript in preparation). The release profile of the interferon was determined over 3 months by quantitating the radioactivity remaining in a mouse at the site of SC implantation of a PLGA–interferon pellet or film (containing [^{125}I]-rHuIFN$_{\beta ser17}$ (Eppstein, 1986*a*). The release profile of the interferon was influenced both by the method of preparation as well as by the geometry of the final implant.

Release profiles were obtained over 2 months by employing different formulation methods as well as different final implant geometries. With some formulations, a very triphasic release profile was obtained, as has been observed with PLGA formulations of small peptides (Sanders *et al.*, 1984). Such a triphasic release profile is believed to represent first an augmented initial release rate of peptide as a result of diffusion from the surface of the PLGA implant or microspheres, followed by a relatively latent period of minimal drug release, while the polymer chains are gradually hydrolyzed to progressively shorter chain lengths, culminating with a high-release rate when the polymer has become hydrolyzed to sufficiently short chain lengths to become solubilized (Sanders *et al.*, 1984). However, by varying parameters of implant geometry, *in vivo* release profiles of interferon could be obtained that were fairly linear over 60–70 days. Extraction of the interferon from implants retrieved up to 5 weeks after implantation in the mouse showed that the rHuIFN$_{\beta ser17}$ retained remarkably good biologic (antiviral) activity. In fact, the PLGA formulation appeared to stabilize the interferon significantly, such that only a 0.3 $\log_{10}$ loss in biologic activity occurred after 5 weeks of implantation in the mouse. More than two $\log_{10}$ values of activity were lost when the same interferon (containing stabilizers) was incubated at 37°C in buffer in a test tube (Eppstein, 1986*a*; Schryver, van der Pas, and Eppstein, manuscript in preparation).

Although sustained-release delivery systems have been shown to act as immunologic adjuvants for certain proteins (Amkraut and Martins, 1984), no antibody formation (neutralizing or non-neutralizing) was detected in the mouse with this human interferon–PLGA formulation. As the actual daily dose of interferon protein can be quite low in a sustained-release system, this may help circumvent the unwanted formation of antibodies against the exogenous interferon.

D. Treatment of Influenza Virus Infection with Squalene-in-Water Preparations

Studies were performed to investigate the ability of bacterial-derived immunomodulators, natural trehalose dimycolate (TDM), or synthetic muramyl dipeptide (MDP) to stimulate nonspecific resistance against respiratory viral infection by a well-defined aerosol model of influenza virus in mice (Table 5). The compounds were prepared in 1% squalene-in-water and adsorbed to the surface or mixed within the oil droplets. Outbred NMRI mice were pretreated with MDP, TDM, or a combination of MDP + TDM. A relatively long period (3–4 weeks) elapsed before aerosol infection with mouse-virulent A/PR/8/34 (H1N1) influenza virus. The results, summarized in Table 5, show that 150 μg MDP in saline was ineffective against a lower dose (LD_{80}) of influenza virus, whereas administration of MDP in 1% squalene-in-water preparation provided borderline protection. The increase in survival was not statistically significant. A dose of 75 μg of TDM in squalene-in-water preparation induced significant protection in 70% of treated mice.

The combination of 150 μg MDP plus 75 μg TDM afforded complete protection

Table 5. Effect of MDP and TDM on Aerogenic Influenza Viral Infection in Mice[a,b]

Pretreatment	Dose (μg)	% Surviving influenza infection	
		LD_{80}	LD_{100}
MDP–saline	150	30	0
MDP–1% squalene-in-water	150	50	0
	300	40	0
TDM–1% squalene-in-water	75	70	15
	150	70	10
MDP + TDM 1% squalene-in-water	150 + 75	100	91
	300 + 150	100	90
Formalinized influenza vaccine	200 HAU	100	100
1% squalene-in-water		20	0

[a] All treated by the IV route.

[b] Groups of 10–20 mice were pretreated 3 weeks before A/PR/8/34 influenza viral infection. Formalin-inactivated A/PR/8/34 influenza virus was used as specific vaccine.

against influenza virus infection (LD_{80}). When given alone, MDP and TDM were ineffective against a completely lethal infection (LD_{100}), even when higher doses of 300 μg or 150 μg, respectively, were used. The combination of MDP + TDM was consistently found to be efficacious against both low and high doses of influenza virus.

The efficacy of MDP + TDM combination was corroborated by tissue culture titration of the virus in the infected lungs. MDP + TDM-pretreated mice showed marked decreased levels of virus 72 hr after infection, and the clearance of detectable infectious virus occurred earlier in this group. Serum hemagglutination-inhibiting antibodies could be detected already on day 5 in the MDP + TDM-pretreated mice, but only on day 7 in the control group. It should be mentioned that all survivors of the pretreated group showed a long-lasting resistance against lethal rechallenge 6 months later, even if a heterologous influenza [A/Port Chalmers/1/73 (H3N2)] virus was used for rechallenge (Fig. 11).

Significant protection against influenza virus mortality was obtained with 6-0-acyl analogues and a ubiquinone derivative of MDP. A greater degree of protection was induced by the combination of TDM with lipophilic derivatives, like the seryl analogue of desmethyl-MDP (Table 6).

Cell walls of gram-negative bacteria contain endotoxic lipopolysaccharide, a potent stimulator of the immune system, even in nanogram quantities. Clinical application of endotoxin has been hampered, as it plays a major role in the pathophysiology of bacterial sepsis. Concerted efforts have been made to modify endotoxin for possible therapeutic use in humans. Investigations on endotoxic lipopolysaccharide have led to the identification of lipid A as an important immunopharmacologic constituent of the endotoxin molecule.

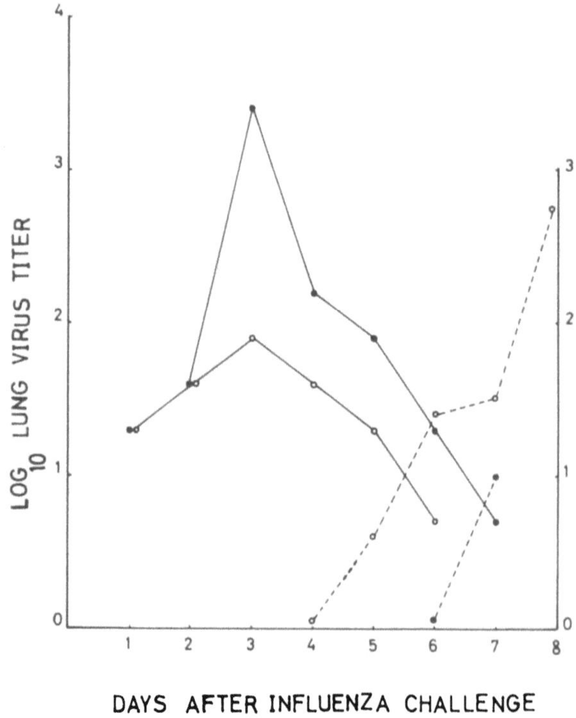

Figure 11. Reduction of A/PR/8/34 (HINI) influenza virus in the lungs of mice treated with 150 μg MDP and 75 μg TDM.

Table 6. MDP Analogues Effective against Influenza Viral Infection in Mice When Combined with TDM

N-acetylmuramyl-L-alanyl-D-isoglutamine (MDP)
6-0-(2-Tetradecylhexadecanoyl)-*N*-acetylmuramyl-L-alanyl-D-isoglutamine (B30-MDP)
6-0-Stearoyl-*N*-acetylmuramyl-L-alanyl-D-isoglutamine (L18-MDP)
6-0-Isopentadecanoyl-*N*-acetylmuramyl-L-alanyl-D-isoglutamine (Iso.15-MDP)
2,3-Dimethoxy-5-methyl-6-(9′-carboxynonyl)-1,4-benzoquinone-*N*-acetylmuramyl-L-valyl-D-isoglutamine-methyl ester (QS-10-MDP-66)
N-Acetylmuramyl-L-seryl-D-isoglutamine
N-Acetyldesmethylmuramyl-L-seryl-D-isoglutamine
6-0-Succinyl-*N*-acetyldesmethylmuramyl-L-seryl-D-isoglutamine
6-0-Capryl-*N*-acetyldesmethylmuramyl-L-seryl-D-isoglutamine
N-Acetylmuramyl-L-α-aminobutyryl-D-isoglutamine
N-Acetyldesmethylmuramyl-L-α-aminobutyryl-D-isoglutamine

Recently, selective reduction of the harmful toxicity and pyrogenicity of lipid A, while retaining its beneficial adjuvant property, have been achieved. Effective chemical treatment of toxic lipid A isolated from refined cell walls (CWS) of polysaccharide-deficient heptose-free Re mutant strains of *Salmonella typhimurium* or *Salmonella minnesota* yielded monophosphoryl lipid A (MPL), which was 1000-fold less toxic than endotoxin. The availability of nontoxic MPL has already led to clinical trials in human patients.

Combinations of MPL with CWS and TDM in squalene-in-water emulsions were investigated for their effect on influenza viral infection. The results of a representative experiment show that the combination of 100 μg of CWS with either 50 μg MPL or 50 μg TDM did not induce significant resistance. By contrast, the combination of 50 μg TDM with MPL afforded complete protection against lethal disease. Even amounts as low as 5 μg MPL combined with 50 μg TDM could induce significant resistance to the viral infection. Combination treatment with CWS, MPL, and TDM was effective when at least 50 μg TDM was used in the preparations (Table 7).

The efficacy of the MPL + TDM combination was checked by the titration of the virus present in the lungs. The resistance-stimulating MPL + TDM combinations significantly reduced lung viral titers, in contrast to the ineffective CWS + MPL combinations or the controls.

Natural killer (NK) cell activity of 3–4-week pretreated animals was not significantly different from controls at the time of infection. In contrast to NK cells, macrophages appear to be among the target cells for the immunopotentiating activity of MPL + TDM combination 2–4 weeks after pretreatment.

To elucidate the role of the macrophage in the resistance induced by MDP + TDM combination, animals pretreated 21 days earlier with the immunostimulants were given silica, dextran sulfate, and carrageenan, and 24 hr later, an aerosol of influenza virus. The results show that the resistance to influenza virus that was induced by MDP + TDM combination was abrogated by treatment with all three agents known to inhibit selectively or impair macrophage function *in vivo* (Table 8).

Phagocytosis stimulates oxidative metabolism of the phagocytic cells, which results in increased hexose monophosphate shunt activity and the generation of activated oxygen

Table 7. Protection against Influenza Viral Infection by Combination of MPL, TDM, and CW

Treatment[a]	Dose (μg)	% Survival	p
MPL	50	0	—
TDM	50	0	—
CWS	100	0	—
CWS + MPL	100 + 50	0	—
CWS + TDM	100 + 50	0	—
MPL + TDM	50 + 50	100	<0.001
	5 + 50	30	<0.05
	50 + 5	0	—
CWS + MPL + TDM	100 + 50 + 50	80	<0.001
	100 + 5 + 50	40	<0.02
	100 + 50 + 5	0	—
Vehicle control	—	0	—

[a] Groups of 10–20 mice were pretreated intravenously with various squalene-in-water emulsions 3 weeks before aerosol infection with A/PR/8/34 influenza virus.

metabolites, such as singlet oxygen, superoxide anion, hydrogen peroxide, and hydroxyl radical with the emission of photons. The release of energy in the form of light (chemiluminescence) can be detected and measured. Spleen cells from animals pretreated with the combination of MDP + TDM exhibited markedly enhanced chemiluminescence activity in response to stimulation by *Staphylococcus aureus;* similar results were obtained with zymosan (Fig. 12).

The effects of lipophilic MDP on the morphology of murine macrophages were investigated by scanning electron microscopy (SEM) and transmission electron microscopy (TEM). In addition, alteration of the phagocytic function was determined by Fc receptor-mediated phagocytosis of ^{51}Cr-labeled sheep erythrocytes. Resident peritoneal macrophages cultured *in vitro* for 24 hr with 100 μg lipophilic MDP underwent a striking change in appearance. Rounding of macrophages and large cytoplasmic vacuoles were

Table 8. Abrogation of Effect of MDP and TDM on Aerogenic Influenza Viral Infection in Mice with Silica, Carrageenan, and Dextran Sulfate[a]

Pretreatment (150 + 75 μg)	Compound	Dose (mg)	% Survival	p
MDP + TDM	—	—	90	—
MDP + TDM	Silica	3	10	0.01
MDP + TDM	Carrageenan	1	10	0.01
MDP + TDM	Dextran sulfate	1	20	0.02

[a] Groups of 10 mice were pretreated with 1% squalene-in-water preparations 3 weeks before aerosol infection with A/PR/8/34 influenza virus. Anti-macrophage compounds were administered intravenously 24 hr before infection.

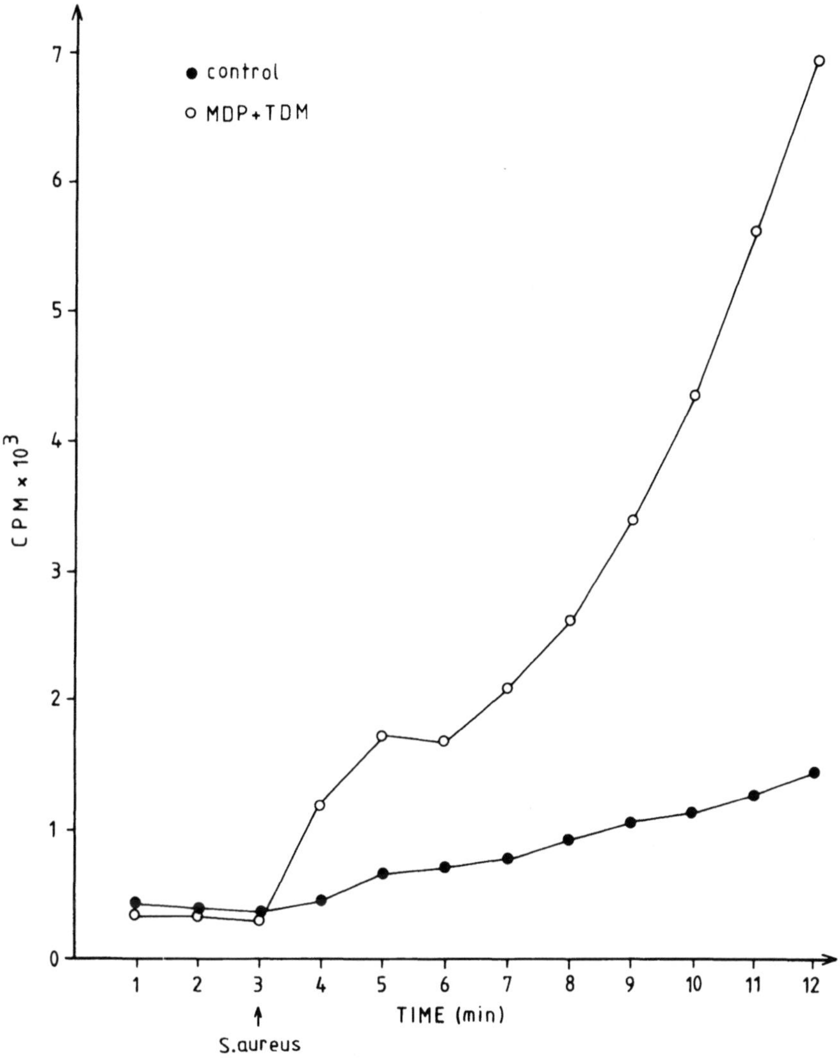

Figure 12. Enhancement of chemiluminescense activity of mouse spleen cells with MDP and TDM combination treatment.

observed. Figure 13 shows SEMs of untreated and MDP-treated macrophages. Untreated macrophages (Fig. 13A) showed spreading elongated forms with many ridges, microvilli, and thin pseudopods. By contrast, MDP-treated macrophages (Fig. 13B) showed rounding and extensive ruffling of the cell surface.

Transmission electron microscopy confirmed the morphologic changes. Large vacuoles were observed in the cytoplasm of macrophages treated with lipophilic MDP (Fig. 13B) but not in the untreated macrophages (Fig. 13A). The morphologic alterations in the treated macrophages could not be ascribed to nonspecific cytotoxic effects, since the viability of macrophages was not affected by lipophilic MDP, as shown by dye exclusion.

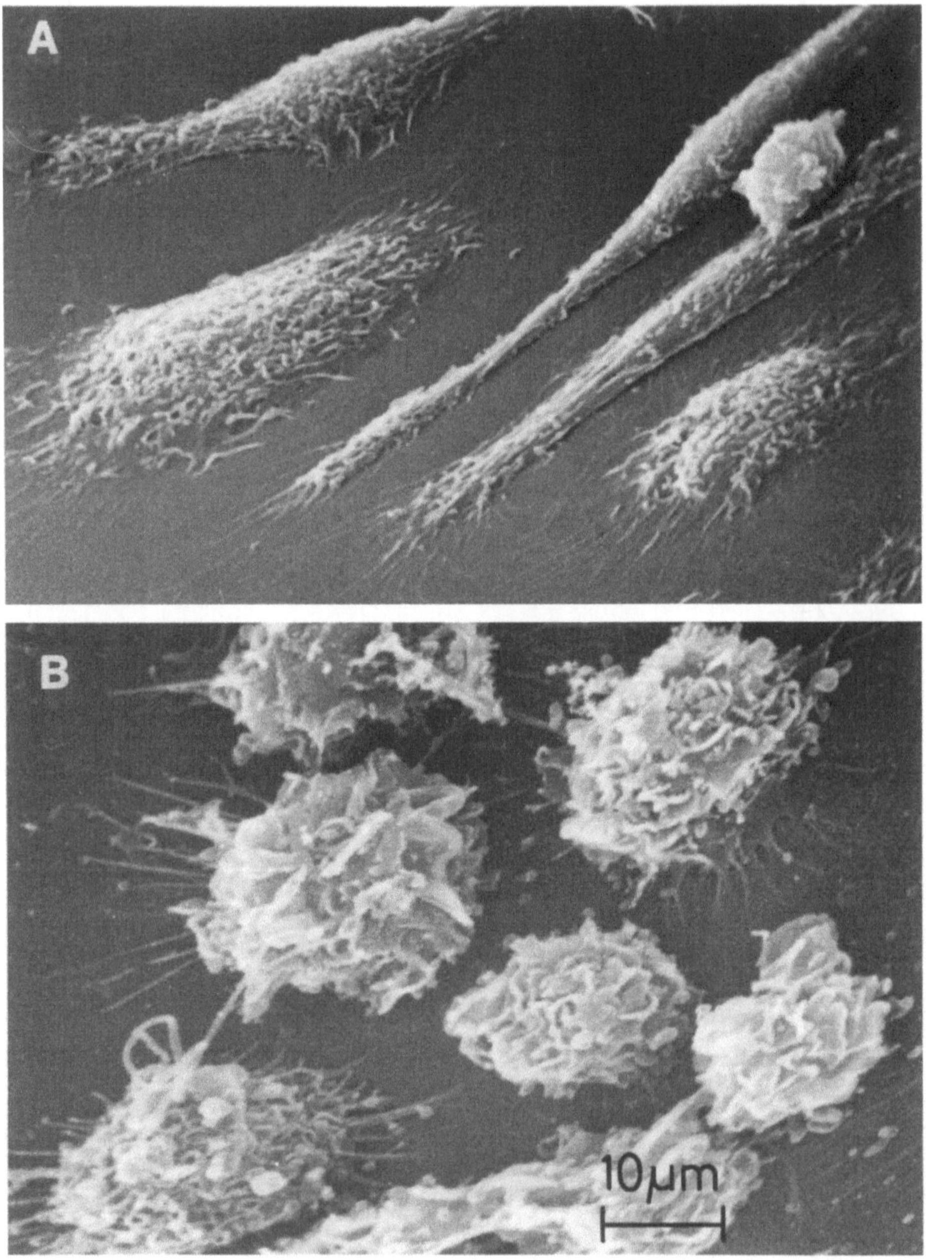

Figure 13. Scanning electron micrograph of untreated (A) and MDP-treated (B) mouse macrophages.

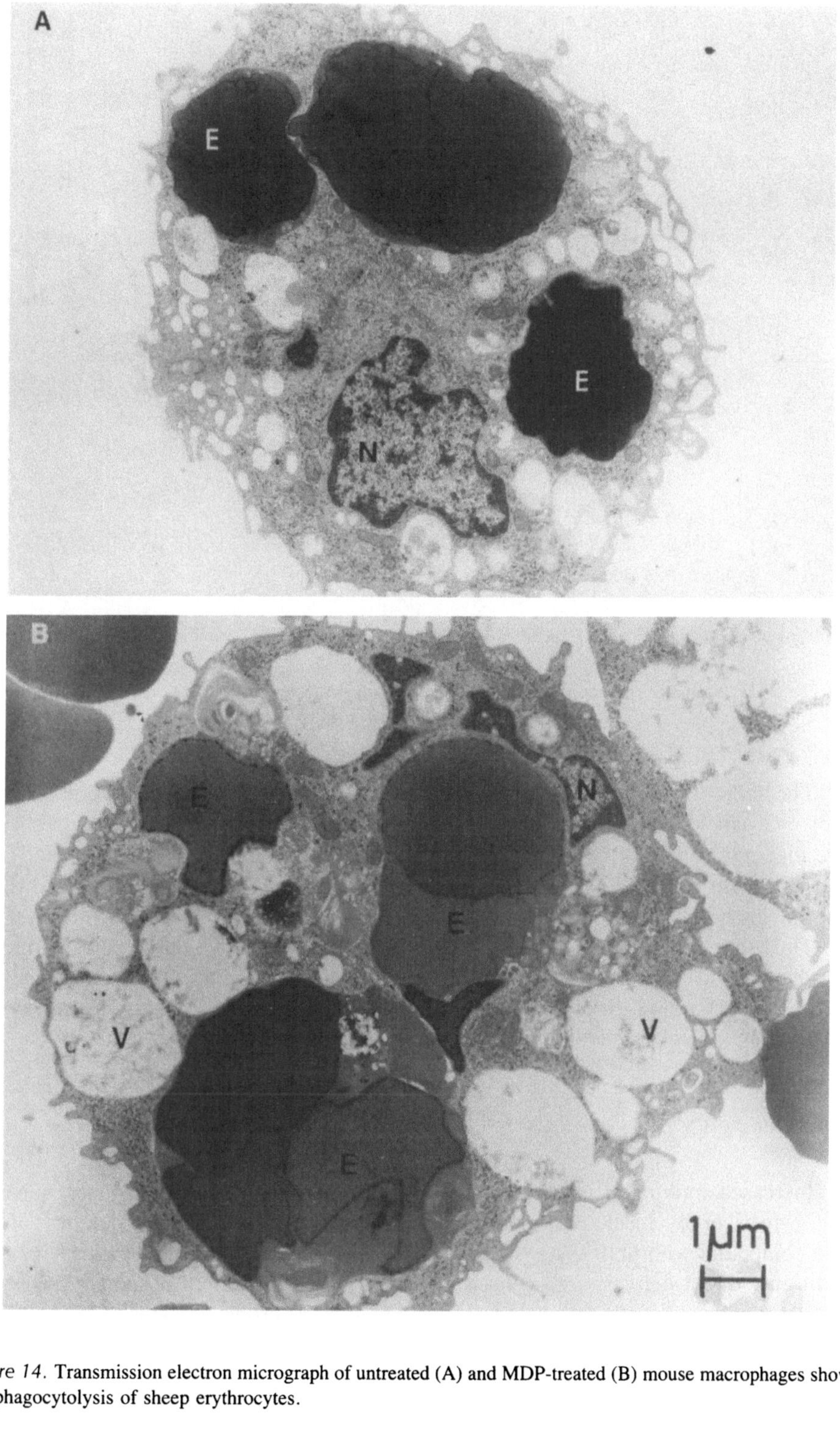

Figure 14. Transmission electron micrograph of untreated (A) and MDP-treated (B) mouse macrophages showing phagocytolysis of sheep erythrocytes.

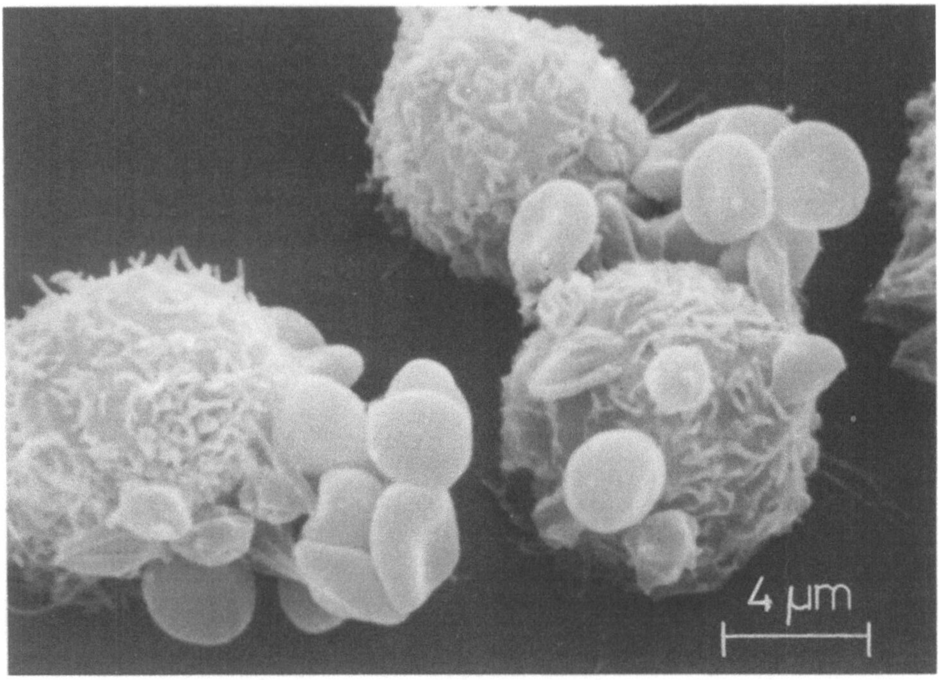

Figure 15. Scanning electron microscopy sheep erythrocytes bound to MDP-treated macrophages.

The ability of treated and control macrophages for Fc receptor-mediated phagocytosis of ^{51}Cr-labeled sheep erythrocytes was tested. A large number of sheep erythrocytes were ingested by lipophilic MDP-treated macrophages, in contrast to the untreated controls, as shown in the TEMs presented in Fig. 14. SEM revealed many opsonized sheep erythrocytes bound to the treated macrophages (Fig. 15).

Thus, the combination of trehalose dimycolate with muramyldipeptides or monophosphoryl lipid A induces long-term resistance, and the mechanisms involved include macrophage activation. Stimulation of nonspecific host-defense mechanisms against viral infections by potent immunomodulators appears feasible.

V. CONCLUSION

Increased antiviral efficacy was achieved with liposomal drug delivery via endocytic phagocytosis or by localized sustained release. Although comparative delivery of the same compound with both systems and the assessment of their antiviral efficacy has not been done, direct delivery into cells is probably more efficient. Liposomal delivery of antiviral compounds into cells is currently limited to macrophages alone, but localized delivery with liposomes provides a viable alternative. Development of targeted delivery of a polymer–matrix–drug complex via receptor-mediated endocytic pinocytosis into cells other than reticuloendothelial origin also appears to be attainable. Furthermore, systemic delivery with polymeric microspheres or with squalene-in-water has also been proved suitable in order to achieve the much-needed control of viral infections. The pitfall of these systems is that they cannot deliver antiviral agents into the CNS. For this purpose, new currently nonexisting technology will be required.

ACKNOWLEDGMENTS

The skillful typing of the manuscript by Ms. Betty J. Martin and the editorial work by Ms. Katheryn F. Kenyon are greatly appreciated. The workshop on Carrier-Mediated Antiviral Therapy was funded in part by grant 86G-6024 from U.S. Army Medical Research and Development Command. The views of the authors, M. Kende and P. Canonico, do not purport to reflect the position of the Department of the Army or the Department of Defense. In conducting the research described in this chapter, the investigators, M. Kende and P. Canonico, adhered to the *Guide for the Care and Use of Laboratory Animals,* as promulgated by the Committee on Care and Use of Laboratory Animals of the Institute of Laboratory Animal Resources, National Research Council. The facilities are fully accredited by the American Association for Accreditation of Laboratory Animal Care.

REFERENCES

Alving, C. R. (1983). *Pharmacol. Ther.* **22,** 407–424.

Alving, C. R., Steck, E. A., Chapman, W. L., Jr., Waits, V. B., Hendricks, L. D., Swartz, G. M., Jr., and Hanson, W. L. (1978). *Proc. Natl. Acad. Sci. USA* **75,** 2959–2963.

Amkraut, A. A., and Martins, A. B. (1984). U.S. Patent #4,484,923.

Beck, L. R., Cowsar, D. R., and Lewis, D. H. (1980). In *Biodegradable Delivery Systems for Contraceptive* (E. S. E. Hafez and W. A. A. Van Os, eds.), pp. 63–81, G. K. Hall, Boston.

Couvreur, P., Kante, P., Roland, M., Gulot, P., Bauduin, P., and Speiser, P. (1979). *J. Pharm. Pharmacol.* **31,** 331–332.

Creque, H. M., Langer, R., and Folkman, J. (19OO). *Diabetes* **29,** 37–40.

De Clercq, E., and Luczak, M. (1976). *J. Infect. Dis.* **133,** A226–A236.

Desiderio, J. V., and Campbell, S. G. (1983). *J. Reticuloendothel. Soc.* **34,** 279–287.

Dietrich, F. M., Lukas, B., and Schmidt-Ruppin, H. H. (1983). *Proc. 13th Int. Congr. Chemother. Vienna* **91,** 50–53.

Durr, F. E., Lindh, H. F., and Forbes, M. (1975). *Antimicrob. Agents Chemother.* **7,** 582–586.

Eppstein, D. A. (1986*a*). In *Drug Targeting with Synthetic Systems* (G. Gregoriadis, ed.), pp. 207–219, NATO ASI Conference, 1985, Plenum, New York.

Eppstein, D. A. (1986*b*). In *Delvery Systems for Peptide Drugs* (S. S. Davis, L. Illum, and E. Tomlinson, eds.), pp. 277–283, Plenum, New York.

Eppstein, D. A., and Felgner, P. L. (1987). In *Liposomes as Drug Carriers: Trends and Progress* (G. Gregoriadis, ed.), Wiley, New York (in press).

Eppstein, D. A., Marsh, Y. V., van der Pas, M. A., Felgner, P. L., and Schreiber, A. B. (1985). *Proc. Natl. Acad. Sci. USA* **82,** 3688–3692.

Fidler, I. J., Sone, S., Fogler, W. E., Smith, D., Braun, D. G., Tarcsay, L., Gisler, R. H., and Schroit, A. J. (1982). *J. BIol. Response Modif.* **1,** 43–55.

Gisler, R. H., Schumann, G., Sackman, W., Pericin, C., Tarcsay, L., and Dietrich, F. M. (1982). In *Immunomodulation by Microbial Products and Related Synthetic Compounds* (Y. Yamamura and S. Kotani, eds.), pp. 167–170, Excerpta Medica, Amsterdam.

Graybill, J. R., Craven, P. C., Taylor, R. L., Williams, D. M., and Magee, W. E. (1982). *J. Infect. Dis.* **145,** 748–752.
Grislain, L., Couvreur, P., Lenaerts, V., Roland, M., Deeprez-Decampeneere, D., and Speiser, P. P. (1983). *Int. J. Pharm.* **83,** 335–345.
Kende, M., Alving, C. R., Rill, W. L., Swartz, G. M., Jr., and Canonico, P. G. (1985). *Antimicrob. Agents Chemother.* **27,** 903–907.
Koff, W. C., Fidler, I. J., Showalter, S. D., Chakrabarty, M. K., Hapar, B., Ceccorulli, L. M., and Kleinerman, E. S. (1984). *Science* **224,** 1007–1009.
Koff, W. C., Showalter, S. D., Hampar, B., and Fidler, I. J. (1985). *Science* **228,** 495–497.
Kopacek, J., Rejmanova, P., Duncan, R., and Lloyd, J. B. (1985). In *Macromolecules as drugs and as Carriers for Biologically Active Materials* (D. A. Tirell, L. G. Donaruma, and A. B. Turek, eds.), pp. 93–104, New York Academy of Science, New York.
Kreuter, J. (1983). *Pharmacol. Acta Helv.* **58,** 217–226.
Kreuter, J., and Liehl, E. (1978), *Med. Microbiol. Immunol. (Berl.)* **165,** 111–117.
Kreuter, J., and Liehl, E. (1981). *J. Pharm. Sci.* **70,** 367–371.
Kreuter, J., and Speiser, P. P. (1976*a*). *Infect. Immun.* **13,** 204–210.
Kreuter, J., and Speiser, P. P. (1976*b*), *J. Pharm. Sci.* **65,** 1624–1627.
Kreuter, J., and Zehnder, H. J. (1978). *Radiat. Effects* **35,** 161–166.
Kreuter, J., Mauler, R., Gruschkau, H., and Speiser, P. P. (1976). *Exp. Cell. Biol.* **44,** 12–19.
Kreuter, J., Nefzger, M., Liehl, E., Czok, R., and Voges, R. (1983). *J. Pharm. Sci.* **72,** 1146–1149.
Langer, R. (1981). *Methods Enzymol.* **73,** 57–75.
Lopez-Berestein, G., Mehta, G. R., Hopfer, R. L., Mills, K., Kasi, L., Mehta, K., Fainstein, V., Luna, M., Hersh, E. M., and Juliano, R., (1983). *J. Infect. Dis.* **147,** 939–945.
Nachtigal, M., and Caulfield, J. B. (1984). *Am. J. Pathol.* **115,** 175–185.
New, R. R., Chance, M. L., and Heath, S. (1981). *J. Antimicrob. Chemother.* **8,** 371–381.
Sanders, L. M., Kent, J. S., McRae, G. I., Vickery, B. H., Tice, T. R., Lewis, D. H. (1984). *J. Pharm. Sci.* **73,** 1294–1297.
Sidwell, R. W., Allen, L. B., Khare, G. P., Huffman, J. H., Witkowski, J. T., Simon, L. N., and Robbins, R. K. (1973). *Antimicrob. Agents Chemother.* **3,** 242–246.
Soike, K. F., Eppstein, D., Gloss, C. A., Cantrell, C., Chou, T. C., and Jerome, P. J. (1987). *J. Infect. Dis.* **156,** (in press).
Tremblay, C., Barza, M., Fiore, C., and Szoka, F. (1984). *Antimicrob. Agents Chemother.* **26,** 170–173.

20

Synthetic Antiphytoviral Substances

G. Schuster

I. INTRODUCTION

Viral diseases affect all important agricultural crops and cause large economic losses. Unfortunately, hitherto only indirect control measures are available and practiced; for instance, isolation or removal of infected plants, resistance breeding, or prevention of virus transmission by insects through treatment with insecticides in the case of persistent viruses or with oils, milk, and so forth, in the case of nonpersistent viruses. As these prophylactic control measurements are not fully effective, the development and use of antiphytoviral compounds interfering with steps of the replication cycle of viruses but not with the nucleic acid and protein replication of the host would constitute a major technical advance.

More work has been done and greater progress achieved in the fundamental research work on substances controlling vertebrate viruses than in the development of antiphytoviral substances. Reasons are better methodical prerequisites for studying the interference of antiviral substances with the replication cycle of animal and human viruses, such as cell culture methods as the basis of agar-diffusion plaque-inhibition tests and one-step growth-cycle experiments. Other difficulties encountered in the development of effective antiphytoviral substances are attributable to the fact that plants in contrast to animals and humans do not generally possess an efficient humoral system for the rapid distribution of xenobiotic and other therapeutic agents. Moreover, plants do not form antibodies against viruses and therefore cannot eliminate residual viruses after treatment with antiviral substances. Thus, reinfection of the whole plant can start from unaffected

Abbreviations for frequently mentioned viruses: AMV, alfalfa mosaic virus; BNYVV, beet necrotic yellow vein virus; BRV, brome mosaic virus; BSMV, barley stripe mosaic virus; BWYV, beet western yellow virus; BYMV, bean yellow mosaic virus; CCMV, cowpea chlorotic mottle virus; CGMV, cucumber green mottle virus; CMV, cucumber mosaic virus; NRSV, necrotic ringspot virus; PLRV, potato leaf roll virus; PNRV, prunus necrotic ringspot virus; PVA, potato virus A; PVM, potato virus M; PVS, potato virus S; PVX, potato virus X; PVY, potato virus Y; RCMV, red clover mottle virus; SBMV, southern bean mosaic virus; TBSV, tomato bushy stunt virus; TEV, tobacco etch virus; TMV, tobacco mosaic virus; TNV, tobacco necrosis virus; TRV, tobacco ringspot virus; TSWV, tomato spotted wilt virus; TYMV, turnip yellow mosaic virus; WMV, watermelon mosaic virus.

G. Schuster • Department of Biosciences, Karl-Marx-University of Leipzig, DDR 7010 Leipzig, German Democratic Republic.

single virions as soon as the concentration of the antiviral substances has dropped below the effective threshold.

Nevertheless, remarkable and promising progress has been achieved during the past three decades. Very often, the development of antiphytoviral substances has started with compounds that had proved active against vertebrate viruses (e.g., Dawson, 1984). Base analogues (see Section II.) have been especially successful against plant viruses. In other groups, such as thioureas and compounds with azine structure, substances exhibiting high activity against animal viruses were only infrequently active against plant viruses and vice versa (Schuster and Vassilev, 1983*b*; Schuster *et al.*, 1984*a*). Thus, the search for antiphytoviral substances has to go its own way, and special strategies must be developed.

Extensive screening will undoubtedly lead to new approaches, but this screening should proceed along specific lines. Wherever large numbers of compounds have to be screened, tests on whole plants or tissue slices may be too expensive. In these cases, screening with phages and their photosynthetically active microbial hosts may provide more economical alternatives. Studies of structure–activity relationships are a suitable tool for a more directed development of antiphytoviral substances (Dove *et al.*, 1982). Besides, elucidation of the mode of action of antiphytoviral substances represents a basis for the improvement of the antiphytoviral chemotherapy. Finally, most promising approaches have been obtained starting from fundamental research into the replication of plant viruses and the pathophysiology of virus infected plants (see Section VIII.).

The following survey is confined to major results and trends of the development of synthetic antiphytoviral substances. It will become clear that a number of substances exhibit remarkable antiphytoviral activities without being phytotoxic. However, only a very limited number fulfill all criteria necessary to keep viral infection in field crops below a critical level or to prevent the virus-induced depressions in growth and yield:

Interference with the replication of a great number of plant viruses but not with host processes

Being nontoxic to humans and animals as well as to the microflora of the soil

Being broken down relatively quickly in plants as well as in the soil so that no residues will be detectable

Being available at a reasonable cost, which would enable farmers to apply them repeatedly, in spite of the relative low prices of plant products

These same criteria apply to natural antiphytoviral substances, including antibiotics, *in vitro* inhibitors of virus (Hirai, 1977), and inductors of resistance. Only those substances are included in the following outline that are closely related to synthetic antiphytoviral substances. Only a very limited number, if any, fulfill these criteria.

II. ANALOGUES OF BASES, NUCLEOSIDES, AND NUCLEOTIDES AND OF ANABOLITES (PRECURSORS) AND CATABOLITES

Only the more important base analogues are discussed herein. More detailed data were reported by Lapierre *et al.* (1971), Hirai (1977), Hecht and Diercks (1978), and

Schuster and Byhan (1980). Of those studied, the purine base analogue 8-azaguanine (8-AG) has been very effective in reducing plant virus multiplication and/or infectivity. Effects against tobacco mosaic virus (TMV) are reported by Lindner *et al.* (1960), Matthews (1955*a*), Chiu and Sill (1962), Harder and Kirkpatrick (1970), and Dawson (1984). Kroell (1985) found only a small reduction of the TMV concentration by 8-AG in inoculated leaves of intact plants and an augmentation of the TMV concentration in secondarily infected leaves. Effects against potato virus X (PVX) were described by Zhuk *et al.*, (1970) and by Pennazio (1973) and in inoculated leaves by Kroell (1985). Effects against potato virus Y (PVY) were reported by Hecht and Diercks (1978), against brome mosaic virus (BRV) by Chiu and Sill (1962), against turnip yellow mosaic virus (TYMV) by Matthews (1955*b*), and against stone fruit ringspot virus by Kirkpatrick and Lindner (1961). 8-AG had no significant effect on red clover mottle virus (RCMV) (Sinha, 1960) or Dolichos enation mosaic virus (Hariharasubramanian, 1968) or upon cycumber mosaic virus (CMV) in inoculated leaves. In secondarily infected leaves, the CMV concentration was augmented (Kroell, 1985) following treatment. The effect of treatments with 8-AG or other antiphytoviral substances on virus multiplication can be influenced by the mode of application and other factors. Harder and Kirkpatrick (1970) reported that at 26°C and 30°C a single application of 8-AG stimulated TMV infectivity. But at lower temperatures or with repeated applications TMV infectivity was reduced. Moreover, treatment with low concentrations of 8-AG stimulated TMV multiplication up to ninefold, whereas higher concentrations caused a 10–1000-fold inhibition (Sander, 1969).

9-(S)-(2,3-Dihydroxypropyl)adenine [(S)-DHPA] (De Clercq *et al.*, 1978), an acyclic nucleoside analogue exhibiting high activity against a broad spectrum of animal and human viruses (De Clercq and Holý, 1985) has also proved inhibitory against TMV and cowpea chlorotic mottle virus (CCMV) (Dawson, 1984).

One of the most important analogues of a precursor of purine nucleosides is 1-β-D-ribofuranosyl-1,2,4-triazole-3-carboxamide, developed by Sidwell *et al.* (1972). This substance, an analogue of 5-amino-4-imidazole-carboxamide-ribotide (AICAR), is known under the tradenames Virazole and Ribamidil. It shows marked inhibitory effects on numerous viruses occurring in humans and has also been found to be active against the plant viruses TMV (Schuster, 1976; Dawson, 1984), PVX (Schuster, 1976; Lerch, 1977, 1978; Shepard, 1977), PVY (Lerch, 1978; Cassells and Long, 1980), belladonna mottle virus (Lerch, 1980), CMV (Kluge and Oertel, 1978; Cassells and Long, 1980), red clover mottle virus (RCMV) (Kluge, 1980), CCMV (Dawson, 1984), apple chlorotic leaf spot virus (Hansen, 1979), TYMV (Kalonji-M'Buyi and Kummert, 1982*a*), tomato white necrosis virus (De Fazio *et al.*, 1978), TSWV (De Fazio *et al.*, 1980*a,b*, 1984), turnip mosaic virus (Kalonji-M'Buyi and Kummert, 1982*b*), and green ring mottle causal agent (Hansen, 1984). However, in some cases only phytotoxic concentrations were effective, which were harmful to plants. For instance, the effective control of TMV in systemically infected plants of *Nicotiana tabacum* Samsun required plant-damaging concentrations of ribavirin (Schuster, 1976). The concentrations of TMV, PVX, CMV, and other viruses were less strongly reduced by ribavirin in inoculated than in secondarily infected leaves (Schuster, 1976; Kluge and Oertel, 1978). The same is true with a great number of other antiphytoviral compounds such as 2-thiouracil or 5-azadihydrouracil. Examples of efforts to make practical use of the antiphytoviral activities of ribavirin are given in Sections X.B. and X.C.

Besides ribavirin, pyrazofurine (4-hydroxy-5-β-D-ribofuranosylpyrazole-3-carboximide) proved a promising analogue of a precursor of purine nucleotides (Cadman and Benz, 1980). Lerch (1980) reported satisfying activities of pyrazofurine against PVS, PVY, and other plant viruses.

Of the analogues of pyrimidine bases, 2-thiouracil (2-TU) is one of the most studied compounds, which has been shown to inhibit a great number of plant viruses effectively. These include TMV (Commoner and Mercer, 1952; Bawden and Kassanis, 1954; Francki, 1962; Yun and Hirai, 1968; Meyer *et al.*, 1977; Dawson and Grantham, 1983; Kroell, 1985), PVX (Bawden and Kassanis, 1954; Bergmann, 1958; Quak, 1961; Oshima and Livingston, 1961; Rao and Raychaudhuri, 1965; Yun and Hirai, 1968; Kroell, 1985), PVY (Bawden and Kassanis; 1954, Mitra and Raychaudhuri, 1973; Hecht and Diercks, 1978), TYMV (Francki and Matthews, 1962; Ralph and Wojcik, 1976), TNV (Bawden and Kassanis, 1954), SBMV (Kuhn, 1977), and CMV (Kroell, 1985). By contrast, CCMV (Dawson and Kuhn, 1972; Dawson, 1984) and AMV (Kuhn, 1973) were not inhibited but often stimulated. When several virus hosts were included in the investigations (TMV, PVX, PVY, TYMV), 2-TU inhibited the viruses in all hosts. However, when the same host was infected once or twice with two different plant viruses, such as cowpea leaves with CCMV and SBMV, 2-TU increased accumulation and infectivity of CCMV and decreased accumulation and infectivity of SBMV (Kuhn, 1977).

Ralph and Wojcik (1976) suggested that 2-TU and the corresponding nucleoside affect virus synthesis by inhibiting the biosynthesis of uridylic acid, possibly by inhibiting orotydilic acid decarboxylase. In a system synchronized by differential temperature inoculation (Dawson *et al.*, 1975; Dawson and Grantham, 1983), 1 mM 2-TU inhibited TMV protein and RNA synthesis totally if treatment began within the first 4 hr after initiation of replication. Contrary to these results, Barta *et al.* (1981) found that 2-TU failed to inhibit TMV multiplication in tobacco protoplasts and attributed the inhibitory effect to prevention of cell-to-cell movement of the virus. This explanation seems unlikely in view of the distinct points of attack in the replication cycle of TMV, identified by Dawson and others. Instead, it seems that in the experiments of Barta *et al.* (1981) the 2-TU concentration was too low to influence TMV (Dawson and Grantham, 1983). Sander (1969) demonstrated that treatments with low concentrations of 2-TU stimulated the TMV synthesis, whereas high concentrations resulted in strong inhibition.

Besides 2-TU, 4-methyl-2-TU inhibited CMV, PVX, and, in part, TMV (Kroell, 1985). 2-Thiocytosine and 2-thiothymine inhibited TMV (Mercer *et al.*, 1953). 5-Fluorouracil (5-FU) was inhibitory against TMV (Staehelin and Gordon, 1960; Kroell, 1985), PVX (Pennazio, 1973; Kroell, 1985), and CMV (Kroell, 1985). 6-Azauracil and 6-azauridin inhibited TYMV (Ralph and Wojcik, 1976). Dijkstra (1969) demonstrated inhibitory activities of 6-azauracil against TMV. TMV was also inhibited by 2-amino-4-methyl-6-chloropyrimidine and 2-amino-4-methyl-6-hydroxypyrimidine (Ulrychová-Zelinková, 1960) as well as by 5,6-diaminouracil sulfate (Schlegel and Rawlins, 1954). 2-Amino-4,6-dichloropyrimidine interfered with TBSV infection (Pennazio *et al.*, 1981).

A promising analogue of a catabolite of pyrimidine bases is 5-azadihydrouracil (2,4-dioxohexahydro-1,3,5-triazine), or 5-ADHU (DHT) (Schuster *et al.*, 1979*a,b*). There is evidence that this substance is processed by the plant in relatively slow bioconversion steps to an active base analogue with a time course of inhibition nearly completely coinciding with that of 2-thiouracil (Schuster and Arenhoevel, 1984). This may explain why relatively high doses of 5-ADHU (DHT) are well tolerated by plants. Since the bioconversion of 5-ADHU (DHT) to an active base analogue apparently does not take place in animals and humans, 5-ADHU (DHT) is not toxic to vertebrates. Moreover, the persistence of 5-ADHU (DHT) in plants and soils is low (Byhan *et al.*, 1981; Liebert and Schuster, 1980), and most soil microorganisms either tolerate 5-ADHU (DHT) or use it (El-Dahtory *et al.*, 1983, 1984). Chemotherapy using 5-ADHU (DHT) is therefore toxicologically and ecologically safe and highly plant compatible (Schuster, 1982*a*). Thus, catabolites and anabolites of compounds that have to be processed by appropriate host or virus activities are of special interest for chemotherapy of plant and human viruses. This may be considered one of the most important trends of the past few years in the field of development of antiviral substances, in welcome contrast to most normal base analogues, which are toxic to the host plant at concentrations just above those required to inhibit viral multiplication. Moreover, most base analoĝues are unsuitable for use on crops because of their costs and toxicity to animals.

5-ADHU (DHT), which is available under favorable economical conditions, is active against a relatively broad spectrum of viruses, e.g., PVX, PVY, PVA, PVS, PVM, TMV, BNYVV, CGMV, WMV, REMV, CMV, PLRV, TRV, and PNRV (Schuster *et al.*, 1979*a,b*; Schuster, 1982*a*; Schuster and Hanzsch, 1981; Kluge, 1980; Kroell, 1982, 1985; Oshima *et al.*, 1984; Hillman, 1984; Bogusch *et al.*, 1985; Borissenko *et al.*, 1985*a,b*; El-Mazaty, 1985), but some viruses, such as PVX, are inhibited better than others, such as TMV.

Following the discovery of the remarkable antiphytoviral activities and advantages of 5-ADHU (DHT), practical open-air trials were performed (for results see Sections X.C. and X.D.). Efforts were also made to improve application methods using 5-ADHU (DHT) as a base. Remarkable progress was obtained by developing granulated materials that release therapeutic quantities of 5-ADHU (DHT) over prolonged periods of time, thus depressing virus titers for more than 3 months (Schuster and Kramer, 1982). The four to five treatments that are usually necessary because of the relatively rapid breakdown of DHT by the host plant can be reduced to one application of granulated material during planting. Moreover, most viruses are inhibited better by granulated 5-ADHU (DHT) even in inoculated leaves of younger plants because 5-ADHU (DHT) is taken up readily by the roots and rapidly transported in an acropetal direction. Following granulated application, the concentration of 5-ADHU (DHT) in leaves is relatively high even a short time after sowing or planting and 5-ADHU (DHT) will be present before the first replication cycle occurs. Under these conditions, the antiphytoviral effects of 5-ADHU (DHT) are highest, since 5-ADHU (DHT) influences a very early event in the replication cycle of viruses (Schuster and Arenhoevel, 1984). Granulated substances will possibly be a prerequisite for obtaining an optimal effect in areas in which virus vectors are present and active throughout the year, such as tropical and some subtropical countries.

Further improvement of the antiphytoviral chemotherapy was obtained by combined treatments with two or more antiphytoviral substances. For instance, the remarkable antiphytoviral activities of ribavirin and 5-ADHU (DHT) were augmented synergistically by combined application of these substances. The concentration of PVX could be reduced by very small quantities, namely 15 g/ha of each active agent, to a level at which it was no longer detectable serologically in secondarily infected leaves. Moreover, the concentration of some difficult-to-inhibit viruses, such as CMV and TMV, was reduced much better by combinations of ribavirin with 5-ADHU (DHT) than by treatments with the single agent (Schuster, 1982*b*).

Besides ribavirin, many other antiphytoviral substances augment the antiphytoviral activities considerably and in many cases synergistically when applied in combination with 5-ADHU (DHT). In this context, some compounds with guanidine structure (Schuster, 1983*a*) as well as substituted ureas and thioureas (Schuster, 1982*d*; Schuster and Rehnig, 1982; Schuster and Vassilev, 1985) are of interest. Moreover, some compounds with noncyclic azine structure (Schuster *et al.*, 1984*a*), some 2,5-substituted thiadiazoles (Schuster *et al.*, 1984*b*), and some 2,4,5-substituted oxazoles (Schuster and Ulbricht, 1985) proved to be excellent synergists. Apparently, some of these plant virus inhibitors will obtain practical importance only in combination with 5-ADHU (DHT). Examples for the practical use of combined treatments are given in Section X.D.

III. HETEROCYCLIC COMPOUNDS EXCEPT BASE ANALOGUES

Among those heterocyclic compounds that cannot be classified as base analogues, benzimidazole derivatives are most active against plant viruses. Most results have been obtained with methyl benzimidazole-2-yl-carbamate (MBC, carbendazim), which, like other benzimidazole derivatives, acts as a systemic fungicide and is the product of the breakdown of methyl-*N*-[1-(butylcarbamoyl)-2-benzimidazole] and similar benzimidazoles. MBC reduces the severity of symptoms produced by TMV infection of tobacco plants and by BWYV infection of lettuce plants (Tomlinson *et al.*, 1976), but tests of TMV-infected tobacco plants treated with MVC showed no marked reduction of virus content (Tomlinson, 1977). This may in part be due to cytokinin-like properties of MBC as well as of a number of other benzimidazole derivatives (Skene, 1972; Thomas, 1974). Measurements of the accumulation of TMV RNA in MBC-treated tobacco plants showed that MBC treatment reduced the TMV accumulation significantly (Fraser and Whenham, 1978*a,b*). These authors conclude from their experiments on time of application, dose, level, and leaf age that MBC inhibits viral RNA synthesis indirectly, by maintaining the host in a state unsuitable for viral multiplication. Reduction of PVX and TYMV concentrations by MBC treatments is described by Menzel and Schuster (1979) and by Hattori and Sarkar (1981).

Benomyl (methyl-1-butylcarbamoyl-1-benzimidazole), which is easily converted to MBC and which exhibits cytokinin-like activity, reduced the content of TMV (Bailiss *et al.*, 1977*a*), PVX (Menzel and Schuster, 1979), and urdbean leaf crinkle virus (Bhardwaj *et al.*, 1982). PVX is also influenced by thiabendazole (thiazole-2-benzimidazole) (Menzel and Schuster, 1979). High activity against TMV is exhibited by 2-(α-hydroxybenzyl)benzimidazole, and a smaller inhibition was observed after treatments with corresponding 2-benzyl-, 2-phenyl-, 2-methyl-, and 2-hydroxymethyl-benzimidazoles (Cassells and Cocker, 1982).

Antiphytoviral activities of several 2,5-substituted 1,3,4-thiadiazoles (TDA) were described by Schuster *et al.* (1984*a*). The highest activities against PVX, in part combined with high activities against TMV, are exhibited by 2-anilino-5-adamantyl-TDA, 2-anilino-5-4′-pyridyl)-TDA, and 2-anilino-5-benzyl-TDA.

Among 99 2,4- and/or 5-substituted 1,3-oxazoles, 18 reduced the concentration of PVX in inoculated and/or secondarily infected leaves of *N. tabacum* Samsun. Some oxazoles also reduced the number of local lesions caused by TMV on leaves of *Nicotiana glutinosa* (Schuster and Ulbricht, 1985). The strongest inhibitory effects against PVX were obtained with 2,4-dihydroxy-acetophenone-(4,5-diphenyl-oxazole-2-yl-hydrazone)-4,5-diphenyl-oxazole-2-yl-hydrazine-adduct and *m*-guanidinoacetophenone(4,5-dipenyl-oxazole-2-yl-hydrazone)hydrochloride. These compounds combine in one molecule the activity of the oxazoles with that of the guanidino, acetophenone, hydrazine, or hydrazone groups. Combining several groups with known antiphytoviral activities within one molecule seems to be an interesting way to obtain substances displaying high antiphytoviral activities, quite similar and, in part, comparable to combining several antiphytoviral substances in one preparation or during treatment of the plants.

Compounds with antiphytoviral effects have also been described among triazines, above all among triazine herbicides. The triazine herbicide simazine (2-chloro-4,6-bis(ethylamino)-1,3,5-triazine) influences TMV and CMV when applied in low doses, but the difference between the curative dose and the phytotoxic dose is very small (Bobyr, 1976; Rajyalakshmi and Raychaudhuri, 1978; Žmurko and Bobyr, 1974). Atrazine (2-chloro-6-ethylamino-4-isopropylamino-1,3,5-triazine), propazine (2-chloro-4,6-isopropylamino-1,3,5-triazine), and prometryne (2-methylmercapto-4,6-isopropylamino-1,3,5-triazine) are reported to reduce local lesions of TMV on *N. glutinosa* (Bobyr, 1976). Bladex (2-ethylamino-4-chloro-6-cyanopropylamino-1,3,5-triazine) reduced BYMV (Sĭndelář *et al.*, 1979). On the other hand, application of atrazine to *N. tabacum* Samsun, a systemic host of TMV, resulted in an increase of TMV in the tissues (Sĭndelář and Makovcová, 1975).

Hitherto it was not possible to decide whether the mentioned triazine effects are obtained by influencing the replication cycle or by inhibition of host processes necessary for virus multiplication. But it is worth mentioning that some triazines proved to be antiphytoviral, even though they showed no herbicidal activity. Discrimination analyses and investigations into structure–activity relationships were helpful in finding such nonherbicidal triazines with considerable antiphytoviral activity (Dove *et al.*, 1982). In this way, the antiphytoviral activities of hydrated triazines have been detected, chiefly those of 2,4-dioxohexahydro-1,3,5-triazine, or DHT, or azadihydrouracil, or 5-ADHU (Schuster *et al.*, 1979*a,b*).

2,3-Dihydroxy-6-bromopyrazino(2,3-β)pyrazine delayed the spread of TSWV in tobacco plants (De Fazio and Kudamatsu, 1983). Among morpholine derivatives the systemic fungicide tridemorph (*N*-tridecyle-2,6-dimethyl-morpholine) inhibited PVX (Schuster, 1977).

IV. ISOCYCLIC COMPOUNDS

The herbicide trifluralin (α,α,α-trifluoro-2,6-dinitro-*N,N*-dipropyl-*p*-toluidine) when incorporated into the soil, remarkably inhibited the multiplication of AMV and TMV in intact primary leaves of *Phaseolus vulgaris* cv. Pinto: "Therefore, trifluralin, used for chemical weed control in beans, may be important not only as a herbicide but probably as a viricide too" (Horváth and Hunyadi, 1973).

The chelating agent salicylaldoxime inactivates the necrotic lettuce yellow virus by degeneration of its structure (Atchison *et al.*, 1969). The same substance inhibits PVY in intact potato plants and reduces the infectivity of this virus (Hecht and Diercks, 1978). Acetylsalicylic acid (aspirin) reduced numbers and size of local lesions of TBSV when applied prior to inoculation. No time interval between acetylsalicylic acid supply and TBSV inoculation was necessary to obtain these reductions (Pennazio and Redolfi, 1980). The size and viral antigen content of non-self-limiting necrotic lesions produced by TMV in detached tobacco leaves were reduced by sodium salicylate at concentrations close to the limits of toxicity (Pennazio *et al.*, 1983). Injection of leaves with acetylsalicylic acid induced resistance to infection with TMV in tobacco cultivars (Antoniw and White, 1980). These results are consistent with the concept that acetylsalicylic acid acts twice, first by interfering with steps of the replication cycle and second by inducting resistance. Naphthalene–acetic acid reduced the concentration of PVY and suppressed the expression of symptoms of this virus completely when applied daily to tobacco (Mitra and Raychaudhuri, 1973). The aromatic amines *p*-coumarylputrescine, di-*p*-coumarylputrescine, and caffeylputrescine inhibited the multiplication of TMV (Martin-Tanguy *et al.*, 1976). Amantadine, which is not a cyclic compound *sensu strictu*, has proved active against chrysanthemum stunt viroid (Horst and Cohen, 1980).

V. NONCYCLIC COMPOUNDS

Substituted thioureas (TU), which are remarkable for a diversity of biologic activities (surveyed by Schuster and Vassilev, 1983*a*), also exhibit remarkable antiphytoviral activities. Using the virus–host combination PVX–*N. tabacum* 'Samsun,' TUs with antiphytoviral activity were found among *N*-alkyl-*N'*-aryl-, *N*-phenyl-*N'*-aryl- as well as *N*-alkyl-*N'*-alkyl-substituted TUs (Schuster, 1978; Schuster and Vassilev, 1983*a,b,* 1985). Those compounds that have the strongest inhibitory effects are *N*-allyl-*N'*-*p*-ethoxyphenyl-TU, *N*-allyl-*N'*-*p*-carboxyphenyl-TU, *N*-ethyl-*N'*-*p*-carboxyphenyl-TU, *N*-phenyl-*N'*, -*o*-, -*m*-, and -*p*-carboxyphenyl-TU, *N*-cyano-S-methyliso-*N'*-*m*-tolyl-TU, *N,N'*-bis-methyl-TU as well as *N*-allyl-*N'*-aminoethyl-TU. *N*-phenyl-*N'*-*p*-carboxyphenyl–urea has been described as a urea compound with remarkable antiphytoviral activities (Rehnig and Schuster, 1982). This substance diminishes the concentration of PVX when sprayed weekly on leaves of tomato plants and reduces yield losses caused by double infection with PVX and TMV. A number of TU as well as *N*-phenyl-*N'* carboxyphenyl–urea are excellent synergists of 5-azadihydrouracil (DHT) (Schuster, 1982*d*; Rehnig and Schuster, 1982; Schuster and Vassilev, 1985).

Compounds with guanidine structure include inhibitors of a number of animal, human, and plant viruses. Varma (1968) described inhibition of TNV in bean leaves by guanidine (GD) carbonate. TMV is inhibited in leaf discs of *N. tabacum* by GD hydrochloride (Dawson, 1976, 1984). GD nitrate, GD carbonate, as well as *N,N'*-diamino-GD-bicarbonate, *N,N'*,-triamino-GD-hydrochloride, *N*-cyano-GD, *N*-acetyl-GD, and guanyl urea sulfate inhibit PVX and, in part, TMV and BRV. On the other hand, only 3 of 18 more extensively substituted guanidines showed a satisfying activity against PVX (Schuster, 1982*c*). A great number of GDs augment the antiphytoviral activities of 5-azadihydrouracil (DHT) (Schuster, 1983*a*).

Remarkable antiphytoviral activities are exhibited by variously substituted noncyclic compounds with azine structure:

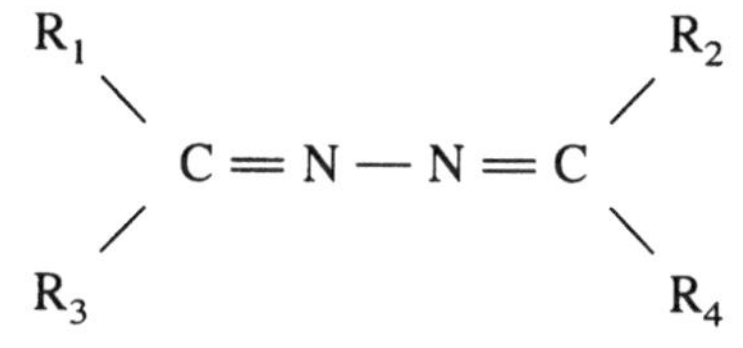

(Schuster *et al.*, 1984*a*). Thus, the azine structure and similar heteroconjugated systems seem to be important not only in cyclic compounds such as thiadiazoles (Schuster *et al.*, 1984*b*) or oxazoles (Schuster and Ulbricht, 1985), but also in noncyclic compounds including isothiosemicarbazones, thiosemicarbazones, thiocarbonylhydrazones, semicarbazones, and guanylhydrazones. Of a total of 90 compounds tested, 42 inhibited the concentration of PVX (Schuster *et al.*, 1984*a*). One of the most effective compounds is pyridine-3-aldehyd-*S*-ethyl-isothiosemicarbazone, which inhibits even TMV in a systemic host. Several compounds proved excellent synergists of 5-ADHU (DHT). When used in combination with 5-ADHU (DHT), the above-mentioned compound and others reduced symptom severity of potato plantlets and increased the yield.

As some noncyclic compounds with azine structure also exhibit activities against animal and human viruses, the corresponding structures were compared with regard to their activities in different virus–host systems. Surprisingly, compounds displaying high antiphytoviral activity were only infrequently active against animal viruses and vice versa. However, the compounds active in these two different virus–host systems are often closely related structurally. Possibly, some specific structures induce the antiviral activity, while others are responsible for the reaction with different types of receptors occurring in plants and animals, respectively (Schuster *et al.*, 1984*a*).

VI. POLYANIONS, POLYCATIONS, AND NONCHARGED POLYMERS

Polycarboxylates with maleic or acrylic acids as the anionic component of the polymer as well as mannans and other polysaccharides are known as inducers of resistance to infection with TMV, TNV, and other necrosis-inducing viruses, above all in hypersensitive hosts (e.g., Stein and Loebenstein, 1972; Gianinazzi and Kassanis, 1974; Stein *et al.*, 1979; Kovalenko *et al.*, 1981). But there is evidence that some of them may also directly affect the interaction between the virus and its host. For instance, polyanions, such as polyacrylic acid or polyaspartic acid would compete directly with the negatively charged virus nucleic acid for the cationic receptor sites on cells. An inhibitor isolated from carnation would initially or primarily interact via its ε-amino groups with anionic host sites (Ragetli and Weintraub, 1974). Likewise, Cassels *et al.* (1978) and Fernandez and Gáborjányi (1976) suggested that polyacrylic acid can protect plants against virus infection by affecting membrane properties. Kassanis and White (1975) and Schuster (1983*b*) applied polyanions to PVX- and PVY-susceptible plants without a time interval between treatment and inoculation. Their results indicate that the virus-inhibiting effect of the treatment was immediate and direct and that no antiviral principle was induced in the host. Vicente *et al.* (1972) reduced numbers of lesions of TMV on leaves of *N. glutinosa* by spraying solutions of ammonium polyphosphate before or after inoculation. Pretreatment with the synthetic polycation polylysine caused a reduction of TMV local lesions compared with the control (Tyihák and Balázs, 1976).

VII. AMINO ACID ANALOGUES

The inhibitory potential of amino acid analogues has only rarely been tested. Dawson (1984) inhibited the multiplication of TMV and CCMV using slightly phytotoxic concentrations of cycloleucine. Apparently, amino acid analogues influence not only processes of virus replication but the formation of host proteins as well. This may be the reason for the infrequent use of amino acid analogues as antiphytoviral substances.

VIII. MEMBRANE LIPID ANALOGUES AND OTHER COMPOUNDS WITH MEMBRANE ACTIVITY

The effect of membrane lipid analogues on replication has been tested because of the close link between membranes and virus replication (Kiho, 1970; De Zoeten *et al.*, 1974; Beachy and Zaitlin, 1975; Zabel *et al.*, 1976; Dawson, 1978; Gaard and De Zoeten, 1979; Ohashi and Shimomura, 1982; Pustowoit *et al.*, 1984, 1985). These substances are supposed to be bound to or incorporated into membrane bilayers, thereby disturbing membrane-bound processes of virus replication or virus-induced alterations of the cell membranes. Lipid extracts and fractions thereof from biomasses of *Lodderomyces elongisporus* markedly reduced the concentration of PVX. A nearly inactive fraction containing a high proportion of phosphatides was converted by partial hydrolysis into a highly virus-inhibitory lysophosphatides-containing fraction (Voigt *et al.*, 1985). This is one of the first examples of the production of antiphytoviral substances in cell culture and for the conversion of inactive into antiphytovirally active compounds.

Kluge *et al.* (1984) reported antiphytoviral activities of fully synthetic lysolecithines. Further fully synthetic substances with high antiphytoviral activity were obtained when the bridge molecule of the membrane lipids was eliminated, while one or two long alkane chains were maintained; the latter were bound directly to an altered hydrophilic end group. The best results were obtained with compounds containing a negatively charged head group, followed by substances with a positively and dually charged group, and finally by substances with an uncharged group. Among the substances with a negatively charged group, a specific preparation of alkane monosulfonates with unbranched alkane chains of C_9–C_{16} is the most active. The longer these alkane chains, the higher is the antiphytoviral activity against PVX and TMV. The antiphytoviral activity of these compounds is higher in inoculated than in secondarily infected leaves (Schuster, 1987). This alkane monosulfonate preparation also inactivates the free virion, but the percentage of *in vitro* inactivation is lower than that of the *in vivo* inactivation. The alkane monosulfonate preparation proved an excellent synergist of ribavirin, 5-ADHU (DHT), of several guanidines, thiourea compounds, and a thiadiazole (Schuster, 1987). Some combinations with these substances may be of practical importance (see Section X.D.).

Substances with a hydrophilic and a hydrophobic end group are considered detergents. Thus, there is a close connection between the substances described above and the inhibition of local lesions of TMV by the anionic detergent sodium dodecyl sulfate (SDS) and the nonionic detergents Triton X 100 and Tween 80 described by Taniguchi (1976). Based on the results of several experiments, Taniguchi (1976) suggested that detergents act by interfering with the adsorption of virus to cell surface and by disruption of unions between virus and the site of infection in cells. This hypothesis is consistent with the action of these substances as membrane lipid analogues.

Not only membrane lipid analogues but the lipids themselves may influence the activity of the free virion and of membrane-bound processes such as *in vivo* uncoating (Kiho *et al.*, 1979). The latter investigators observed *in vitro* disassembly of TMV virions in the presence of lecithine and of a lipid fraction isolated from tobacco leaf membrane.

IX. PLANT HORMONES AND SUBSTANCES WITH PLANT HORMONE ANALOGUE ACTIVITY

A. Auxin and Auxin Analogues

The concentration of PVX and PVY was considerably reduced in secondarily infected older leaves after treatment of intact *N. tabacum* 'Samsun' with (β-indoleacetic acid (IAA), 2,4-dichlorophenoxyacetic acid (2,4-D), 4-chloro-2-methylphenoxyacetic acid (MCPA), and 4-chloro-2-methylphenoxybutyric acid (MCPB) (Schuster, 1971). Šindelář and Makovcová (1975) noted an increase in TMV content after application of 2,4-D and MCPA. In tobacco protoplasts, the multiplication of TMV was enhanced by 2,4-D (Loebenstein *et al.*, 1980). The concentration of CMV was augmented after treatment with low concentrations of IAA, indole-3-butyric acid (IBA), naphtoxy acetic acid (NOAA), and 2,4-D and was reduced after treatments with high concentrations (Seth and Raychaudhuri, 1975). The concentration of BSMV in wheat plants increased after treatment with 2,4-D (Makovcová *et al.*, 1976). Likewise, the number of local lesions of TMV on *N. glutinosa* increased after treatment with IAA, 2,4-D, MCPA, and MCPB (Schuster, 1975*a*). Thus, no consistent approach to virus inhibition seems to be possible using auxin and auxin analogue substances.

B. Gibberellic Acids

Gibberellin concentrations progressively decline in plants stunting by viral infections (Bailiss, 1977). Numerous attempts have been made to overcome this stunting by application of giberellic acids (GA). The stunting caused by CMV, AMV, TMV, corn stunt virus, wound tumor virus, and other viruses either disappeared or became insignificant after treatments with GA_3 (Chessin, 1958; Maramorosch, 1957; Nariani, 1963; Chant *et al.*, 1963; Singh, 1964; Mukherjee and Raychaudhuri, 1966; Bailiss, 1968; Selman and Arulpragasam, 1970; Fernandez and Gáborjányi, 1976). However, virus concentrations are only rarely reduced by treatments with GA (Selman and Arulpragasam, 1970; Schuster, 1974). The effect of GA is related to cell elongation rather than to antiphytoviral activity. On the other hand, treatment of potato tubers with GA_3 and thiourea increased the yields and resulted in pratically virus-free tubers (Zherebchuk, 1967). The transmission of cabbage ringspot virus to turnip plants by *Myzus persicae* (Sulz.) was reduced by foliar sprays of GA together with ammonium nitrate, but no reduction occurred if each substance was applied separately (Selman and Kandiah, 1971). The number of local lesions of TMV on leaves of hypersensitive tobacco varieties was increased by application of GA_3 (Schuster, 1975*b*).

C. Cytokinins and Cytokinin Analogues

There are conflicting reports as to whether exogenously supplied cytokinins, such as kinetin or benzyladenine, increase or decrease virus replication and local lesion formation (Király and Szirmai, 1964; Selman, 1964; Daft, 1965; Pozsár and Király, 1965; Aldwinckle and Selman, 1967; Soans, 1967; Mukherjee *et al.*, 1967; Király *et al.*, 1968). The effect of treatments with cytokinins proved to be dependent on the time, frequency of treatments as well as on concentration (Aldwinckle and Selman, 1967; Schuster, 1967, 1972; Bailiss *et al.*, 1977*b*). There is evidence that cytokinins reduce virus synthesis indirectly by maintaining the host in a state unsuitable for virus multiplication, or by promoting the attachment of host messengers to ribosomes. This latter mechanism may influence the competition for ribosomes between host mRNA and virus RNA in favor of the host. On the other hand, virus multiplication may be indirectly promoted in lower leaves by cyctokinins because of their ability to rejuvenate older leaves or tissues. Finally, kinetin seems to inhibit a relatively early event in the replication cycle of PVX (Huber and Schuster, 1987).

It is remarkable that several antiphytoviral benzimidazole derivatives (see Section III.) as well as thioureas (see Section V.) exhibit cytokinin activities (Fraser and Whenham, 1978*b*; Vassilev and Jonova, 1976). Similar effects are induced by catabolites and anabolites of base analogues, above all 5-ADHU (DHT) and ribavirin (see Section II.). The optimum cytokinin activity of 5-ADHU (DHT) and ribavirin exceeds that of reference cytokinins considerably. However, it is impossible to say at this time whether there is a causal relationship between antiphytoviral and cytokinin-like activities. Cytokinins may be considered a special kind of base analogue, and there is some probability they may influence virus multiplication like base analogs (Schuster, 1983*c*; Hoehne *et al.*, 1983).

D. Abscisic Acid

The abscisic acid (ABA) concentration of host plants increased progressively after systemic or local infection with TMV (Balázs *et al.*, 1973; Bailiss *et al.*, 1977*a*; Whenham and Fraser, 1981), contributing to the senescence of virus infected plants. Pretreatment of tobacco leaves with ABA increased the number of local lesions of TMV and induced senescence (Bailiss *et al.*, 1977*a*). By contrast, treatment with ABA (10^{-7} mol/liter) reduced the concentration of PVX in secondarily infected tobacco leaves. Moreover, ABA potentiated the antiviral effect of ribavirin substantially and counteracted ribavirin phytotoxicity (Schuster, 1979). Treatment with ABA may therefore improve the chances of virus inhibition, especially in situations where near-phytotoxic concentrations of ribavirin are needed for complete chemotherapy.

E. Ethylene and Ethylene-Releasing Substances

Increased ethylene production during formation of local lesions and the involvement of ethylene in virus localization and host necrosis are well documented (see Bailiss *et al.*, 1977*b*; Gold and Faccioli, 1972). An important effect of treatments with ethylene or ethylene-releasing synthetic substances such as 2-chloroethylphosphonic acid (Ethrel, Camposan) is dwarfing and hastening of senescence of plants. Treatment with Ethrel increased the number and size of local lesions and the virus multiplication in systemically TMV-infected leaves of *N. tabacum* 'Xanthi' (Balázs and Gáborjányi, 1974; Bailiss *et al.*, 1977*b*). On the other hand, treatment with high doses of Ethrel or Camposan decreased the concentration of PVX, above all in combination with ribavirin or 5-ADHU (DHT) (Schuster, 1974, 1975*b*, 1980). Treatment with Ethrel can also induce a systemic resistance in hypersensitive hosts comparable to that induced by viral infections (van Loon, 1977).

X. PRESENT AND PROSPECTIVE FIELDS OF APPLICATION OF ANTIPHYTOVIRAL SUBSTANCES

A. General Remarks

Antiphytoviral substances that influence the steps of the replication cycle generally have no effect on the replicated virion. Once the first replication cycle is completed, eradication of the virus is scarcely possible. It is restricted to some exceptional cases. One is described by Hansen (1984). As a rule, virus replication will start again from the unaffected virions once the antiphytoviral substance concentration has decreased beneath the curative dose as a consequence of natural breakdown. The application of antiphytoviral substances should occur earlier than the virus transmission to avoid, or at least diminish, the probability of the formation of virions, which could intitiate an infection. In other words, treatments should be preventive and should be repeated several times. The quicker the breakdown of the antiphytoviral substance used, the shorter the interval between the various treatments.

The practical application of antiphytoviral chemotherapy is limited in many cases by the relative narrow range of plant viruses, which are influenced by a single antiphytoviral substance. This is a consequence of the highly specific interaction between compounds and virus species or even virus strains. However, more and more new antiphytoviral substances have been and will be developed with other spectra of efficacy. Appropriate combinations of several of these substances will widen the range of the affected viruses and will increase the antiphytoviral activity because of the synergistic effects. This means that an appropriate combination therapy will improve the effectiveness of antiphytoviral chemotherapy. Examples are given in Section X.D.

B. Regeneration of Virus-Free Plants from Explants

Antiphytoviral substances are generally most active in plant parts in which virus multiplication is naturally limited, such as meristematic tissues, cell and callus cultures, and protoplasts. Application to these plant parts and subsequent regeneration of plants from meristems should make virus-free plants easier to obtain. This was first demonstrated by Quak (1961) with 2-thiouracil in potato meristems. Shepard (1977) obtained tobacco plants free of PVX from infected tobacco leaves by regenerating plants from proliferating calli in the presence of ribavirin. Likewise, adventitious shoots regenerated from CMV and PVY infected tobacco explants were virus free when the explants were cultured in the presence of ribavirin, whereas the controls were completely virus infected (Cassells and Long, 1980). Growing tops of PVX-infected potato plants could be kept virus free by treatment with ribavirin so that healthy material could be obtained by regeneration of plants from these tops (Lerch, 1979). Amantadine incorporated into a tissue culture medium for chrysanthemums was effective in obtaining chrysanthemum plantlets free of chrysanthemum viroid (Horst and Cohen, 1980).

A high percentage (66% and 72%, in two cultivars) of potato plantlets free of PVX, PVM, and PVS was obtained from fully virus-infected stocks by a combination of meristem (tip) culture with 5-ADHU (DHT) chemotherapy. Treatment with 2-thiouracil, with cyanoguanidine, and with Imanine and Novoimanine (extracts from *Hypericum perforatum*) resulted in 46.6%, 28.6%, 43%, and 21.6%, respectively, of plantlets in which neither PVX, PVM, nor PVS was serologically detectable (Borissenko *et al.*, 1985*a*). No PVY was detectable in 98% of explants of the cultivar Priobskij rannija after application of 5-ADHU (DHT) and in 99% of explants after application of cyanoguanidine (Borissenko *et al.*, 1985*b*).

When trees of four necrotic ringspot virus-infected sweet and sour cherry cultivars were injected three times with 5-ADHU (DHT), 99% of the tested buds proved virus free. By contrast, only 20–32% of the buds from solvent-injected control plants were found to be free of virus. About 65% of the fruit trees grown from buds of the DHT-treated plants were still virus-free 2 years later (Bogusch *et al.*, 1985). These examples indicate that the application of antiphytoviral substances can lead to a higher percentage of virus-free explants and can become one of the first practical uses of this method.

C. Freeing Plants from Viruses

Weekly applications of ribavirin as a foliar spray to 2-year-old *P. serrulata* Kwanzan trees infected with the green mottle causal agent (GRM) prevented symptom development on newly developing foliage and gradually eliminated the infective principle from the previously infected older wood. Back-indexing confirmed that new shoots were free of GRM and demonstrated that GRM in older wood gradually disappeared. One year after the ribavirin treatments were discontinued, no GRM could be detected in these trees. By contrast, weekly foliar applications of ribavirin to peach trees infected with NRSV prevented neither symptom development nor virus replication (Hansen, 1984). Complete plant virus chemotherapy by treatments with a single antiphytoviral substance is so far restricted to few examples.

D. Keeping Seed Potatoes and Other Plants under a Critical Level of Virus Infection

According to Hecht and Diercks (1978), the use of antiphytoviral agents should be useful in keeping the virus level in seed potato fields under a critical level of infestation by viruses, even when these fields are immediately adjacent to infected commercial potato areas. In one such case, treatments with a combination of 5-ADHU (DHT), alcane sulfonic acid, and cyanoguanidine reduced the infections with PLRV from 35.1 to 11.7% (Schuster, 1984). In other experiments, a completely healthy potato stock was cultivated for 2 years in a region with high natural infection pressure. This combination of three substances prevented most infections and reduced the potato leaf roll incidence from 20 to 5.6%. Treatments with a combination of DHT, alcane sulfonic acid, and anilinoadamantylthiadiazole led to reductions from 20 to 2.4%, respectively (Kleinhempel *et al.*, 1984). On the other hand, treatments with a single substance did not result in a significant reduction of virus infection. Experiments are being carried out to keep other vegetatively propagated crops, including bananas, under a critical level of virus infection.

E. Prevention or Reduction of Virus-Induced Depressions of Growth and Yield

In a model experiment, CMV-induced growth retardation of *N. glutinosa* was virtually eliminated by application of 5-ADHU (DHT), even though the virus concentration was only reduced by 62% (Schuster *et al.*, 1979*a*). Yields of naturally infected dwarf beans could be increased by 8% following treatment with DHT (Schuster, 1983*b*). Substantial yield increases can therefore be considered an objective to be accomplished by the systematic use of antiphytoviral substances.

Several years' potato plot tests showed that repeated treatments with DHT increased the yield in 57 virus-infected stocks with an average infestation of 24.6%. The yield of tubers rose by 6.5% following the treatment. On the other hand, 49 identically treated healthy stocks with an average 4.8% virus infestation showed no changes in yield whatsoever. Thus yield increase may be regarded as being due to the antiviral effects produced by DHT rather than to nonspecific growth-regulating or -fertilizing effects, for the latter would have been observed in healthy as well as in virus-infected plants. In a 20-ha field test, yields increased by 12.6% following DHT treatment of 19 potato stocks with an average viral infection of 10–35%. However, DHT only reduced and did not fully suppress the development of symptoms of viral diseases (Schuster *et al.*, 1979*c*).

The yield of PVX- and TMV-infected tomato plants was stabilized by preplanting application of granulated DHT. Untreated virus-infected plants showed strong growth depressions and heavy virus symptoms. They had a yield of 19.6% compared with that of the healthy control. On the other hand, the virus-infected plants that had received granulated DHT were of normal growth and yielded 84% compared with the healthy control. Symptoms appeared relatively late and were light. Subsequent field trials gave similar results (Schuster, unpublished observations). Consequently, stabilization of yields in virus-infected field plants by appropriate measures will be an interesting objective of the use of antiphytoviral substances. The present trends promise that applications of antiphytoviral compounds will soon support the battle against viruses and will be economically feasible and profitable.

REFERENCES

Aldwinckle, H. S., and Selman, I. W. (1967). *Ann. Appl. Biol.* **60,** 49–58.

Antoniw, J. F., and White, R. F. (1980). *Phytopathol. Z.* **98,** 331–341.

Atchison, B. A., Francki, R. I. B., and Crowley, N. C. (1969). *Virology* **37,** 396–403.

Bailiss, K. W. (1968). *Ann. Bot.* **32,** 543–551.

Bailiss, K. W. (1977). *Current Topics in Plant Pathology. Proceedings of a Symposium, Budapest, June 24–27, 1975, Hung. Acad. Sci.* 361–373.

Bailiss, K. W., Cocker, F. M., and Cassells, A. C. (1977*a*). *Ann. Appl. Biol.* **87,** 383–392.

Bailiss, K. W., Balázs, E., and Király, Z. (1977*b*). *Acta Phytopathol. Acad. Sci. Hung.* **12,** 133–140.

Balázs, E., and Gáborjányi, R. (1974). *Z. Pflanzenkr. Pflanzenschutz* **81,** 389–393.

Balázs, E., Gáborjányi, R., and Király, Z. (1973). *Physiol. Plant Pathol.* **3,** 341–346.

Barta, A., Sum, I., and Foeglein, F. J. (1981). *J. Gen. Virol.* **56,** 219–222.

Bawden, F. C., and Kassanis, B. (1954). *J. Gen. Microbiol.* **10,** 160–173.

Beachy, R. S., and Zaitlin, M. (1975). *Virology* **63,** 84–97.

Bergmann, L. (1958). *Phytopathol. Z.* **34,** 209–220.

Bhardwaj, S. V., Dubey, G. S., and Sharma, I. (1982). *Phytopathol. Z.* **105,** 87–91.

Bobyr, A. D. (1976). In *Chemophrophylaxis and Chemotherapy of Viruses* Naukowa dumka, Kiew.

Bogusch, L. J., Verderevskaja, T. D., and Schuster, G. (1985). *Phytopathol. Z.* **114,** 189–191.

Borissenko, S., Schuster, G., and Schmygla, W. (1985*a*). *Phytopathol. Z.* **114,** 185–188.

Borissenko, S., Schmygla, W., and Schuster, G. (1985*b*). *Dokl. VASCHINIL Okt. 1985,* 10–12.

Byhan, O., Liebert, A., and Schuster, G. (1981). *Biochem. Physiol. Pflanzen* **176,** 577–583.

Cadman, E., and Benz, C. (1980). *Biochim. Biophys. Acta* **609,** 372–382.

Cassells, A. C., and Cocker, F. M. (1982). *Z. Naturforsch.* **37c,** 390–393.

Cassells, A. C., and Long, R. D. (1980). *Z. Naturforsch.* **35c,** 350–351.

Cassells, A. C., Barnett, A., and Barlass, M. (1978). *Physiol. Plant Pathol.* **13,** 13–21.

Chant, S. R., Kimmins, W., Stevens, W., and Wrigley, T. C. (1963). *Phyton* **20,** 105–114.

Chessin, M. (1958). *Proceedings of the Third Conference on Potato Virus Diseases, Lisse-Wagenigen,* Veenman & Zonen, Wageningen.

Chiu, R.-J., and Sill, W. H., Jr. (1962). *Phytopathology* **52,** 432–438.

Commoner, B., and Mercer, F. L. (1952). *Arch. Biochem. Biophys.* **35,** 275–289.

Daft, M. J. (1965). *Ann. Appl. Biol.* **55,** 51–56.

Dawson, W. O. (1976). *Intervirology* **6,** 83–89.

Dawson, W. O. (1978). *Intervirology* **9,** 304–309.

Dawson, W. O. (1984). *Phytopathology* **74,** 211–213.

Dawson, W. O., and Grantham, G. L. (1983). *Intervirology* **19,** 155–161.

Dawson, W. O., and Kuhn, C. W. (1972). *Virology* **47,** 21–29.

Dawson, W. O., Schlegel, D. E., and Lung, M. C. Y. (1975). *Virology* **65,** 565–573.

De Clercq, E., and Holý, A. (1985). *J. Med. Chem.* **28,** 282–287.

De Clercq, E., Descamps, J., De Somer, P., and Holý, A. (1978). *Science* **200,** 563–565.

De Fazio, G., and Kudamatsu, M. (1983). *Antiviral Res.* **3,** 109–113.

De Fazio, G., Caner, J., and Vicente, M. (1978). *Arch. Virol.* **58,** 153–156.

De Fazio, G., Caner, J., and Vicente, M. (1980*a*). *Arch. Virol.* **63,** 305–309.

De Fazio, G., Kudamatsu, M., and Vicente, M. (1980*b*). *Fitopatol. Bras.* **5,** 343–349.

De Fazio, G., Kudamatsu, M., and Vicente, M. (1984). *Arq. Inst. Biol. Sao Paulo* **51,** 27–34.

De Zoeten, G. A., Assink, A. M., and Van Kammen, A. (1974). *Virology* **59,** 341–355.

Dijkstra, J. (1969). *Neth. J. Plant Pathol.* **75,** 321–328.

Dove, S., Franke, R., Kern, E., Schuster, G., Kochmann, W., and Steinke, W. (1982). *Wiss. Z. Karl-Marx-Univ. Leipzig, Math.-Naturwiss. R.* **31,** 313–320.

El-Dahtory, Th. A. M., Stenz, E., and Schuster, G. (1983). *Zentrabl. Mikrobiol.* **138,** 99–106.

El-Dahtory, Th. A. M., Stenz, E., and Schuster, G. (1984). *Zentrabl. Mikrobiol.* **139,** 375–382.

El-Mazaty, M. A. A. (1985). Dissertation A, Karl-Marx-University, Leipzig, Math.-Naturwiss. Faculty of Sciences.

Fernandez, T. F., and Gáborjányi, R. (1976). *Acta Phytopathol. Acad. Sci. Hung.* **11,** 271–275.

Francki, R. I. B. (1962). *Virology* **17,** 9–21.

Francki, R. I. B., and Matthews, R. E. F. (1962). *Virology* **17,** 367–380.

Fraser, R. S. S., and Whenham, R. J. (1978*a*). *J. Gen. Virol.* **39,** 191–194.

Fraser, R. S. S., and Whenham, R. J. (1978*b*). *Physiol. Plant Pathol.* **13,** 51–64.

Gaard, G., and De Zoeten, G. A. (1979). *Virology* **96,** 21–31.

Gianinazzi, S., and Kassanis, B. (1974). *J. Gen. Virol.* **23,** 1–9.

Gold, A. H., and Faccioli, G. (1972). *Phytopathol. Medit.* **11,** 145–153.

Hansen, A. J. (1979). *Plant Dis. Rep.* **63,** 17–20.

Hansen, A. J. (1984). *Plant Dis. Rep.* **68,** 216–218.

Harder, D. E., and Kirkpatrick, H. C. (1970). *Phytopathology* **60,** 1255–1258.

Hariharasubramanian, V. (1968). *Phytopathol. Z.* **62,** 92–99.

Hattori, T., and Sarkar, S. (1981). *Mittlg. BBA* **203,** 242.

Hecht, H., and Diercks, R. (1978). *Bayer. Landwirtsch. Jahrb.* **55,** 433–457.

Hillmann, U. (1984). *Diss. Justus Liebig Univ. Giessen.*

Hirai, T. (1977). *Plant Dis.* **1,** 285–306.

Hoehne, C., Frommhold, I., and Schuster, G. (1983). *Potsdamer Forschungen-Wiss. Schriftenreihe Paedagog. Hochsch. Potsdam, Reihe B* **35,** 153–157.

Horst, R. K., and Cohen, D. (1980). *Acta Hort.* **110,** 315–319.

Horváth, J., and Hunyadi, K. (1973). *Acta Phytopathol. Acad. Sci. Hung.* **8,** 347–350.

Huber, S., and Schuster, G. (1987). Symposium *Neue Erkenntnisse und Trends in der Pflanzenphysiologie,* Leipzig, June 10–12, (abstr.), p. 39.

Kalonji-M'Buyi, G., and Kummert, J. (1982*a*). *Med. Fac. Landbouww. Rijksuniv. Gent.* **47,** 1053–1059.

Kalonji-M'Buyi, G., and Kummert, J. (1982*b*). *Parasitica* **38,** 75–84.

Kassanis, B., and White, R. F. (1975). *Ann. Appl. Biol.* **79,** 215–220.

Kiho, Y. (1970). *Jpn. J. Microbiol.* **14,** 291–301.

Kiho, Y., Abe, T., and Ohashi, Y. (1979). *Microbiol. Immunol.* **23,** 1067–1076.

Király, Z., and Szirmai, J. (1964). *Virology* **23,** 286–288.

Király, Z., El Hammady, M., and Pozsár, B. I. (1968). *Phytopathol. Z.* **63,** 47–63.

Kirkpatrick, H. C., and Lindner, R. C. (1961). *Phytopathology* **51,** 727–730.

Kleinhempel, D., Schuster, G., Vogel, J., and Falkenhagen, D. (1984). *Abst. Symp. Biotechnol.* 10.–14. Sept. 1984, Leipzig (GDR), 115.

Kluge, S. (1980). Report I of the *Seventh International Congress on Infectious and Parasitic Disease, Varna, Bulgaria, October 2–6, 1978.*

Kluge, S., and Oertel, C. (1978). *Arch. Phytopathol. Pflanzenschutz. (Berl.)* **14,** 219–225.

Kluge, S., Nuhn, P., Kertscher, N.-P., and Dobner, B. (1984). *Acta Virol.* **28,** 128–133.

Kovalenko, A. G., Barkalova, A. A., Bobyr, A. D., and Votselko, C. K. (1981). *Mikrobiol. J.* **43,** 627–632.

Kroell, J. (1982). *Arch. Phytopathol. Pflanzenschutz. (Berl.)* **18,** 367–368.

Kroell, J. (1985). *Zentralbl. Mikrobiol.* **140,** 225–230.

Kuhn, C. W. (1973). *Phytopathology* **63,** 1235–1238.

Kuhn, C. W. (1977). *Intervirology* **8,** 37–43.

Lapierre, H., Kusiak, C., Albouy, J., Macquaire, G., Férault, A. C., and Staron, T. (1971). *Ann. Phytopathol.* **3,** 3–18.

Lerch, B. (1977). *Phytopathol. Z.* **89,** 44–49.

Lerch, B. (1978). *Biol. Bundesanst. Land. Forstwirtsch., Berl. Braunschweig, Jahresber.,* 135.

Lerch, B. (1979). *Ber. Landwirtschaft* **57,** 555–558.

Lerch, B. (1980). Abstracts of the Fourth International Symposium of Socialist Countries: Antiviral Substances, Szeged, Hungary, 1980.

Liebert, A., and Schuster, G. (1980). *Zentralbl. Bakteriol. Mikrobiol. Hyg. [II]* **135,** 691–695.

Lindner, R. C., Cheo, P. C., Kirkpatrick, H. C., and Govindu, H. C. (1960). *Phytopathology* **50,** 884–890.

Loebenstein, G., Gera, A., Barnett, A., Shabtai, S., and Cohen, J. (1980). *Virology* **100,** 110–115.

Makovcová, O., Sĭndelář, L., and Polák, Z. (1976). *Biol. Plant. (Praha)* **18,** 190–194.

Maramorosch, K. (1957). *Science* **126,** 651–652.

Martin-Tanguy, J., Martin, C., Gallet, M., and Vernoy, R. (1976). *C.R. Acad. Sci. Paris [D]* **282,** 2231–2234.

Matthews, R. E. F. (1955*a*). *Atti VI Congr. Int. Microbiol.* **3,** 363–372.

Matthews, R. E. F. (1955*b*). *Virology* **1,** 165–175.
Menzel, G., and Schuster, G. (1979). *Abh. Akad. Wiss. DDR, Abt. Math., Naturwiss., Technol.* **2,** 353–362.
Mercer, F. L., Landhorst, T. E., and Commoner, B. (1953). *Science* **117,** 558–559.
Meyer, M., Foeglein, F. J., and Farkas, G. L. (1977). *Phytopathol. Z.* **90,** 139–146.
Mitra, D. K., and Raychaudhuri, S. P. (1973). *Proc. Indian Nat. Sci. Acad. [B] Biol. Sci.* **39,** 194–198.
Mukherjee, A. K., and Raychaudhuri, S. P. (1966). *Plant Dis. Rep.* **50,** 88–90.
Mukherjee, A. K., Soans, L. C., and Chessin, M. (1967). *Nature (Lond.)* **216,** 1344–1345.
Nariani, T. K. (1963). *Indian Phytopathol.* **16,** 101–102.
Ohashi, Y., and Shimomura, T. (1982). *Physiol. Plant Pathol.* **20,** 125–128.
Oshima, N., and Livingston, C. H. (1961). *Am. Potato J.* **38,** 294–299.
Oshima, N., Sholara, K., Takahashi, Y., Araki, T., Ko, K., and Misato, T. (1984). *Abstracts of the Sixth International Congress of Virology, Sendai, Japan,* pp. 46–51.
Pennazio, S. (1973). *Riv. Pat. Veg. Ser. 4* **9,** 3–10.
Pennazio, S., and Redolfi, P. (1980). *Microbiologica* **3,** 475–479.
Pennazio, S., Luisoni, E., and La Colla, P. (1981). *Microbiologica* **4,** 353–357.
Pennazio, S., Roggero, P., and Lenzi, R. (1983). *Antiviral Res.* **3,** 335–346.
Pozsár, B. I., and Király, Z. (1965). *Tagungsber. DAL* (Berl.), **74,** 333–343.
Pustowoit, B., Pustowoit, W., and Schuster, G. (1984). *Biochem. Physiol. Pflanz.* **179,** 507–516.
Pustowoit, B., Pustowoit, W., and Schuster, G. (1985). *Biochem. Physiol. Pflanz.* **180,** 423–429.
Quak, F. (1961). *Adv. Hort. Sci.* **1,** 144–148.
Ragetli, J. P. H., and Weintraub, M. (1974). *Contrib. Res. Sta. Agri. Can. Vancouver, B.C.,* **339,** 1–13.
Rajyalakshmi, R., and Raychaudhuri, S. P. (1978). *Natl. Acad. Sci. Lett.* **1,** 125–126.
Ralph, R. K., and Wojcik, S. J. (1976). *Biochim. Biophys. Acta* **444,** 261–268.
Rao, D. G., and Raychaudhuri, S. P. (1965). *Indian J. Microbiol.* **5,** 9–12.
Rehnig, A., and Schuster, G. (1982). *Wiss. Z. Karl-Marx-Univ. Leipzig, Math.-Naturwiss. Reihe* **31,** 331–340.
Sander, E. (1969). *J. Gen. Virol.* **4,** 235–244.
Schlegel, D. E., and Rawlins, T. E. (1954). *J. Bacteriol.* **67,** 103–109.
Schuster, G. (1967). *Flora Abt. A.* **158,** 325–343.
Schuster, G. (1971). *Arch. Pflanzenschutz* **7,** 171–187.
Schuster, G. (1972). *Arch. Pflanzenschutz* **8,** 89–102.
Schuster, G. (1974). *Biochem. Physiol. Pflanz.* **166,** 393–400.
Schuster, G. (1975*a*). *Arch. Phytopathol. Pflanzenschutz* **11,** 9–17.
Schuster, G. (1975*b*). *Arch. Phytopathol. Pflanzenschutz* **11,** 255–262.
Schuster, G. (1976). *Ber. Inst. Tabakforsch.* **23,** 21–36.
Schuster, G. (1977). *Arch. Phytopathol. Pflanzenschutz* **13,** 229–239.
Schuster, G. (1978). *Zentralbl. Bakteriol. Mikrobiol. Hyg. [B]* **133,** 686–689.
Schuster, G. (1979). *Phytopathol. Z.* **94,** 72–79.
Schuster, G. (1980). *Tagungsber. Akad. Landw.-Wiss. DDR* **179,** 347–353.
Schuster, G. (1982*a*). *Wiss. Z. Karl-Marx-Univ. Math.-Naturwiss. Reihe* **31,** 295–312.
Schuster, G. (1982*b*). *Phytopathol. Z.* **103,** 323–328.
Schuster, G. (1982*c*). *Phytopathol. Z.* **103,** 77–86.
Schuster, G. (1982*d*). *Wiss. Z. Karl-Marx-Univ. Leipz. Math.-Naturwiss. Reihe* **31,** 321–330.
Schuster, G. (1983*a*). *Phytopathol. Z.* **106,** 262–271.
Schuster, G. (1983*b*). *Zeszyty Problemowe Postepów Nauk Rolniczych.* **291,** 275–289.
Schuster, G. (1983*c*). *Potsdamer Forschungen-Wiss. Schriftenreihe Paedagog. Hochsch. Potsdam, Reihe B* 127–142.
Schuster, G. (1984). *Symp. Biotechnol. (GDR),* (10–19, Sept.) 87–88 (abst.).
Schuster, G. (1987). *J. Phytopathol.* **118,** 109–122.
Schuster, G., and Arenhoevel, C. (1984). *Intervirology* **21,** 134–140.
Schuster, G., and Byhan, O. (1980). *Tagungsber. Akad. Landw.-Wiss. DDR* **184,** 125–136.
Schuster, G., and Hanzsch, C. (1981). *Phytopathol. Z.* **100,** 226–236.
Schuster, G., and Kramer, W. (1982). *Arch. Phytopathol. Pflanzenschutz* **18,** 77–87.
Schuster, G., and Rehnig, A. (1982). *Wiss. Z. Karl-Marx-Univ. Leipzig Math-Naturwiss.* **31,** 331–340.
Schuster, G., and Ulbricht, H. (1985). *J. Plant Dis. Prot.* **92,** 27–36.
Schuster, G., and Vassilev, G. N. (1983*a*). *J. Plant Dis. Prot.* **90,** 194–199.

Schuster, G., and Vassilev, G. N. (1983*b*). *J. Plant Dis. Prot.* **90,** 500–504.
Schuster, G., and Vassilev, G. N. (1985). *Arch. Phytopathol. Pflanzenschutz* **21,** 41–51.
Schuster, G., Hoeringklee, W., Winter, H., Esser, G., Steinke, U., Kochmann, W., Kramer, W., and Steinke, W. (1979*a*). *Zentralbl. Bakteriol. Mikrobiol. Hyg. [II]* **134,** 64–69.
Schuster, G., Hoeringklee, W., Winter, H., Esser, G., Steinke, U., Kochmann, W., Kramer, W., and Steinke, W. (1979*b*). *Acta Virol.* **23,** 412–420.
Schuster, G., Kochmann, W., Kramer, W., Lehmann, K., Steinke, W., Hoeringklee, W., Esser, G., Steinke, U., and Winter, H. (1979*c*). *Potato Res.* **22,** 279–288.
Schuster, G., Heinisch, L., Schulze, W., Ulbricht, H., and Willitzer, H. (1984*a*). *Phytopathol. Z.* **111,** 97–113.
Schuster, G., Just, H., and Schwarz, J. (1984*b*). *Z. Pflanzenkr. Pflanzenschutz* **91,** 569–579.
Selman, I. W. (1964). *Ann. Appl. Biol.* **53,** 67–76.
Selman, I. W., and Arulpragasam, P. V. (1970). *Ann. Bot.* **34,** 1107–1114.
Selman, I. W., and Kandiah, U. (1971). *Bull. Entomol. Res.* **60,** 359–365.
Seth, M. L., and Raychaudhuri, S. P. (1975). *Z. Pflanzenkr. Pflanzenschutz* **82,** 593–603.
Shepard, J. F. (1977). *Virology* **78,** 261–266.
Sidwell, R. W., Huffman, J. H., Khare, G. P., Allen, L. B., Witkowski, J. T., and Robins, R. K. (1972). *Science* **177,** 705–706.
Šindelář, L., and Makovcová, D. (1975). *Biol. Plant. (Praha)* **17,** 371–373.
Šindelář, L., Makovcová, O., and Polak, Z. (1979). *Sb. Uvtiv-Ochrana Rostlin* **15,** 89–94.
Singh, K. B. P. (1964). *Diss. Abstr.* **25,** 19–20.
Sinha, R. C. (1960). *Ann. Appl. Biol.* **48,** 749–755.
Skene, K. G. M. (1972). *J. Hort. Sci.* **47,** 179–182.
Soans, L. C. (1967). *Diss. Abstr.* 27-12-1, B.P. 4210-1.
Staehelin, M., and Gordon, M. P. (1960). *Biochim. Biophys. Acta* **38,** 307–315.
Stein, A., and Loebenstein, G. (1972). *Phytopathology* **62,** 1461–1466.
Stein, A., Loebenstein, G., and Spiegel, S. (1979). *Physiol. Plant Pathol.* **15,** 241–255.
Taniguchi, T. (1976). *Phytopathol. Z.* **86,** 246–251.
Thomas, T. H. (1974). *Ann. Appl. Biol.* **76,** 237–241.
Tomlinson, J. A. (1977). *Proceedings of the 1977 British Crop Protection Conference: Pests and Diseases,* pp. 807–813.
Tomlinson, J. A., Faithfull, E. M., and Ward, C. M. (1976). *Ann. Appl. Biol.* **84,** 31–41.
Tyihák, E., and Balázs, E. (1976). *Acta Phytopathol. Acad. Sci. Hung.* **11,** 11–16.
Ulrychová-Zelinková, M. (1960). *Biol. Plant. (Praha)* **3,** 240–243.
van Loon, L. C. (1977). *Virology* **80,** 417–420.
Varma, J. P. (1968). *Virology* **36,** 305–308.
Vassilev, G. N., and Jonova, P. A. (1976). *Biochem. Physiol. Pflanz.* **170,** 517–529.
Vicente, M., Barradas, M. M., and Noronha, A. (1972). *Biológico* **38,** 400.
Voigt, B., Mueller, H., and Schuster, G. (1985). *Acta Biotechnol.* **5,** 313–317.
Whenham, R. J., and Fraser, R. S. S. (1981). *Physiol. Plant Pathol.* **18,** 267–278.
Yun, T. G., and Hirai, T. (1968). *Ann. Phytopathol. Soc. Jpn.* **34,** 109–113.
Zabel, P., Jongen-Neven, I., and van Kammen, A. (1976). *J. Virol.* **17,** 679–685.
Zherebchuk, L. K. (1967). *Mikrobiol. J.* **29,** 361–364.
Zhuk, I. P., Didenko, L. F., and Gorbarenko, N. J. (1970). *Biol. Nauki (Mosc.)* **13,** 108–110.
Žmurko, L. I., and Bobyr, A. D. (1974). *Izd-vo VASCHNIL* **38.**

21

Chemotherapy of Plant Virus Infections

A. J. Hansen

I. INTRODUCTION

Plant virus chemotherapy, in the strict sense, can be defined as the production of virus-free plants from infected stock by chemical means. Currently available plant virus chemotherapeuticals are not virucidal but inhibit replication and therefore prevent the spread of infections in the plant. The basic principles of human, animal, and plant virus chemotherapy are very similar. The major differences lie in the physiology and growth characteristics of the hosts and in the goals of the therapy. Plants have no true immune system, and plant viruses are always in the active state. Plant virus chemotherapy therefore aims at complete inhibition of virus replication, so that plant parts which develop during treatment (seeds, shoots, bulblets, corms, tubers, buds, suckers) are absolutely virus free and can be used as nuclear stock for mass propagation and eventual distribution to the industry. The cost of plant virus chemotherapy is generally too high to justify direct curative treatment of producing field plants or trees.

Attempts to implement plant virus chemotherapy are probably as old as plant virology itself, yet there have been very few positive results until the last decade. The major reason for this lack of success seems to have been an incomplete understanding of the difference between compounds that inhibit the infection process and sometimes depress the virus titer in inoculated leaves, and the true chemotherapeuticals which are capable of completely preventing the replication of viruses in systemically infected host plants. The functional difference between these two groups of compounds is probably more apparent than real, since the success or failure of plant virus chemotherapy often depends more on the application method than on the choice of compound (Bogusch *et al.*, 1985). The work with the first group, the general plant virus inhibitors, is summarized in Chapter 20. The present review is concerned only with the compounds and methods that have been used successfully for complete plant virus chemotherapy.

Traditionally, plants free of virus have been obtained from infected stocks by three major methods: seed propagation, thermotherapy, and selection of plants or plant parts that had become virus free more or less by chance (scale bulblets, tip buds of trees, meristematic domes) (Simpkins *et al.*, 1981; Blom-Barnhoorn and van Aartrijk, 1985; Blom-Barnhoorn *et al.*, 1986; Hansen, 1985). Plant virus chemotherapy promises to

---

A. J. Hansen • Agriculture Canada, Research Station, Summerland, British Columbia VOH 1Z0, Canada.

become a fourth method for obtaining virus-free plant material. In some cases, it will replace the traditional methods; in others, it will supplement them.

Any attempt to eliminate viruses from infected plants has to be followed by detailed tests to confirm the virus-free status. This evaluation is usually carried out by attempts to transmit residual virus to specific highly susceptible test plants or by serology in its various forms, e.g., Ouchterlony and enzyme-linked immunosorbent assay (ELISA). Post-treatment evaluation tends to be a major cost factor in the production of virus-free plant material, especially with viruslike agents (VLAs) which are poorly understood, are not sap transmissible, and can therefore generally not be detected by serology or on herbaceous hosts.

II. CHEMOTHERAPEUTIC METHODS

Any attempt to summarize and evaluate the methodology used in plant virus chemotherapy will be incomplete and inaccurate, since the research work is scattered through many laboratories and countries and has been directed more to an understanding of the process than to actual chemotherapy. Consequently, no effort at standardization has been made. However, some major approaches are emerging and an initial evaluation is possible.

A. Application

Four major methods are being used to introduce potential chemotherapeuticals into virus-infected plants: foliar application, root drench, injection or wick application, and incorporation into solid or liquid artificial medium for meristems or shoots. Callus culture, protoplasts, and floating leaf discs are used in experimental systems (Simpkins *et al.*, 1981; Bové *et al.*, 1982; Dawson, 1984; Lerch, 1984; Lozoya-Saldaña *et al.*, 1984).

Foliar application (De Fazio *et al.*, 1978, 1980*b*; Hansen, 1979; Akius, 1983) with or without surfactants (Kalonji-M'Buyi and Kummert, 1982) to inoculated leaves or systemically infected plants is simple, quick and very useful where quantitative data are needed for a first evaluation of a new compound. Data obtained from foliar application on inoculated leaves are a compound measure of virus inhibition due to (1) prepenetration inhibition, (2) incomplete epidermal penetration, and (3) inhibition of uncoating and of the first replicative steps. These data do not necessarily reflect the effect of a compound upon systemic replication. Foliar application is used most frequently for an initial screening of viruses that induce local lesions on the inoculated leaf and for an evaluation of the inhibitory effect on established infections. For model tests of the second kind, young plants are inoculated at an early stage, are allowed to develop systemic infection, and are then treated with the potential chemotherapeuticals (Lerch, 1977; De Fazio *et al.*, 1978, 1980*a*). Alternately, vegetatively propagated shoots from infected mother plants or trees can be used for these experiments (Hansen, 1984; Shepard, 1977; Rohloff and Lerch, 1978; Simpkins *et al.*, 1981). The success of the treatment is measured by the percentage of shoots that are virus free by the end of the treatment period. Tests with systemically infected plants take more time and effort but have the crucial advantage of resembling operational conditions more closely than any other test. Foliar application to systemically infected plants will likely become the standard routine method of plant virus chemotherapy of the future.

Root drench application likely leads to excellent absorption, even distribution and good systemic movement of compounds throughout the treated plant, although this effect has never been critically evaluated for chemotherapeuticals. Root drench has been used by several working groups to evaluate the effect of potential candidate compounds on various viruses in systemically infected herbaceous host plants. The main drawback of drench application is the potential of the growing medium for absorbing some compounds; this has never been tested experimentally but could influence the results. Plant culture in sterile and chemically nonabsorbing quartz sand would overcome this objection only partially, since the required nutrient medium might interact with some compounds.

Injection of plants with candidate compounds or application by wick (Secor and Nyland, 1978; Akius, 1983; Meyer, 1985) has been used only rarely but quite successfully. Bogusch *et al.* (1985) recently showed that injection of cherry trees can lead to complete and permanent elimination of *Prunus* necrotic ringspot virus (NRSV) in cases in which application of the same compound with other methods (Meyer, 1985) had failed. More comparative work is needed to determine whether this is an isolated incident or whether injection in general is a more promising method. Injection methods are slightly more inconvenient and time consuming than other methods, especially with small plants, but they have the important advantage that quantitative data on compound uptake can be obtained.

In meristem or shoot treatment, short shoot tip pieces or axillary buds are aseptically removed from infected plants and are grown on a sterile solid or liquid medium to which candidate compounds have been added (Hansen and Lane, 1985). The growth of shoots or roots and rapid proliferation can be manipulated by addition of growth hormones to the medium (Cassells and Long, 1980; Klein and Livingston, 1982; Wambugu *et al.*, 1985). Transfers to fresh medium are carried out every 4–6 weeks. Virus inhibition can then be evaluated by counting the number of virus-free shoots following treatment. For laboratories equipped with meristem facilities, this is one of the simplest and most flexible methods, having the additional advantage that a large number of virus-free plants are produced with very little extra effort. One drawback with meristems and shoot cultures is the appearance of spontaneously virus-free shoots that arise at unpredictable rates (Hansen, 1985). This affects primarily basic and experimental studies, because it is impossible to distinguish between spontaneous and compound-induced freedom from virus infection. In routine chemotherapy, in which the ultimate goal is not an understanding of the process but the production of virus-free plants, such spontaneously virus-free shoots can only improve the result.

Floating leaf discs have frequently been used to evaluate candidate compounds or to study the mode of action of these compounds (Turner and Dawson, 1984; Dawson, 1984; Dawson and Lozoya-Saldaña, 1984). They offer the great advantage of providing a large number of uniform replicates, facilitating quantitative evaluation and statistical analysis. However, results obtained with leaf discs are not fully applicable to differentiated plants. Similar differences have been encountered in comparison of virus–host interaction between animal cell cultures and differentiated animals (Streeter *et al.*, 1973).

Callus cultures have been used for the study of viruses but apparently only rarely for that of virus–chemotherapeutant interactions. Callus cultures do not offer any specific advantage over meristem cultures, especially since callus reactions to virus infections and to chemotherapeuticals are not comparable to that of intact plants (Hansen, 1974; Shepard, 1977; Dawson, 1984; Lozoya-Saldaña *et al.*, 1984).

Similar objections apply to the use of protoplasts for plant virus chemotherapy (Shepard, 1977). However, the use of protoplasts as experimental models for studies of virus replication and of basic aspects of virus–chemotherapeuticals interaction has several advantages and will likely increase in the future.

B. Timing

When routine efficacy tests are carried out by the foliar application method (Hansen, 1979; De Fazio *et al.*, 1980*a,b*; Akius, 1983), applications are generally timed in such a way that the compound is being applied to the leaf surface before inoculation takes place. Thus, the efficacy of the compound can be evaluated at all stages of inoculation, subsequent uncoating and first replication.

Depending on the purpose of the experiment and the host–virus–compound combination, plants are treated once or repeatedly. Solid and liquid culture and wick application (Meyer, 1985) are the only examples of continuous treatment. Repeated daily or weekly treatments have been used only rarely, yet they may hold the key to more effective applied chemotherapy with intact plants. Especially promising candidates are the many compounds that have been shown to reduce virus replication after one or two applications without resulting in complete virus elimination (Caner *et al.*, 1985). Treatment with some of these compounds would likely have resulted in complete virus elimination if the applications had been continued beyond the initial period. Recent evidence suggests that some compounds have a mode of action that differs from that of ribavirin (1-β-D-Ribofuranosyl-1,2,4-triazole-3-carboxamide) (Dawson, 1984). Meyer (1985) showed, in very detailed studies with three viruses in herbaceous hosts and in *Prunus* branches and young trees, that 5-azauracil and 6-azauracil affect a very early stage of replication but have less effect on systemic infection and do not eliminate the viruses completely even when applied continuously and in near-phytotoxic doses.

When tests are being carried out on systemically infected plants or shoots, it is important to remember that intact virions are resistant to chemotherapeuticals and that any effect can therefore be observed only in actively growing plants. In other words, the effect is not virucidal but is inhibitory to the replicative process (Akius, 1983). Application intervals should be short enough to maintain relatively constant chemotherapeutical levels in the plant. Evaluations should be carried out only after the compound has had sufficient time to affect replication in the growing tissues.

Timing is especially critical if the goal is not chemotherapy but rather determination of the stage of virus replication that is affected. However, no clear data are available regarding the time course of inactivation of ribavirin in treated plants. Using foliar inhibition tests, Akius (1983) and Hansen (1979) found that the major inhibiting effect lasts approximately 3 days in *Chenopodium quinoa* Willd. if local lesions or systemic movement of apple chlorotic leaf spot virus (ACLSV) were used as a criterion of effectiveness. Others have arrived at the same order of magnitude, using different hosts and viruses. The actual time course of inactivation in each experiment likely depends upon variables such as original virus concentration, growth stage of the plants, temperature, and light. More detailed and elegant methods have been developed for specific purposes by Dawson's group (Dawson and Lozoya-Saldaña, 1984) and by Meyer (1985).

C. Dosage

The optimal dosage is generally the highest dosage that does not induce severe phytotoxicity. The actual dosage level depends strongly on the experimental system, the application method, the light conditions and the type and age of the test plant. The published data are based chiefly on ribavirin in potato material and a few other host plants, at about 20°C and daylight conditions. Commonly used ribavirin dosages can be summarized as follows: protoplasts, 1–10 ppm; floating leaf discs, 1–50 ppm; meristems, 20–80 ppm; calli, 5–400 ppm; herbaceous plants, 50–1000 ppm; and field trees, 500–1000 ppm. Frequency and interval of application have also been highly variable, depending on the purpose of the experiment. We have ample (Hansen, unpublished results) evidence that two half-doses applied twice in a short time (1 or 2 hr) induce much less phytotoxicity than a whole dose given in one single application, but no systematic study of this aspect has been carried out. We have also found that host plants growing under higher light intensities and higher temperatures permit application of higher doses of ribavirin without showing phytotoxicity.

The influence of the host on virus chemotherapy has not been systematically explored. Scattered observations indicate that chemotherapeutical dosages required for successful therapy can vary depending upon genus, species or even cultivar. In one of the few direct comparisons, De Fazio *et al.* (1980*a*) found that 500 ppm ribavirin was sufficient to prevent systemic movement of tomato spotted wilt virus in tobacco, but not in tomato. A similar difference between two ribavirin-treated virus-infected lily species was observed by Blom-Barnhoorn and van Aartrijk (1982). Jordán *et al.* (1983) found that 14 days of treatment with ribavirin were sufficient to obtain virus-free shoots from one potato cultivar, while other cultivars required 3 months. Bogusch *et al.* (1985) showed that some cultivars of sour cherries can be cured of NRSV by injection with a chemotherapeutant, while the virus status of another cultivar remained unaffected by an identical treatment. The reasons for such differences are not clear.

Phytotoxic effects have been observed in numerous experiments. They generally consist of reduced growth, misshapen foliage and reduced capacity for rooting (Kluge and Oertel, 1976; Simpkins *et al.*, 1981; Jordán *et al.*, 1983). The level at which phytotoxicity becomes severe depends on the medium and growth stage of the protoplasts, leaf discs, calli, shoots, or plants. The upper effective levels for chemotherapy listed in Table 1 are generally the levels at which phytotoxicity begins to have a strong and inhibiting effect.

D. Virus Detection

Virus detection following treatment is a time-consuming yet essential step. Methodologic details depend on the goal of the experiment. In many cases, sap transmission from treated plants to susceptible herbaceous indicators is the most reliable method. For those viruses and VLAs that are not sap transmissible, graft or insect transmission to

Table 1. Virus Inhibition and Development of Fully Virus-Free Plants from Infected Stock by Chemotherapy with Ribavirin (MW 244.2)

Virus or viruslike agent	Host	Treatment and minimum concentration needed	Comments	Reference
Apple chlorotic leafspot virus	Apple	Meristems, 20 ppm for 4 weeks	All shoots virus free	Hansen and Lane (1985)
	Apple trees	Weekly spray, 500 ppm for 6 months	All new shoots virus free	Hansen (unpublished observations)
Apple stem grooving	*Chenopodium quinoa*	Daily spray, 100 ppm for 18 days	Most plants virus free	Hansen (unpublished observations)
Cucumber mosaic and potato virus Y	Tobacco	Meristem, 50 ppm prolonged	Virus free after 22 weeks	Cassells and Long (1980)
Cucumber mosaic virus	Tobacco	Meristem, 100 ppm prolonged	Many virus free after 3 weeks	Simpkins *et al.* (1981)
Green ring mottle	*Prunus serrulata*	Weekly spray, 500 ppm for 8 weeks	Minimum needed for shoot tip	Hansen (1984)
Potato virus M	Potato	Meristems, 40 ppm	All plants virus free	Cassells and Long (1982)
Potato viruses S and Y	Potato	Axillary buds, 20 ppm	90% of shoots virus free	Wambugu *et al.* (1985)
Potato virus X	Tobacco	Nutrient solution 0.5 ppm, prolonged	Top leaves virus free	Lerch (1977); Rohloff and Lerch (1978)
	Potato	Protoplast-derived shoots, 100 ppm	94% virus free; similar callus not	Shepard (1977)
	Potato	Liquid medium, meristems	80% of plants in 10 ppm were virus free	Klein and Livingston (1982)
	Potato 7 cultivars	Nutrient solution, up to 100 ppm	Shoots treated with 25 ppm or more were virus free	Jordán *et al.* (1983)
Tomato spotted wilt virus	Tomato, tobacco	Plants, 500 ppm	Partial effect in tomato; full cure in tobacco	De Fazio *et al.* (1980*a*)
Tomato white necrosis virus	Tomato	Plants, 500 ppm	60% virus free; 40% reduced concentration	De Fazio *et al.* (1978)
Turnip yellow mosaic virus	Chinese cabbage	Protoplasts, 10 ppm	Virions reduced by 90%	Bové *et al.* (1982)

specific test plants is the only known method. Serology can be used under some circumstances but has the inherent disadvantage that it detects infective as well as noninfective virions and does not distinguish between them. Kluge and Marcinka (1978) showed a good example of one such case in which virus infectivity had dropped significantly following application of polyacrylic acid and ribavirin, while the corresponding serologic titer remained unaffected.

III. CHEMOTHERAPEUTICALS

Virus-inhibiting characteristics have been found, usually by foliar tests, in such diverse groups of compounds as plant constituents, fungal constituents, skim milk, salts, aspirin, and others. The effect of plant extracts and natural substances has been studied in great detail by Verma *et al.* (1985), while synthetic substances have been tested by Schuster (Chapter 20), Dawson and Lozoya-Saldaña (1984), Lozoya-Saldaña *et al.* (1984), and by our own investigations. Most of these compounds were capable of reducing the number of local lesions or of depressing the virus titer in inoculated leaves; many also affected systemic movement or replication, but only a few were effective enough to inhibit replication to the point where virus-free plants could be recovered following treatment.

Those that eventually were found to be capable of completely inhibiting virus replication in systemically infected plants had certain characteristics in common: water solubility, ability to penetrate the epidermis, translocation at least upward in the plant, and not being overly phytotoxic at levels that inhibited virus replication. These four characteristics seem to be the basic requirements for any effective plant virus chemotherapeutical.

The search for plant virus chemotherapeuticals has been a long one. In recent years, it has received major stimuli and direction from the voluminous work with human and animal virus chemotherapeuticals. By the time the first success was achieved with inhibitors of mammalian viruses, most plant virus researchers had given up hope for plant virus chemotherapeuticals and had shifted their attention to other fields. As a consequence, very little effort is being devoted to research on chemical plant virus inhibition, compared with other fields of plant virology, and most mammalian virus inhibitors have never been tested against plant virus. However, three inhibitors have been shown to be very effective.

A. Ribavirin

Ribavirin, a guanosine analogue, has been by far the most successful of the compounds tested. It has not only been effective in model tests but has been used to create permanently virus-free plants from infected stock in 13 virus–host combinations involving 10 different viruses and VLAs (Table 1). In another combination (rose ring pattern VLA in rose), symptoms disappeared following a single application of ribavirin by injection but reappeared several months later (Secor and Nyland, 1978). This treatment would likely have been successful if the applications had been continued or if the shoot tips that grew during the application had immediately been budded onto healthy understock. In principle, this treatment was very similar to the successful elimination of the green ring mottle VLA from *Prunus serrulata* Lindl (Table 1), except for the fact that treatment was continued in the latter case.

Negative results with chemotherapy are not always reported yet are just as important as the successes for an understanding of the mode of action. Ribavirin applied by one or several of the above mentioned methods is apparently not inhibitory to hyacinth mosaic virus (Blom-Barnhoorn *et al.*, 1986), several ILAR and at least one PVY virus (Hansen, 1984), red clover mottle virus (Kluge and Marcinka, 1978), lily symptomless and tulip-breaking virus in 'Enchantment' (Blom-Barnhoorn and van Aartrijk, 1985), tobacco mosaic virus (TMV), and cowpea chlorotic mottle virus (Lozoya-Saldaña, 1981).

In many virus–host–treatment combinations, a virus titer reduction occurs but is not followed by complete elimination of the virus from growing tips during the course of the experiment. In some of these cases, prolonged exposure or repeated treatment probably would have led to success. In other cases, treatments were applied to recently inoculated plants, and it was not always clear whether the observed effects were due to inhibition of incipient infections in the inoculated leaf or of truly systemic infections. As long as the test methodology is not standardized, there will always be a degree of subjective judgement in classifying viruses as fully, moderately, or not susceptible to treatment with a given compound. This bias is inevitable and is reflected in Table 1 as well as in the text.

The evidence for susceptibility of plant viruses to compounds other than ribavirin is much less documented, which is in part a reflection of the lesser amount of research that has been conducted with these compounds.

Partial success has been obtained with polyacrylic acid against red clover mottle virus (Kluge and Marcinka, 1978) and against TMV (Kassanis and White, 1978), with (S)-9(2,3-dihydroxypropyladenine (DHPA) or vidarabine against grapevine leafroll (Monette, 1984), with various benzimidazole compounds against cucumber mosaic virus (Simpkins *et al.*, 1981), and with adenine arabinoside against TMV (Lozoya-Saldaña *et al.*, 1984).

Many other compounds delay systemic movement or reduce virus titer in inoculated plants (Dawson, 1984; Schuster *et al.*, 1984; Caner *et al.*, 1985; see also Chapter 20). Changes in methodology, such as repeated application of lower doses or evaluation in systemically infected instead of recently inoculated plants, may show that some of these compounds can be very useful in chemotherapy.

The effect of ribavirin on plant viroids has not been explored in detail. Cheplick and Agrios (1982) found reduction in symptom severity with apple scar skin and dapple apple, both of which are suspected to be caused by a viroid or viroids. Bellés *et al.* (1986) found a similar effect in *Gynura aurantiaca* DC. plants affected with citrus exocortis viroid; they also noted a reduced viroid level. N. S. Wright and A. J. Hansen (unpublished observations) found no effect on potato spindle tuber viroid concentrations in potato after prolonged treatment.

Safe, practical, long-term use of any chemotherapeutical depends in part on the likelihood that the target pathogen will not develop resistant mutants. Theoretical considerations indicate that it is unlikely that mutants will develop that are resistant to ribavirin. Huffman *et al.* (1977) showed in a detailed experiment that no such mutants developed in herpes simples virus (HSV) during four to five cell-culture passages under heavy selective pressure for mutants, and general experience confirms the view that such mutants do not develop (Oxford, 1977). The only report on mutant development (Indulen and Feldblum, 1982) is based on one experiment only and has apparently not been repeated.

In order to clarify the situation for plant viruses, we conducted an experiment with two isolates of ACLSV which were kept under heavy selective ribavirin pressure for 25 weekly passages and were then compared with untreated lines of the same isolates in regard to ribavirin susceptibility (A. J. Hansen, unpublished observations). The results show that the treated and untreated lines were equally susceptible to ribavirin at the end of the 25-week period. If ribavirin-resistant mutants had developed during the course of this experiment, they would likely have had such a competitive advantage during subsequent replication that the whole population of the treated line or lines would have consisted of the resistant mutant or mutants.

B. Tiazofurin

Prevention of systemic movement of tomato spotted wilt virus in recently inoculated tomato plants has been achieved by two applications of 200 ppm tiazofurin (2-β-D-ribofuranosylthiazole-4-carboxamide) (Caner *et al.*, 1984). The effect of this newly released compound on other plant viruses has not yet been explored.

C. DHT

Schuster *et al.* (1984) showed that 2,4-dioxo-hexahydro-1,3,5-triazine (DHT) can be used to reduce the titer of potato virus X (PVX) in potato plants to the point where yields increase significantly. However, no fully PVX-free plants were obtained during this work. Meyer (1985) observed similar reductions in NRSV titers in recently inoculated cucumber plants without obtaining fully virus-free plants.

In a recent study, Bogusch *et al.* (1985) injected DHT at various concentrations into sweet and sour cherry trees infected with NRSV. Buds were taken from the treated trees and were used to create new trees. When these trees were tested for virus content 2 years later, 65% were free of NRSV.

The results obtained by Meyer and by Bogusch are not necessarily contradictory, since different experimental systems and application methods were used. Similar changes of application methods may permit chemotherapy with uracil analogues and with other compounds that so far have only had titer-depressing effects.

D. Combination Therapy

Combinations of antiviral drugs are being used increasingly against mammalian virus diseases (Galasso, 1984) but have only rarely been tried against plant viruses under field conditions. Schuster (1982) showed under experimental conditions that the concentration of PVX, TMV, and cucumber mosaic virus (CMV) in growing tobacco plants was significantly reduced by a combination treatment with ribavirin and DHT. With PVX, no virus could be detected by serology 2 weeks after inoculation. Treatments with one compound alone reduced the titer with all three viruses but did not lead to complete inhibition. Lozoya-Saldaña *et al.* (1984) showed that combination therapy in callus culture with ribavirin and adenine arabinoside against TMV can be very successful, even though neither compound alone was capable of eliminating the virus. This result is especially promising because it is more difficult to eradicate viruses from callus than from differentiated tissue. If similar effects can be obtained under field conditions, combination therapy will lead to considerable cost saving and will greatly expand the range of viruses that can be treated by chemotherapy.

IV. MODE OF ACTION

As the major proven plant virus chemotherapeutical, this discussion concentrates on ribavirin. Considerable effort has been devoted to a clarification of the mode of action of ribavirin and other virus inhibitors against human and animal viruses. The inhibitory action in plants has been less thoroughly investigated, but some similarities and differences are becoming apparent. The implicit assumption in many studies has been that viral replication and chemotherapy take place via very similar or identical pathways in animal and plant hosts, even though the two host groups are phylogenetically unrelated. The assumption may be correct, but so far too little is known about plant virus replication to permit any definite comparison. The hypothesis should be tested in more detail regarding the various chemotherapeuticals, for both practical and theoretical reasons. Ribavirin is not a very suitable tool for this purpose, since it is a broad-spectrum chemotherapeutical (Sidwell *et al.*, 1972; Witkowski *et al.*, 1972) that inhibits some viruses in a nonspecific way. Unfortunately most of the pertinent work has been carried out not with intact plants, but with model systems, such as leaf discs floating on nutrient culture, protoplasts, or callus cultures (Lerch, 1984). There is ample evidence (Dawson, 1984) that biochemical reactions in model systems can be qualitatively different from those in plants. The conclusions drawn from such work do therefore not necessarily apply to the chemotherapeutic action in differentiated plants.

The fact that ribavirin is a broad-spectrum plant virus chemotherapeutant, yet highly selective for some virus groups (Sidwell, 1984; Huggins *et al.*, 1984), indicates that its effect is not on the host metabolism in general but rather against specific virus-directed functions. Ribavirin is translocated rapidly in growing plants; the active antiviral form in plants and animals seems to be the 5′-triphosphate, which is relatively immobile. Interference with the capping mechanism at the 5′-end of the messenger RNA (mRNA) is most likely one of the major modes of action against mammalian viruses (Goswami *et al.*, 1979; Smith, 1984). Many of the highly ribavirin-susceptible plant viruses are also known to have 5′-capped mRNAs and a similar mechanism of inhibition seems likely; however, TMV has the capped mRNA, yet is only mildly susceptible to ribavirin (Lozoya-Saldaña *et al.*, 1984).

Some detailed work has been carried out in model systems by Dawson's group (Dawson and Lozoya-Saldaña, 1984; Dawson, 1984) and by Byhan *et al.* (1978). Their results indicate that ribavirin affects an early step in TMV replication by preventing a function that is necessary to initiate viral RNA synthesis following inoculation. Both ssRNA and dsRNA were inhibited, to approximately the same degree, but host RNA synthesis was much less affected. Later replicative steps were not influenced by ribavirin treatment. Inhibition in other host–virus combinations may follow a different pattern by affecting a later replicative step. Simpkins *et al.* (1981) showed that systemic infections of alfalfa mosaic and CMV in meristem cultures can be strongly inhibited or even eliminated by treatment with ribavirin. This was likely due to inhibition at such a later step. The exact nature of this step will remain unclear as long as the details of local and systemic plant virus replication are unknown.

One possibly significant difference between the action of ribavirin against mammalian and plant viruses is the response of ribavirin-inhibited host–virus systems to application of guanosine. In animal cell cultures, an excess of guanosine acts as a neutralizer of the inhibitory action of ribavirin (Streeter *et al.*, 1973). Similar tests in PVX-infected ribavirin-treated leaf discs have shown that the guanosine level has very little effect on the interaction, while thymidine in 10-fold molar concentration blocked the inhibitory action of ribavirin (Lerch, 1984), probably by delaying the phosphorylation of ribavirin to its antiviral form, the 5′-triphosphate (Goswami *et al.*, 1979; Smith, 1984). We had no success in trying to reproduce this effect by foliar application to intact plants (A. J. Hansen, unpublished data); however, this may have had more to do with the rate of absorption in the two systems rather than with any basic principle. The fact that guanosine did not counteract the antiviral effect of ribavirin on PVX indicates that the mode of action against plant viruses may be different from that against animal viruses. Ultimate proof regarding the mode or modes of action will likely come from work with antiviral compounds that have a more limited range of target viruses and a more specific mode of action.

Experience in several laboratories has shown that the degree of ribavirin-induced virus inhibition in locally infected and systemically invaded leaves can be different. The reason for this is unclear, and any explanation would be speculative. There has also been some discussion (Simpkins *et al.*, 1981) as to whether inhibition of systemic infection is due to ribavirin-induced restriction of virus movement or to restriction of virus replication. However, inhibition of virus movement and of virus replication are likely different manifestations of the same process, in view of increasing indications that viruses move systemically not as intact virions but in some other form. Any treatment affecting the systemic replicative process should therefore affect replication and movement in the same way.

The results obtained with ribavirin therapy of *Prunus* green ring mottle VLA (Hansen, 1984) provide a rough indication of the time frame during which inactivation of systemic infection takes place: immediately following treatment of systemically infected trees there were no further leaf symptoms on the developing foliage and virus tests showed that no VLA could be recovered from these parts. Simultaneous tests from lower parts of the tree that had developed prior to treatment showed that the VLA could be recovered 1 month later, but not after 2 months. This finding is consistent with the hypothesis that treatment immediately and completely inhibits virus replication and that the VLA gradually loses its infectivity due to the normal process of viral degradation (Simpkins *et al.*, 1981). The same process presumably takes place in virus-infected protoplasts and calli, which gradually lose their infectivity following treatment with ribavirin.

V. PRACTICAL APPLICATIONS OF PLANT CHEMOTHERAPY

Our increasing understanding of plant virus disease epidemiology (Thresh, 1982) has led to a clear distinction between the two major modes of field spread: vertical and horizontal. Vertical spread takes place with vegetatively propagated planting stock, such as flower bulbs, corms, fruit trees, orchids, potatoes, many ornamental shrubs, and with crops affected by seed-transmitted viruses. Horizontal transmission from infected to healthy plants is dependent on vectors such as insects, athropods, nematodes, nursery, and greenhouse workers or soil fungi.

Economic considerations restrict the potential use of the present generation of plant virus chemotherapeuticals to those situations in which the high cost of the compounds and the treatments can be economically justified because of the high value of the treated plants. These plants are of three kinds:

1. *Mother plants selected as elite stock for virus-free propagation systems:* In these cases, the high value of the treated plants results from the fact that they provide the initial and crucial virus-free source from which all future propagation material is derived. This applies especially to crops such as fruit trees, potatoes, and strawberries, perennials and ornamental bulbs, and shrubs and trees in which successful virus-free propagation systems already exist (Hansen, 1985). With these and other vegetatively propagated crops, chemotherapy will usually be the fastest and cheapest method, provided the virus in question is susceptible to the compound used. Chemotherapy may also provide a solution in those cases in which propagation systems are hampered by the difficulty encountered in eliminating one or several specific viruses that are highly resistant to the traditional methods of virus elimination (Lozoya-Saldaña and Merlin-Lara, 1984; P. Friedlund, personal communication, 1986).
2. *Crops in which individual plants are of a high enough value to warrant direct curative virus chemotherapy:* This may be the case with orchids, which generally suffer from a high degree of virus infection.
3. *Crops affected by seed-transmitted viruses:* In crops such as lettuce, in which even a low degree of seed transmission can lead to rapid horizontal virus spread under field conditions, almost total yield loss can result (Grogan, 1983). Chemotherapy of seed-transmitted viruses has apparently never been tried, but it should present no major obstacles as long as the virus in question is susceptible to the chemotherapeuticals used. Applications at the dormant stage may be unsuccessful, since viruses presumably do not replicate at this stage. However, treatment of germinating seeds or presowing treatment should be successful if the effect of the compound lasts long enough to inhibit the replicating stage of the virus.

Because of the many biologic, agronomical, economical, and regulatory variables inherent in each field situation, each case must be assessed on its own merit. Relative advantages and disadvantages of chemotherapy over other methods of virus elimination vary from crop to crop and from laboratory to laboratory. The first large-scale field application will most likely be in one of the above-mentioned three groups. As costs for present and future chemotherapeutants decrease, other uses may become apparent. However, they are likely to face increasing competition from cheaper compounds, such as DHT, which prevent most virus-induced crop losses without eliminating the virus completely. They may also have to compete with new or improved nonchemical control methods, such as cross-protection.

Whenever chemotherapy with reasonably priced compounds can be applied to routinely grown plants it will most likely be the most economical method of virus elimination. This is because chemotherapy requires virtually no special equipment or laboratory, and the end product is a full-grown plant that can be used immediately for further propagation.

The technical aspects of chemotherapeutic virus elimination are relatively simple, since new shoots and buds developing during treatment can be cut from the source plant or meristem, can be rooted or budded onto virus free understock, can be tested for freedom from viruses, and can then be grown into mother plants or trees.

Chemotherapy has the further advantage that in many cases all plants are virus free following treatment (Hansen, 1984, 1985). This is especially important with viruses and VLAs that cannot be detected by simple means, such as serology or visual evaluation, but that require lengthy post-treatment testing. With tree fruit viruses and VLAs, where the percentage of virus free shoots following thermotherapy can be especially low, this post-treatment testing has to be extensive, can take several years and is the major cost factor in the whole process of virus elimination.

The problem of environmental contamination with plant virus chemotherapeuticals has apparently not yet been addressed. Existing regulations have been devised to fit the requirements of human and veterinary chemotherapy, and of fungicide and insecticide applications to plants. Development of appropriate guidelines for the use of plant virus chemotherapeuticals under field conditions is presently under consideration.

A frequent argument against plant virus chemotherapy has been the occasional observation that ribavirin treatment can temporarily suppress symptoms, but that virus activity can later be recovered from the symptom-free parts (Secor and Nyland, 1978; Stace-Smith, 1985). However, such observations are the exception rather than the rule. In most cases of successful chemotherapy, frequent post-treatment checks have been carried out by various methods and for up to 4 years (Cassells and Long, 1980; Hansen, 1984; Blom-Barnhoorn *et al.*, 1986). The results have shown that no residual virus infection could be detected. This parallels the situation following thermotherapy, wherein tests for presence of the virus generally give identical results when carried out soon after treatment or a year later. However, plants with initially undetectable levels of virus have been found occasionally in several virus–host combinations following thermotherapy (P. Fridlund, personal communication, 1986; M. F. Welsh, personal communication, 1986). Similar cases will undoubtedly occur following chemotherapy as well.

Combinations of chemotherapy and meristem methods have sometimes been recommended, even though little supporting evidence for efficiency or economy was offered. Such combinations are advantageous only when chemotherapy or meristem therapy alone do not give the desired result, when meristem methods are already an established routine method, or when large numbers of propagules are required soon after virus elimination.

A more general acceptance of plant virus chemotherapy will depend on the availability of a wider range of broad-spectrum chemotherapeuticals. These will likely be found among the compounds being developed for use with animal viruses. Our lack of understanding of the details of the mode of action against plant viruses and the lack of an immediate market make it unlikely that specific plant virus chemotherapeuticals will be developed.

REFERENCES

Akius, M. 91983). *Vaxtskyddsnotiser (Stockh.)* **17,** 39–45.

Awasthi, L. P., Pathak, S. P., and Gautam, N. C. (1985). *Indian J. Plant Pathol.* **3,** 59–63.

Bellés, J. M., Hansen, A. J., Granell, A., and Conejero, V. (1986). *Physiol. Plant Pathol.* **28,** 61–65.

Blom-Barnhoorn, G. J., and van Aartrijk, J. (1982). Acta Hort. **164,** 163–168.

Blom-Barnhoorn, G. J., van Aartrijk, J., and van der Linde, P. C. G. (1986). *Acta Hort.* **177** (in press).

Bogusch, G. J., Verderevskaja, T. D., and Schuster, G. (1985). *Phytopathol. Z.* **114,** 189–191.

Bové, J. M., Renaudin, J., and Mouchez, C. (1982). In *La sélection des plantes pour la résistance aux maladies,* pp. 7–18, INRA, France.

Byhan, O., Kluge, S., Pustowoit, B., Pustowoit, W., and Schuster, G. (1978). *Biochem. Physiol. Pflanz.* **173,** 521–535.

Caner, J., Amelia, M., Alexandre, V., and Vicente, M. (1984). *Antiviral Res.* **4,** 325–331.

Caner, J., De Fazio, G., Vaz Alexandre, M. A., Kudamatsu, M., and Vicente, M. (1985). *Biologico, Sao Paulo,* **57** (in press).

Cassells, A. C., and Long, R. D. (1980). *Z. Naturforsch. Sect. C Biosci.* **35,** 350–351.

Cassells, A. C., and Long, R. D. (1982). *Potato Res.* **25,** 165–173.

Cheplick, S. M., and Agrios, J. N. (1982). *Phytopathology* **72,** 259. (abst).

Dawson, W. O. (1984). *Phytopathology* **74,** 211–213.

Dawson, W. O., and Lozoya-Saldaña, H. (1984). *Intervirology* **22,** 77–84.

De Fazio, G., Caner, J., and Vicente, M. (1978). *Arch. Virol.* **58,** 153–156.

De Fazio, G., Caner, J., and Vicente, M. (1980*a*). *Arch. Virol.* **63,** 305–309.

De Fazio, G., Kudamatsu, M., and Vicente, M. (1980*b*). *Fitopatol. Bras.* **5,** 343–349.

Galasso, G. J. (1984). *J. Antimicrob. Chemother.* **14,** 127–136.

Goswami, B. B., Borek, E., Sharma, O. K., Fujitaki, J., and Smith, R. A. (1979). *Biochem. Biophys. Res. Commun.* **89,** 830–836.

Grogan, R. G. (1983). *Seed Sci. Technol.* **11,** 1–7.

Hansen, A. J. (1974). *Virology* **57,** 387–391.

Hansen, A. J. (1979). *Plant Dis. Rep.* **63,** 17–20.

Hansen, A. J. (1982). *Acta Hort.* **130,** 183–184.

Hansen, A. J. (1984). Plant Dis. **68,** 216–218.

Hansen, A. J. (1985). *HortScience* **20,** 852–859.

Hansen, A. J., and Lane, W. D. (1985). *Plant Dis.* **69,** 134–135.

Hoffmann, C. E., Togo, Y., Otter, C. W., and Swallow, D. L. (1977). *Ann. NY Acad. Sci.* **248,** 289–293.

Huffman, J. H., Allen, L. B., and Sidwell, R. W. (1977). *Ann. NY Acad. Sci.* **284,** 233–237.

Huggins, J. W., Jahrling, P., Kende, M., and Canonico, P. G. (1984). In *Clinical Applications of Ribavirin* (R. A. Smith, V. Knight, and J. A. D. Smith, eds.), pp. 49–63, Academic, Orlando, Florida.

Indulen, M. K., and Feldblum, R. L. (1982). *Acta Virol.* **26,** 109.

Jordán, M., Montenegro, G., and Apablaza, G. (1983). *Cienc. Invest. Agr.* **10,** 249–256.

Kalonji-M'Buyi, and Kummert, J. (1982). *Parasitica* **38,** 75–84.

Kassanis, B., and White, R. F. (1978). *Ann. Appl. Biol.* **79,** 215–220.

Klein, R. E., and Livingston, C. H. (1982). *Am. Potato J.* **59,** 359–365.

Kluge, S., and Marcinka, K. (1978). *Acta Virol.* **23,** 148–152.

Kluge, S., and Oertel, C. (1976). *Arch. Phytopathol. Pflschutz* **14,** 219–225.

Lerch, B. (1977). *Phytopathol. Z.* **89,** 44–49.

Lerch, B. (1984). *University of Leipzig, DDR,* Biotechnology Symposium 1984 (abst).

Lozoya-Saldaña, H. (1981). Ph.D. thesis, University of California, Riverside.

Lozoya-Saldaña, H., and Merlin-Lara, O. (1984). *Am. Potato J.* **61,** 735–739.

Lozoya-Saldaña, H., Dawson, W. O., and Murashige, T. (1984). *Plant Cell Tissue Organ Cult.* **3,** 41–48.

Meyer, S. (1985). Ph.D. thesis, University of Hannover, Germany.

Monette, P. L. (1984). *Proc. Int. Plant Prop. Soc.* **33,** 90–100.

Oxford, J. S. (1977). *Ann. NY Acad Sci.* **284,** 289–293.

Rohloff, H., and Lerch, B. (1978). *Mitt. Biol. Bundesanst. Land-Forstwirtsch (Braunschweig).* **178,** 181.

Schuster, G. (1982). *Phytopathol. Z.* **103,** 323–328.

Schuster, G., Heinisch, L., Schulze, W., Ulbricht, H., and Willitzer, H. (1984). *Phytopathol. Z.* **111,** 97–113.

Secor, G. A., and Nyland, G. (1978). *Phytopathology* **68,** 1005–1010.

Shepard, J. F. (1977). *Virology* **78,** 261–266.

Sidwell, R. W. (1984). In *Clinical Applications of Ribavirin* (R. A. Smith, V. Knight, and J. A. D. Smith, eds.), pp. 19–32, Academic, Orlando, Florida.

Sidwell, R. W., Huffman, J. H., Khare, G. P., Allen, L. B., Witkowski, J. T., and Robins, R. K. (1972). *Science* **177,** 705–706.

Simpkins, I., Walkey, D. G. A., and Neely, H. A. (1981). *Ann. Appl. Biol.* **99,** 161–169.

Smith, R. A. (1984). In *Clinical Applications of Ribavirin* (R. A. Smith, V. Knight, and J. A. D. Smith, eds.), pp. 1–18, Academic, Orlando, Florida.

Stace-Smith, R. (1985). In *Comprehensive Biotechnology,* Vol. 4 (C. W. Robinson, ed.), pp. 169–179, Pergamon, London.

Streeter, D. G., Witkowski, J. T., Khare, G. P., Sidwell, R. W., Bauer, R. J., Robins, R. K., and Simon, L. N. (1973). *Proc. Natl. Acad. Sci. USA* **70,** 1174–1178.

Thresh, J. M. (1982). *Annu. Rev. Phytopathol.* **20,** 193–218.

Turner, M. S., and Dawson, W. O. (1984). Intervirology **21,** 224–228.

Verma, H. N., Khan, M. M. A. A., and Dwivedi, S. D. (1985). *Indian J. Plant Pathol.* **3,** 13–20.

Wambugu, F. M., Secor, G. A., and Gudmestad, N. C. (1985). *Am. Potato J.* **62,** 667–672.

Witkowski, J. T., Robins, R. K., Sidwell, R. W., and Simon, L. N. (1972). *J. Med. Chem.* **15,** 1150.

Index